Effect of Spaceflight and Spaceflight Analogue Culture on Human and Microbial Cells

Cheryl A. Nickerson • Neal R. Pellis
C. Mark Ott
Editors

Effect of Spaceflight and Spaceflight Analogue Culture on Human and Microbial Cells

Novel Insights into Disease Mechanisms

Springer

Editors
Cheryl A. Nickerson, Ph.D.
School of Life Sciences
The Biodesign Institute, Center for Infectious Diseases and Vaccinology
Arizona State University
Tempe, AZ, USA

Neal R. Pellis, Ph.D
Division of Space Life Science
Universities Space Research Association
Houston, TX, USA

C. Mark Ott, Ph.D
Biomedical Research and Environmental Sciences Division
NASA/Johnson Space Center
Houston, TX, USA

ISBN 978-1-4939-3276-4 ISBN 978-1-4939-3277-1 (eBook)
DOI 10.1007/978-1-4939-3277-1

Library of Congress Control Number: 2016931436

Springer New York Heidelberg Dordrecht London

Printed on acid-free paper

Springer Science+Business Media LLC New York is part of Springer Science+Business Media (www.springer.com)

This book is dedicated to all members of the spaceflight community who have worked tirelessly to support microgravity and microgravity analogue research efforts, especially those crewmembers who risked their lives for the scientific advancement that contributes to the health and quality of life of the astronauts and the general public on Earth. The editors would especially like to acknowledge the patience and support of their families during the writing of this book.

Foreword

We have been sending humans off the planet for half a century. A fortunate few were able to travel to our nearest neighbor, the moon, and experience a fractional gravity environment. Most, however, have spent their time in space hovering several hundred miles above the Earth living and working in microgravity. When we first conceived of sending our species off the planet, no one had any idea what would happen to the human organism. Was it safe? Would the lack of gravity cause physical problems? Lead to an inability of humans to function? Psychological problems? Cause mental reason and judgment to deteriorate?

There were many unanswered questions, and because of this, the selection process for the first astronauts (and I daresay cosmonauts) was very grueling and comprehensive. Even after finding the "best of the best"—those deemed most equipped to handle the unknown effects of those initial encounters with space—other mammals were still sent forth first to gather preliminary data (the USA sent monkeys and chimpanzees and the USSR experimented with dogs.) Men soon followed: Yuri Gagarin on April 12, 1961, and John Glenn on February 20, 1962. Both survived with no ill effects, and thus we learned that humans could indeed venture safely into space. But our voyage of discovery about life off of planet Earth had just started, and we have been learning ever since. The journey continues today on the International Space Station, more than 60 years after those first baby steps.

We have not been idle in the intervening decades, however. The Soviets built ever increasingly complex space stations and sent people to live there for various amounts of time. After closing down the lunar program, the USA built and occupied for a short while the Skylab orbital platform, our first venture into longer duration, low Earth orbit (LEO) missions. The space shuttle program provided us with the capability to conduct a multitude of missions in LEO ranging from satellite deployment and Earth observation to technology demonstration and, of course, science.

While the shuttle only remained on orbit for short periods, on the order of two weeks or less, that was more than enough time to initiate a wide range of science investigations. They included studies on combustion, fluids and colloidal systems, materials processing, and life science, the study of living things. There was much to learn about the effects of microgravity, not only on human beings, of which the astronauts remain the prime "guinea pigs," but also on other organisms ranging from very small-scale microcellular to other species.

Everything was new, and the scope of "we don't know what we don't know" was incredibly broad. We went from asking questions about whether humans could survive being in space to questions such as: "What are the long-term effects on human beings after being exposed to a microgravity environment? Can we mitigate those effects? Are those effects permanent? What happens to some of our cellular level functions in a microgravity environment? How do our individual systems respond? What is the effect of radiation? These are only a sample of the multitude of questions that were raised.

The fact that humans could exist in a near zero gravity environment gave us the opportunity to expand our curiosity and drive for knowledge in a whole new way. New fields of study opened up. New ways of thinking had to be created. Areas of specialty that had never overlapped were now fused together to address the myriad of interesting questions that appeared before us. The excitement and energetic sense of inquiry that our initiation as a space fairing civilization engendered in the life sciences remains today a palatable presence in the community.

The commitment of a group of nations to build and operate together an International Space Station, a permanent outpost in LEO, allowing people to live and work there for long periods of time (a typical mission to the ISS is six months) was remarkable on many levels. For the life science community, it provided an enormous opportunity to expand the search for knowledge into complete new directions. The ISS made it possible to continuously gather useful data on the effects on microgravity on human beings as they stayed in space longer and longer. Also, since the station was intentionally designed with extensive laboratory facilities, including several freezers to preserve specimens, the scope of potential research projects opened up.

The biological research community, recognizing an opportunity when they saw it, rallied to meet that potential, and since the arrival of the Expedition 1 crew to the ISS in November 2001, life science experiments and projects have been ongoing. All told, to date, that is almost 15 years of continual science and 15 years of asking questions, conducting experiments to get answers, analyzing those answers, and, in many cases, being able to tweak hypotheses and repeat the whole process. To say that we have learned a lot is an understatement. To say that we have a lot yet to learn is also an understatement. Like all scientific endeavors, the more you learn, the more you learn that you don't know. The research community continues to tackle this challenge, and life science investigations remain and will continue to remain an important part of the activities on board the ISS.

This book is a compilation of some of what we have learned in spaceflight and spaceflight analogue biological investigations. It would take several weighty tomes to cover all of the interesting and unexpected discoveries made through microgravity experiments. This book instead concentrates on one area, human and microbial cellular processes and subsequent insights into disease mechanisms. But on that topic, it is comprehensive and will walk the reader through some of the principles involved in microgravity and microgravity analogue research. In the process, it will highlight some of the technology developments that have been involved. Before diving into cellular and microbial issues and their relationship to disease mechanisms, the subjects at the heart of the book, it provides an excellent explanation of the suppression of the human immune system in microgravity.

Like all research done in space, the increase in fundamental knowledge and the ability to take gravity out of the equation lead to a clearer understanding of how things work here on Earth in 1 g. In the end, that leads to direct benefits for all mankind—the billions of people who have not been to space.

As I have stated, we have much yet to learn, but with each piece of data we gather, we learn to ask better questions, focus our investigations more appropriately, and increase the body of knowledge about the complexities of the human organism. Our journey, even over 60 years, is still at its beginning. This book is a snapshot documenting where we are at this point in our travels. I look forward to the second edition as I am certain there are more discoveries just around the corner!

Executive Director, American Institute of Aeronautics and Astronautics (AIAA)
Former NASA Astronaut

Dr. Sandra H. Magnus

Contents

Chronology of Key Spaceflight and Spaceflight Analogue Culture Experiments on Human and Microbial Cells

1973	Konstantinova et al. report reduced reactivity of lymphoid blood cells from crew members of "Soyuz" space missions as compared to the preflight status [1].
1975	Dietlein describes reduction in muscle volume and function associated with extended spaceflight during Apollo missions [2].
1982	*Staphylococcus aureus* and *Escherichia coli* that were cultured during the Cytos 2 experiment aboard Salyut 7 display an increased resistance to antibiotics compared to ground controls [3].
1984	Activation of human T lymphocytes is profoundly depressed during in vitro culture in spaceflight microgravity [4].
1986	Taylor et al. show that leukocyte distribution pattern changes in Space Shuttle crew members at landing [5].
1991	The rotating wall vessel (RWV) bioreactor as a model system for simulated microgravity culture of cells is developed at NASA-Johnson Space Center. It remains one of the most frequently used and widely accepted ground-based model systems in microgravity research for culturing microbial and human cells [6].
1994	Kulesh et al. fly the first experiment aboard the Space Shuttle utilizing skeletal muscle cells resulting in the creation of a new, non-fusogenic myoblast variant known as the L8SF cell line, which retained its altered phenotype even on return to Earth [7].
1995–2002	Ground-based experiments in the RWV bioreactor suggest that microgravity may facilitate engineering advanced three-dimensional (3-D) human surrogate tissue models from individual cells. To test this hypothesis, a spaceflight experiment aboard the Mir space station shows that cells formed cartilage that was significantly more compressible than the control on Earth, due in part to a decrease in production of glycosaminoglycan [8–10].
1997	Functional cardiac tissues engineered in the RWV bioreactor [11].

1997 Primary myoblasts cultured in the RWV bioreactor demonstrate myotube formation [12].

1997 Pellis et al. use the RWV bioreactor to analyze microgravity-induced inhibition of lymphocyte locomotion and investigate mechanisms related to blunted lymphocyte movement [13].

1998 Cooper and Pellis use the RVW bioreactor to characterize suppression of T cell activation observed in microgravity and microgravity analogue culture and demonstrate that signaling pathways upstream of protein kinase C are sensitive to these conditions [14].

1998 Lewis et al. document spaceflight-induced alteration of the cytoskeleton on a molecular level when they observe that microtubules of Jurkat T lymphocytes are shortened and extended from poorly defined organizing centers [15].

1998 First infection of RWV-derived 3-D cell culture models with any pathogen (rhinovirus) [16].

1999 3-D skeletal muscle organoids constructed from myoblasts demonstrate that microgravity exposure can induce muscle atrophy even without the removal of extrinsic mechanical load [17].

1999 Based upon Space Shuttle investigations of *Bacillus subtilis* and *E. coli* using liquid and semisolid growth media, Kacena et al. report that differences observed in the growth of these microorganisms are likely the result of external physical forces or factors, such as fluid dynamics or extracellular transport [18].

2000 Crucian et al. identify altered cytokine production by human peripheral blood cell subsets following short duration spaceflight [19].

2000 Nickerson et al. report increased virulence and stress resistance in *Salmonella enterica* serovar Typhimurium cultured in the RWV bioreactor compared to identical cultures grown in a reoriented control [20].

2001 Ground and rocket flight studies of cartilage formation from chondrocytes [21–23].

2001 Permanent and nonrandom phenotypic and genotypic changes observed in prostate cancer epithelial cells upon co-culture in 3-D in the RWV with either prostate or bone stromal cells [24].

2001 Nickerson et al. report first infection of RWV-derived 3-D cell culture models with a bacterial pathogen (*Salmonella enterica* serovar Typhimurium) [25].

2001 Use of the RWV in quantitative studies of cartilage healing [26].

2001 Cells cultured in the microgravity environment of spaceflight demonstrate alterations in cytoskeletal gene expression in T lymphocytes [27].

2002 Wilson et al. report the first use of whole genome microarray analysis of any spaceflight analogue cell culture, identifying 163 differentially expressed genes in *S.* Typhimurium cultured in the RWV bioreactor compared to identical cultures grown in a reoriented control [28].

2003 Fluid-mechanic analysis of cartilage development in the RWV [29].

2005	Wang et al. report 3-D co-culture of prostate cancer epithelial cells and bone stromal cells in space [30].
2005	Pierson et al. demonstrate that Epstein-Barr virus shedding increases in astronauts during Space Shuttle missions [31].
2006	Wilson et al. report the first study to examine the effect of spaceflight on microbial virulence and obtain the entire gene expression profile (transcriptomic and proteomic) of any microorganism to spaceflight. This spaceflight study on Shuttle mission STS-115 also confirmed the increased virulence and select changes in gene expression in *S.* Typhimurium observed in the RWV bioreactor [32].
2006–2013	Cartilage formation from bone marrow cells in the RWV bioreactor [33–35].
2006	RWV bioreactor culture of human bone marrow stem cells produces 3-D tissue constructs resembling those of trabecular bone [36].
2007	Nauman et al. report a correlation between fluid shear levels experienced by microbial pathogens in the RWV and those naturally encountered in the infected host [37]. Different microbial responses in *Salmonella* are observed as a function of incremental changes in fluid shear.
2008	Cohrs et al. show asymptomatic reactivation and shedding of infectious varicella zoster virus in astronauts during Space Shuttle mission [38].
2008	Independent validation of increased *S.* Typhimurium virulence when cultured during spaceflight is reproduced aboard Space Shuttle mission, STS-123. This mission also demonstrates that the altered virulence is dependent on media ion concentration [39].
2008	First infection of RWV-derived 3-D cell culture models with a parasite (*Cryptosporidium parvum*) [40].
2008	Sung et al. use the RWV bioreactor to demonstrate the coevolution of cancer and the interacting stromal cells during prostate cancer progression and metastasis [41].
2010	Radtke et al. report the first study using RWV-derived 3-D intestinal cell culture model that mechanistically confirmed clinical data showing that the *Salmonella* Type Three Secretion System (T3SS) is not required for enteric infection in humans [42] .
2010	Crabbé et al. report the first RWV-derived 3-D immunocompetent co-culture model used for studying infectious disease [43].
2011	Cardiac differentiation of mouse embryonic cells in the RWV bioreactor [44].
2012	First study to demonstrate that commensal bacteria can protect against bacterial infection in vitro in RWV-derived 3-D tissue constructs [45].
2013	Brinley et al. report that RWV bioreactor culture decreased DNA repair in cells infected with Epstein-Barr virus (EBV) and caused greater DNA damage from radiation compared to EBV-negative cells [46].

2013	Crucian et al. demonstrate that immune system dysregulation occurs during short duration spaceflight on board the Space Shuttle [47].
2014	First transcriptomic study to reveal molecular differences in epithelial cells grown in 2-D versus RWV-derived 3-D cell cultures, before and after infection [48].
2014	Crucian et al. report a cytokine dysfunction pattern in crew members on the International Space Station (ISS) during long-duration spaceflight, suggesting that multiple physiological adaptations persist during flight, including inflammation, leukocyte recruitment, angiogenesis, and thrombocyte regulation [49].
2015	Barrila et al. report first human cell infection conducted in spaceflight aboard Space Shuttle mission STS-131 [50].

References

1. Konstantinova, I. V., et al. (1973). Reactivity of lymphoid blood cells in the crew of "Soiuz-6", "Soiuz-7" and "Soiuz-8" spacecraft before and after flight. *Kosmicheskaia Biologiia i Meditsina, 7*(6), 35–40.
2. Johnson, R. S., Dietlein, L. F., & Berry, C. A. (Eds.). (1975). *Biomedical results of Apollo, NASA SP-368*. Washington, DC: National Aeronautics and Space Administration.
3. Tixador, R., et al. (1985). Study of minimal inhibitory concentration of antibiotics on bacteria cultivated in vitro in space (Cytos 2 experiment). *Aviation, Space, and Environmental Medicine, 56*(8), 748–751.
4. Cogoli, A., Tschopp, A., & Fuchs-Bislin, P. (1984). Cell sensitivity to gravity. *Science, 225*(4658), 228–230.
5. Taylor, G. R., Neale, L. S., & Dardano, J. R. (1986). Immunological analyses of U.S. Space Shuttle crewmembers. *Aviation, Space, and Environmental Medicine, 57*(3), 213–217.
6. Schwarz, R. P., & Wolf, D. A. (1991). *Rotating bio-reactor cell culture apparatus.*
7. Kulesh, D. A., et al. (1994). Space shuttle flight (STS-45) of L8 myoblast cells results in the isolation of a nonfusing cell line variant. *Journal of Cellular Biochemistry, 55*(4), 530–544.
8. Freed, L. E., et al. (1998). Chondrogenesis in a cell-polymer-bioreactor system. *Experimental Cell Research, 240*(1), 58–65.
9. Freed, L. E., et al. (1997). Tissue engineering of cartilage in space. *Proceedings of the National Academy of Sciences of the United States of America, 94*(25), 13885–13890.
10. Freed, L. E., & Vunjak-Novakovic, G. (1995). Cultivation of cell-polymer tissue constructs in simulated microgravity. *Biotechnology and Bioengineering, 46*(4), 306–313.
11. Carrier, R. L., et al. (1999). Cardiac tissue engineering: Cell seeding, cultivation parameters, and tissue construct characterization. *Biotechnology and Bioengineering, 64*(5), 580–589.
12. Molnar, G., et al. (1997). Skeletal muscle satellite cells cultured in simulated microgravity. In Vitro *Cellular & Developmental Biology. Animal, 33*(5), 386–391.
13. Pellis, N. R., et al. (1997). Changes in gravity inhibit lymphocyte locomotion through type I collagen. In Vitro *Cellular & Developmental Biology. Animal, 33*(5), 398–405.
14. Cooper, D., & Pellis, N. R. (1998). Suppressed PHA activation of T lymphocytes in simulated microgravity is restored by direct activation of protein kinase C. *Journal of Leukocyte Biology, 63*(5), 550–562.
15. Lewis, M. L., et al. (1998). Spaceflight alters microtubules and increases apoptosis in human lymphocytes (Jurkat). *The FASEB Journal, 12*(11), 1007–1018.
16. Long, J. P., Pierson, S. S., & Hughes, J. H. (1998). Rhinovirus replication in HeLa cells cultured under conditions of simulated microgravity. *Aviation, Space, and Environmental Medicine, 69*, 851–856.

17. Vandenburgh, H., et al. (1999). Space travel directly induces skeletal muscle atrophy. *The FASEB Journal, 13*(9), 1031–1038.
18. Kacena, M. A., et al. (1997). Low gravity and inertial effects on the growth of *E. coli* and *B. subtilis* in semi-solid media. *Aviation, Space, and Environmental Medicine, 68*(12), 1104–1108.
19. Crucian, B. E., Cubbage, M. L., & Sams, C. F. (2000). Altered cytokine production by specific human peripheral blood cell subsets immediately following space flight. *Journal of Interferon and Cytokine Research, 20*(6), 547–556.
20. Nickerson, C. A., et al. (2000). Microgravity as a novel environmental signal affecting *Salmonella enterica* serovar Typhimurium virulence. *Infection and Immunity, 68*(6), 3147–3152.
21. Conza, N., et al. (2001). Tissue engineering in space. *Journal of Gravitational Physiology, 8*(1), P17–P20.
22. Barzegari, A., & Saei, A. A. (2012). An update to space biomedical research: Tissue engineering in microgravity bioreactors. *Bioimpacts, 2*(1), 23–32.
23. Lewis, M. L., & Hughes-Fulford, M. (1997). Response of living systems to spaceflight. In S. Churchill (Ed.), *Fundamentals of space life sciences* (pp. 21–39). Malabar: Krieger Publishing Company.
24. Rhee, H. W., et al. (2001). Permanent phenotypic and genotypic changes of prostate cancer cells cultured in a three-dimensional rotating-wall vessel. In Vitro *Cellular & Developmental Biology. Animal, 37*(3), 127–140.
25. Nickerson, C. A., et al. (2001). Three-dimensional tissue assemblies: Novel models for the study of *Salmonella enterica* serovar Typhimurium pathogenesis. *Infection and Immunity, 69*(11), 7106–7120.
26. Obradovic, B., et al. (2001). Integration of engineered cartilage. *Journal of Orthopaedic Research, 19*(6), 1089–1097.
27. Lewis, M. L., et al. (2001). cDNA microarray reveals altered cytoskeletal gene expression in space-flown leukemic T lymphocytes (Jurkat). *The FASEB Journal, 15*(10), 1783–1785.
28. Wilson, J. W., et al. (2002). Microarray analysis identifies *Salmonella* genes belonging to the low-shear modeled microgravity regulon. *Proceedings of the National Academy of Sciences of the United States of America, 99*(21), 13807–13812.
29. Lappa, M. (2003). Organic tissues in rotating bioreactors: Fluid-mechanical aspects, dynamic growth models, and morphological evolution. *Biotechnology and Bioengineering, 84*(5), 518–532.
30. Wang, R., et al. (2005). Three-dimensional co-culture models to study prostate cancer growth, progression, and metastasis to bone. *Seminars in Cancer Biology, 15*(5), 353–364.
31. Pierson, D. L., et al. (2005). Epstein-Barr virus shedding by astronauts during space flight. *Brain, Behavior, and Immunity, 19*(3), 235–242.
32. Wilson, J. W., et al. (2007). Space flight alters bacterial gene expression and virulence and reveals a role for global regulator Hfq. *Proceedings of the National Academy of Sciences of the United States of America, 104*(41), 16299–16304.
33. Ohyabu, Y., et al. (2006). Cartilaginous tissue formation from bone marrow cells using rotating wall vessel (RWV) bioreactor. *Biotechnology and Bioengineering, 95*(5), 1003–1008.
34. Sakai, S., et al. (2009). Rotating three-dimensional dynamic culture of adult human bone marrow-derived cells for tissue engineering of hyaline cartilage. *Journal of Orthopaedic Research, 27*(4), 517–521.
35. Wu, X., et al. (2013). The effect of the microgravity rotating culture system on the chondrogenic differentiation of bone marrow mesenchymal stem cells. *Molecular Biotechnology, 54*(2), 331–336.
36. Marolt, D., et al. (2006). Bone and cartilage tissue constructs grown using human bone marrow stromal cells, silk scaffolds and rotating bioreactors. *Biomaterials, 27*(36), 6138–6149.
37. Nauman, E. A., et al. (2007). Novel quantitative biosystem for modeling physiological fluid shear stress on cells. *Applied and Environmental Microbiology, 73*(3), 699–705.
38. Cohrs, R. J., et al. (2008). Asymptomatic reactivation and shed of infectious varicella zoster virus in astronauts. *Journal of Medical Virology, 80*(6), 1116–1122.

39. Wilson, J. W., et al. (2008). Media ion composition controls regulatory and virulence response of *Salmonella* in spaceflight. *PLoS One, 3*(12), e3923.
40. Alcantara Warren, C., et al. (2008). Detection of epithelial-cell injury, and quantification of infection, in the HCT-8 organoid model of cryptosporidiosis. *The Journal of Infectious Diseases, 198*(1), 143–149.
41. Sung, S. Y., et al. (2008). Coevolution of prostate cancer and bone stroma in three-dimensional coculture: Implications for cancer growth and metastasis. *Cancer Research, 68*(23), 9996–10003.
42. Radtke, A. L., et al. (2010). Analysis of interactions of *Salmonella* type three secretion mutants with 3-D intestinal epithelial cells. *PLoS One, 5*(12), e15750.
43. Crabbe, A., et al. (2011). Alveolar epithelium protects macrophages from quorum sensing-induced cytotoxicity in a three-dimensional co-culture model. *Cellular Microbiology, 13*, 469–481.
44. Rungarunlert, S., et al. (2011). Enhanced cardiac differentiation of mouse embryonic stem cells by use of the slow-turning, lateral vessel (STLV) bioreactor. *Biotechnology Letters, 33*(8), 1565–1573.
45. De Weirdt, R., et al. (2012). Glycerol supplementation enhances *L. reuteri's* protective effect against *S.* Typhimurium colonization in a 3-D model of colonic epithelium. *PLoS One, 7*(5), e37116.
46. Brinley, A. A., et al. (2013). Characterization of Epstein-Barr virus reactivation in a modeled spaceflight system. *Journal of Cellular Biochemistry, 114*(3), 616–624.
47. Crucian, B., et al. (2013). Immune system dysregulation occurs during short duration spaceflight on board the space shuttle. *Journal of Clinical Immunology, 33*(2), 456–465.
48. David, J., Sayer, N. M., & Sarkar-Tyson, M. (2014). The use of a three-dimensional cell culture model to investigate host-pathogen interactions of *Francisella tularensis* in human lung epithelial cells. *Microbes and Infection, 16*(9), 735–745.
49. Crucian, B. E., et al. (2014). Plasma cytokine concentrations indicate that in vivo hormonal regulation of immunity is altered during long-duration spaceflight. *Journal of Interferon and Cytokine Research, 34*(10), 778–786.
50. Barrila, J., et al. (2015). Microgravity uniquely alters the host-pathogen interaction between human intestinal epithelial cells and *Salmonella enterica* serovar Typhimurium. In *115th American Society for Microbiology General Meeting*, New Orleans, LA.

Part I
The Principles and Translational Impact of Space Life Sciences Research

Chapter 1
Overview and Translational Impact of Space Cell Biology Research

Neal R. Pellis, Alexander Chouker, B. Yic, Svantje Tauber, Oliver Ullrich, and A. Sundaresan

1.1 Introduction

Space life sciences research is critical to preparation for long-duration space exploration, increases our knowledge of basic biological processes, and provides insight into the mechanisms and treatment of various medical conditions here on Earth. Although the study of terrestrial organisms and cells in space poses many challenges, it offers opportunities that may be infrequent or nonexistent in ground-based research.

Cell biology and microbiology provide windows into terrestrial living systems for much of the basic and applied research in biological science today. As we proceed into the next phase of space and exploration wherein humans will travel beyond low Earth orbit and even beyond the solar system, it is essential that we understand the impact of hypogravity on life. Cells and microbes offer an expeditious and cost effective tool to begin addressing the critical questions on the adaptation of terrestrial life to space and planetary environments. Animal experiments have extensive design and hardware requirements and can be physiologically complex to control

N.R. Pellis (✉)
Division of Space Life Sciences, Universities Space Research Association,
3600 Bay Area Blvd, Houston, TX 77058, USA
e-mail: pellis@dsls.usra.edu

A. Chouker • B. Yic
Department of Anesthesiology, Stress and Immunology Lab,
Hospital of the University of Munich, 81366 Munich, Germany

S. Tauber • O. Ullrich
Division of Cell- and Neurobiology, Institute of Anatomy, University of Zürich,
Winterthurerstr. 190, CH-8057 Zürich, Switzerland

A. Sundaresan
Department of Biology, Texas Southern University, Houston, TX 77004, USA

C.A. Nickerson et al. (eds.), *Effect of Spaceflight and Spaceflight Analogue Culture on Human and Microbial Cells*, DOI 10.1007/978-1-4939-3277-1_1

and interpret. Additionally, having live animals demands more crew time and close attention to their health and care. While resources are scarce, we can avail ourselves of the technology to study mammalian cell and tissue culture models as well as microbial cell cultures. This chapter will discuss the implementation of cell and tissue models and microbial cell culture technology on space biology.

1.2 Effect of Microgravity on Mammalian and Microbial Cells

For many years, it was presumed that cells were too insignificant in mass to be affected by a decrease in gravity (*g*). In the basic theory of relativity, the component "mass" has a dominant effect, and the mass of the cell is so small that essentially it should have little impact on relativity. The reality is that space travel, and in particular microgravity, does affect cells. Many of the experiments from the past 30 years demonstrate that spaceflight significantly changes the morphology and function of human and microbial cells in culture [1–9]. This section provides a top level summary of pertinent findings from both microbial and human cell culture experiments conducted in space and in microgravity analogue systems. Details are the subject of subsequent chapters.

1.3 Relationship of Terrestrial Life to Microgravity

At the onset, there are specific considerations required for the discussion of terrestrial life and its relationship to microgravity. As life evolved on Earth, a multiplicity of changes in physical and chemical factors invoked adaptations and selections that induced the complicated evolutionary pressures. Life as we know it today was selected by a complex series of changes in the gaseous components of the atmosphere, the chemical nature of the earth and water bodies that cover the planet, competitive species, and the many physical forces that impact living organisms across the phylogenetic scale. For many of these factors, there are clear examples of the role of changing physical forces in evolution. Living systems on Earth, be they bacterial, fungal, plant, or animal, all in some way or another are affected by forces such as hydrodynamic/fluid shear, hydrostatic pressure, stretch, and other mechanical stresses. The same living systems rely on gravity-driven convective events that take place both in the aqueous and the gaseous environments on the planet. Gravity itself has been the same on Earth practically since its inception. The acceleration due to gravity is a function of the mass of the planet where the acceleration is 9.8 m/s^2. Essentially, the influence of gravity remained constant as evolution proceeded. Living systems did experience transient excursions into hypergravity. These living systems adapted to accelerations, which give momentary exposures to higher *g* forces than the base force on the surface of the planet. In contrast, there is little opportunity on the planet for low gravity. Therefore, hypogravity is likely not an influential selective pressure on the evolving species on Earth. As we examine the response suites of the genes expressed in cells and microbes cultured in space, we observe that unlike many of the other selective influences, there are few defined gene suites that have been identified for response to the microgravity environment; however, common evolutionarily conserved responses have been reported in microorganisms [10–13]. Furthermore, gene array analyses of mammalian cells [14], plant cells [15–17], and microbial cells [10–13] have shown that there are specific classes of genes (e.g., stress) that are affected. Thus, investigation of the response of cells to microgravity is a new window through which we may observe physiological processes relevant to normal homeostasis or disease development that can be unveiled by cells attempting to adapt, survive, and thrive in microgravity. These processes are directly relevant to the natural life cycles of cells on Earth, but may be masked by the force of gravity [3].

1.4 Unique Aspects of Microgravity and the Effect on Microbial and Human Cells

Key aspects of microgravity culture that could impact cellular responses are: (1) significant abatement of sedimentation, (2) gravity-driven convection is no longer present, (3) hydrodynamic/fluid shear is substantially decreased, and (4) hydrostatic pressure gradients akin to those on Earth are greatly reduced. These gradients have significant impact on cellular function in aqueous environments (e.g., cell cultures, lakes, and oceans) as well as in organ systems and tissues. Given the characteristics associated with microgravity, it becomes apparent that in this environment, living systems could experience a significant change in the mass transfer that provides nutrients to cells and facilitates dissipation of waste products. Moreover, the absence of a significant gravity vector renders cells responsive to otherwise subtle physical forces, like a reduction in fluid shear. It is important to keep in mind that any or all of these forces, alone or in combination, may be responsible for mediating the diverse cellular and molecular responses observed during spaceflight culture.

Forces active in biological systems. A brief survey of the forces that are active in biological systems reveals four major categories: (1) gravity, (2) electromagnetic, (3) strong sub-molecular forces, and (4) weak sub-molecular forces. Of these, gravity is actually the weakest, yet still has the greatest sphere of influence (from cells to planets to galaxies). In contrast to gravity, intermolecular forces can be extraordinarily strong over short (nm) ranges. Removal of gravity from the system may evoke a new order of priority among the remaining forces. Thus, newly dominant force(s) may activate mechanosensitive channels or other force transducers differently than in 1 *g*. The result will be an adaptation response that will address the ability of the living system to survive and thrive in the new environment. At present, we have a series of actual and presumed effects that are in various levels of

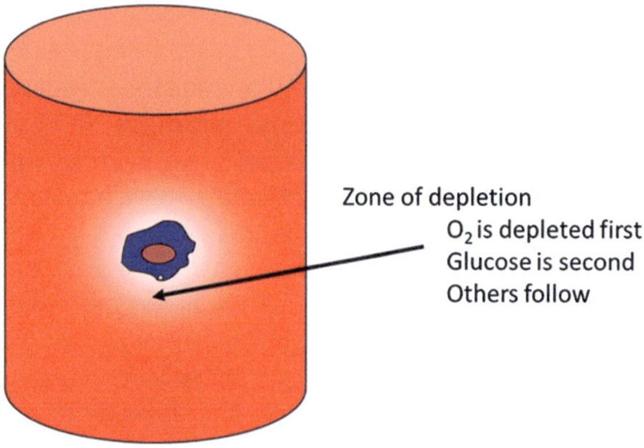

Fig. 1.1 Impact of microgravity environment on cells. The theoretical expectations for the fate of a cell in microgravity are illustrated herein where a metabolizing cell assimilates nutrients in the absence of gravity-driven convection. Thus there is no mixing occurring in the culture vessel. Nutrient transfer to the cell and dissipation of waste from the cell is reduced to the rate of diffusion. As such, the cells should become anoxic and succumb. Actual observations suggest otherwise. In most instances, the cells do not die. It is hypothesized that the more subtle non-gravity-driven convections facilitate a lower rate of mass transfer so the cells adapt and survive. The observation raises the question: Is the cell responding to microgravity or is it adapting to a new mode (rate) of mass transfer?

documentation through many diverse biological experiments conducted in microgravity. The subtle convection that occurs in space is largely due to secondary physical forces; namely, those of the surface tension-driven, convective type. In theory, cells in microgravity should become anoxic rapidly in space cell culture. Figure 1.1 illustrates the loss of gravity facilitated mass transfer, showing a zone of depletion surrounding the cell that should reduce mass transfer to the rate of diffusion. In general, theoretical expectations indicate that mammalian cell culture in space should eventually become anoxic. Nevertheless, the cells appear to adapt and proceed on to a metabolically different profile. Furthermore, this concept does not necessarily apply to all cells. For example, *Salmonella* cultures showed little-to-no induction of mRNA or protein expression profiles related to decreased oxygen content during spaceflight culture [10, 11]. While the present discussion focuses on oxygen, the same can be true for the soluble nutrients, vitamins, and inorganics.

A number of very important fundamental questions are raised that are part of the portfolio of experiments to be done in microgravity over the next decades (see insert in the Significance in 1.13). What is the basis of the response to microgravity? Is this an intrinsic cellular response to microgravity per se (i.e., direct effect), or is it a response to a condition created by microgravity (i.e., indirect effect)? The intrinsic response invokes the existence of sensory systems within the cell that actually can respond to a decrease in gravity and then transduce an adaptive signal to the cell. In contrast, the response to microgravity-induced environmental changes may be the basis of many cellular observations (in microbial and human cells) and human physiological responses. Specifically, perhaps an adaptation to the loss of a hydrostatic pressure gradient, a decrease in hydrodynamic fluid shear, a change in mass transfer and reduced contact with the container because of the absence of sedimentation, may be the manifestation of a reordering of forces following diminution of gravity. The results may show that both direct and indirect mechanisms may be at work in many biological systems [18, 19]. It is clear that some plants cells have direct response to gravity, especially in root systems [20].

1.5 How Can Cells Sense Mechanical Forces?

Despite the fact that there is broad knowledge about the consequences of altered gravity on cellular and molecular responses in immune cells, it is far less understood which mechanism(s) are being used by these cells to sense gravity or gravitational changes. Comparing the mass of a single cell with the other physical forces, it is unlikely that the cell can directly detect the vector of gravity. In contrast, it is more likely that cells respond indirectly to gravitational changes by sensing alterations in the physical microenvironment that result from gravitational changes [21, 22].

During the last decade, research in the interdisciplinary field of mechanobiology has revealed that mammalian cells sense and respond to physical forces in their microenvironment [19, 23–25]. They convert these forces into biochemical signals and finally into cellular responses, a process called mechanotransduction. Ingber et al., posited and expanded the tensegrity model to include how extracellular forces influence the morphology, function, and interaction of cells by stabilizing the three-dimensional form and channeling from the macro- to the nano-scale [19]. Lastly, in mammals, all nucleated cells possess a single, primary cilium [26, 27] at some time during their life. It is not a motile organelle, but rather a sensitive mechanotransducer. Unlike motile cilia, the primary cilium has a 9–0 microtubule configuration, thus missing the control pair essential to motility. Perturbation of the primary cilium is achieved by fluid and mechanical stimulators, hydrostatic pressure gradients, load, and possibly changes in gravity [18].

The engagement of cell surface extracellular matrix (ECM) receptors subsequently results in transmission of the external mechanical force from the cell membrane through the actin cytoskeleton and into the nucleus, which results in changes in gene expression that in turn remodel the ECM [28, 29]. This dynamic and reciprocal cross talk between the ECM and cell nucleus is critical to maintain the balance between normal tissue homeostasis and transition to disease. Indeed, this concept of “dynamic reciprocity” was proposed in cancer research by Mina Bissell and her team to explain the link between tissue form and function [30–33].

The ECM receptors are multimolecular assemblies where transmembrane adhesion receptors, the integrins, provide the mechanical link between the ECM and the cytoskeleton. Integrins bind to a broad spectrum of ECM molecules and thus function as mechanoreceptors [34]. Intracellularly, they are coupled to the actin cytoskeleton directly or indirectly via actin binding proteins. A famous member of these actin binding proteins is talin, which couples forces sensed via integrins to actin assembly in a mechanosensitive manner, and thus functions as a mechanosensor [23, 35]. The ability of talin and other mechanosensory proteins to convert forces into biochemical signals is based on force-dependent conformational changes. Three common mechanisms of mechanosensing are: (1) force-induced exposure of peptide sequences that are otherwise hidden, (2) conformational changes resulting in opening of ion channels, and (3) protein interactions that get stronger with strain [36]. It has turned out that signaling via mechanotransduction does crucially contribute to the control of essential cellular processes including cell proliferation, apoptosis, gene expression, and differentiation [19, 37]. One of the first discovered cellular mechanotransduction responses was the fluid shear stress-induced activation of K+ currents in vascular endothelium [38]. Additional examples of mechanotransduction are evidenced by findings that the rigidity of cell culture matrices determine lineage specification in mesenchymal stem cell cultures [39] and that stretch-activated ion channels contribute to mechanosensitive autoregulation of the heartbeat [40].

A well-established idea that is often discussed in connection with mechanotransduction and sensing of gravitational changes is that the cell cytoskeleton is built in tensegrity architecture. According to this model, the cytoskeleton is in a state of tension (pre-stress) which provides the cell's stability and shape. This tension is induced by the actin-myosin network, focal adhesions, and cell–cell adhesions [21, 41]. In the context of mechanotransduction, the tensegrity nature of the cytoskeleton is interpreted as a prerequisite because it balances the forces that act on cells via the ECM and enables long-distance force propagation via elastic waves along pre-stressed cytoskeletal elements [19]. Based on these mechanisms of mechanotransduction and the model of cellular tensegrity architecture, it has been postulated that alteration of gravity could influence cellular processes by changing the mechanical force environment of cells [21].

In this regard, it is worth noting that prokaryotic cells also have actin cytoskeleton homologues that could potentially be used to sense and translate mechanical force changes due to alterations in gravity (indirectly or directly) into cell-signaling pathways that regulate molecular genetic and phenotypic responses. Indeed, models as to how bacterial cells "might" sense the quiescent microgravity environment have been postulated, and include a possible role for mechanosensitive channel (Msc) proteins, fimbriae, and actin cytoskeletal proteins (Mbl/MreB and FtsZ) [3].

1.6 Investigating Cells in Space

Our experimental methods for investigating the effect of microgravity on biological systems are limited. Space is the most reliable environment in which to conduct these experiments, however, there is currently limited access to the spaceflight research platform. In the absence of routine spaceflight opportunities, there are several approaches that provide brief periods of analogue conditions; namely, drop towers, parabolic flight, and suborbital missions (Fig. 1.2). More recently, a

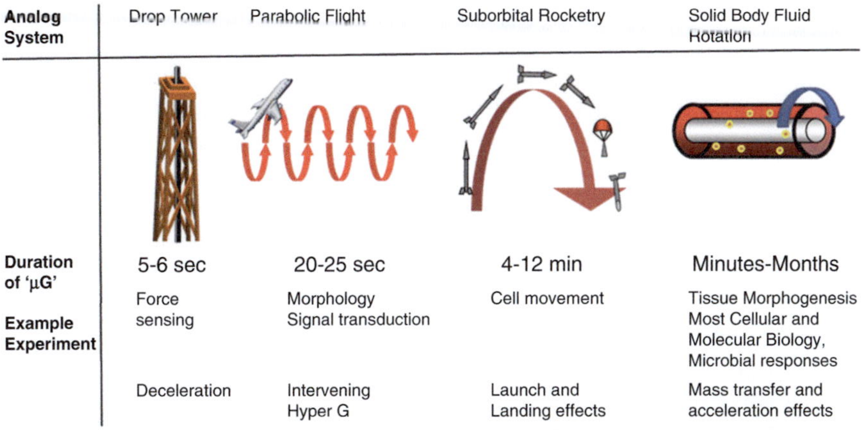

Fig. 1.2 Microgravity analogue systems applicable to mammalian and microbial cells. Analogues are systems and/or devices that mimic some of the characteristics of microgravity. Since microgravity opportunities are infrequent, we rely on ground-based analogues to initiate investigations, profile the actual spaceflight experiment, and design cell culture equipment for flight. There are at least eight analogue systems used for microbial and human cells, and tissue specimens. Four of these systems are represented here

significant number of experiments are being performed worldwide that employ ground-based spaceflight analogue culture in rotating wall vessel (RWV) bioreactor systems (which are covered extensively throughout the book). Lastly, for those few characteristics for which we have valid equations with a *g* term, we may begin computational modeling of many of the experimental options before actually testing them in space. As we elucidate these physical relationships in the context of gravity, the opportunity for computational models will increase, thereby facilitating the success of spaceflight experiments.

Therefore, the observations may or may not be relevant to the actual events occurring in tissue or to what happens in the intact host. Additionally, the observed cellular responses could be the result of combined direct and indirect effects. Investigation of cells in microgravity is often conducted in a stylized model, e.g. cell culture. As such, it may be necessary to include tissue constructs that better emulate the in vivo environment.

The demands on life invoked by microgravity and the space environment have introduced a new age in the study of the physics of life. The influence of mechanical stresses on life as cells, tissue, organs, systems, and complex organisms has risen to a new level of importance as a consequence of the space program. With the diminution of gravity in space, other forces which are in effect (contractile, intermolecular, and electromagnetic) may gain a new prominence and are reflected both in alterations in cellular morphology and physiology.

Elucidation of these critical links between biomechanics and cellular biochemistry hold the potential to lead to novel strategies for the development of new drugs and engineered tissues, as well as biomimetic micro devices and nanotechnologies that more effectively function within the context of living tissues. Cells respond to their environment within a three-dimensional (3-D) tissue structure and context. The number of adhesive structures in a cell and its resulting cell morphological polarity change the way that the cell responds. For example, cells that do not polarize are more likely to undergo apoptosis [42, 43]. Apoptosis may be affected when the cell cytoskeleton is disorganized and may increase or decrease depending on the origin of the cell [44]. Abnormal orientation or cell interactions may cause either premature cell death or the inhibition of death and sometimes at an inappropriate time [43]. Additionally, mitochondrial clustering may cause an increase in the amount of glucose consumed by cells due to crowding, which can also lead to an increase in cell apoptosis [45]. Microgravity-induced changes are reflected in both cell shape and cytoskeleton [46]. Changes in intra- and intercellular architecture can alter the way cells respond to their environment. Thus, the overall alterations in this architecture may occur due to changes in the way that microtubules and other cytoskeletal components organize in microgravity [47].

Changes in the position of cell structures such as mitochondria have been noted [45] in mitochondria clustering and the area around the nuclear envelope during spaceflight analogue culture. These are likely to be caused by changes in the cytoskeleton [45]. Since a cell is confined to a specified diameter, if it is stretched too flat, it will undergo apoptosis and if it is too round, it will undergo cell division [48]. Most mammalian cells exhibit cytoskeletal changes very soon after exposure to microgravity. Some cell types restore their cytoskeleton after approximately 3 days, but often changes are translated to the nucleus and they adapt to a new cell shape. Microtubules and F-actin are affected when cells are first exposed to microgravity, and cells often have increased apoptosis [49]. The reactions of cells to changing gravitational fields are dependent upon cell type and length of exposure. Lymphocyte activation and locomotion through interstitium are diminished in microgravity [50–52]. Thus, there is a complex series of interactions among the cytoskeletal components adapting to microgravity. One driver may be changes in ion flux induced indirectly by the environment or directly by stimulation of cell surface force transducers. Calcium ion flux changes the polymerization of microtubules as well as other cell physiology may be altered in microgravity [53, 54]. It is conceivable that mechanosensitive channels [3] and even the primary cilium may be part of the spaceflight-mediated response [55].

1.7 Partial Gravity (*g*) Experiments

Based upon the profound responses of biological systems to microgravity, where the gravity ranges from 10^{-6} to $10^{-5}g$, we are compelled to further elucidate the relationship of biological responses through partial gravity experiments to define the biological responses at forces between microgravity and normal gravity on Earth. Such studies will be of paramount importance for future crewed

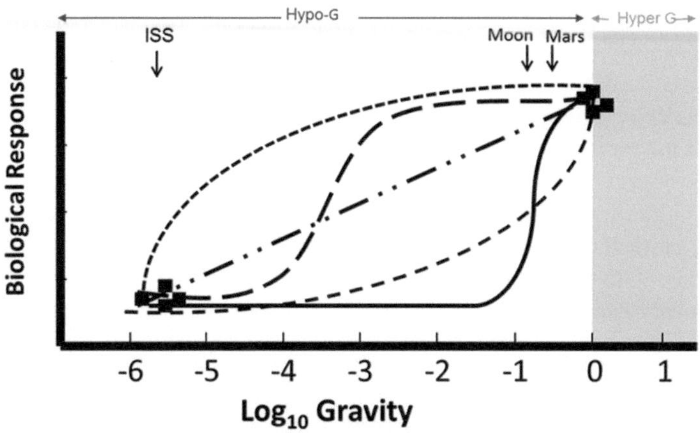

Fig. 1.3 Theoretical relationship between gravity and biological response. Very little has been done to understand the relationship between gravity and a given biological response. We have copious data at $10^0 \times g$ (Earth) and at $10^{-5} - 10^{-6}$ (ISS and Shuttle), but essentially none in between. Each of the *broken line curves* illustrates a potential relationship, but all are unlikely. The most probable may be the *solid line*, since most dose responses in living systems are sigmoidal and occur within 1–1.5 logs of "normal" (Earth gravity)

missions to the Moon (1/6*G*) and Mars (3/8*G*), and as for microgravity studies, hold exciting potential to advance our fundamental understanding of how cells behave normally or transition to disease-causing phenotypes. Our present experience for a given biological response has a robust portfolio of data at 10°*g* (i.e., $1 \times g$, Earth gravity). We also have some, albeit less, data obtained from biological microgravity studies over years of orbital missions. However, an important question in space cell biology and microbiology, where minimal data is available, is the relationship between various levels of hypogravity and the biological response of interest. With Earth gravity and microgravity as reference points, we can envision the theoretical set of graphic relationships shown in Fig. 1.3. Although data that support the concept shown are scant at best, this figure is useful to begin a discourse on what this relationship may be. The space program is critically interested in understanding the threshold of gravity is necessary to sustain a given function. The reality is that without partial gravity studies (by necessity in microgravity using a centrifuge or in hypogravity environments) it will be difficult to support any of the relationships shown in Fig. 1.3. However, there may be critical clues from our general experience in the terrestrial biosciences. Most biological dose responses occur in a sigmoidal relationship relative to the inducer stimulus (in this case, hypogravity). That being assumed, we expect that at some point in the relationship, there will be a flexure in the curve. It is at this point that we should see the minimum gravity necessary for a near normal response. Taking this concept one step farther, our understanding of dose response relationships in biological systems indicates that most of these sigmoidal relationships occur within about 1–1.5 logs of normal [56, 57]. Earth gravity being at 10°*G*

($1 \times g$) on the x axis suggests that the starting point for the low end of the relationship would be within 1–1.5 logs of the Earth's gravitational force. That means somewhere between 10^{-2} and 10^{0}, at best. Thereafter, the response is asymptotically minimal. Under this assumption, there would be no difference in the response from 10^{-2} to 10^{-6} G. If the threshold is much closer to Earth gravity than previously presumed, the implications are profound. The Moon and Mars would provide important partial g points in that interval. One might expect tremendous value in having biological sciences, particularly at the cellular level, conducted during the Mars and Moon expeditions. Otherwise, onboard centrifuges for orbital missions, as in the International Space Station (ISS), remain of value to obtain insight into partial g.

1.8 Microbiological and Human Cell Science Investigations in Space

Efficient use of orbital microgravity research facilities requires a variety of mammalian and bacterial cell culture equipment and frequent access to the laboratory. The first consideration is to conduct focused research programs that address specific hypothesis-driven questions in a manner consistent with an opportunity to make modifications to experiments in real time and to be able to iterate protocols for a 120-day increment in microgravity. Thereby a series of experiments may be completed that result in answers to specific questions which may not be answered within the context of a single experiment. The latter is the result of the payload strategy, where research experiments were flown in space, brought back, and analyzed on Earth. The new ISS strategy affords multiple experiments and real-time modifications. In combination with ongoing improvements in flight hardware that incorporate experimental flexibility, modularity, and advanced capabilities to enable a broad range of biological experiments, the potential for innovative biomedical and biotechnological applications is greatly enhanced. Together with onboard real-time analytical capabilities, there can be meaningful return of data long in advance of retrieving the specimens. Additionally, valuable microgravity analogues can be used on the ground to profile experiments for spaceflight to identify specific biological characteristics to be observed in microgravity. Judicious application of the scientific method requires that we provide the appropriate controls so that the forthcoming answers are valid. Not surprisingly, microgravity experiments offer great challenge for the design of appropriate controls. Some investigators believe that the most appropriate control is an onboard centrifuge wherein 1 g is continuously applied. Others contend that an Earth-based control is the most appropriate. Indeed, design of the control is dictated by the nature of the question. The most difficult aspect of control design is posed by whether the question sought hypothesizes a direct response to microgravity or an indirect response to environmental conditions created by microgravity.

1.9 Flight and Analogue Experiments

Gravitational cell biology may have a traceable history that goes back to the nineteenth century [58]. The origin of gravitational biology may be set in the nineteenth century when Julius Sachs (1863), Charles Darwin (1880), and Wilhelm Pfeiffer (1904) investigated the influence of gravity on plants, with the downward orientation of root caps (Reviewed in the Encyclopedia of Astrobiology) [58]. The experiments actually used hyper *g* to investigate the response of plants and oocytes in accelerated fields. A compilation made in 1991 [59] shows an extensive summary of experiments that indicate both response and non-response to microgravity. The early satellites of the 1960s had some bacterial, plant, and animal experiments aboard. Often these experiments or surveys were confounded by the operational priorities, logistical and technical constraints of these missions. In contrast, Skylab allowed scientists to embark on a new era in microgravity research. These early studies set the stage for investigation of the effects of microgravity on red blood cell shape, bone loss, muscle deconditioning, metabolic changes, and immune performance [60]. Some early findings were conducted using human cell lines, normal human lymphocytes, common bacteria, protozoans, and fungi. Early assessments of microbial populations on crew and in spacecraft were limited. However, the Skylab program hosted a well-ordered monitoring protocol of bacteria and fungi before, during, and after all missions [60]. The crew was monitored for skin, nasal, throat, urine, and feces flora. Air, water, and surface samples were monitored in the Skylab environment. Results showed gross contamination of the environment, and there were several crew infections, none of which affected the mission. Two organisms of note were *Serratia marcescens* and *Propionibacterium acnes*. The longer missions had apparently generated a bacterial load in the environment. Lastly, these studies showed that there was inter-crew transmission of organisms.

More recent reports indicate that spaceflight affects growth [5], morphology [6, 12], metabolism [11], and gene expression and virulence [10] in microorganisms (see Chaps. 10–14). Basic mechanisms indicate that spaceflight culture of the bacterial pathogens *Salmonella* and *Pseudomonas* alters selected gene expression in these organisms [10, 12], with the RNA chaperone Hfq identified as a global regulator of their spaceflight-induced molecular genetic responses.

Following Skylab, the Shuttle era opened new vistas for scientific experiments. It was during this time, and in the follow-on Shuttle Program that some of the key findings with regard to mammalian cells in space took place. These findings include: (1) Inhibition of lymphocyte activation, cytokine production, and locomotion [61–63]; (2) Interference with transmembrane signaling [52, 95, 97]; (3) Changes in metabolism [64–66]; and (4) Tissue morphogenesis [67, 68]. The objective of future experiments will be to delineate the responses and effects that are due to indirect environmental changes and those that are actually a direct response to microgravity from stimulation of force sensing entities within the cell. When mammalian cells arrive in microgravity, if unattached to a substrate, or to each other in multiple cell–cell tissue-like arrangements, within seconds they minimize

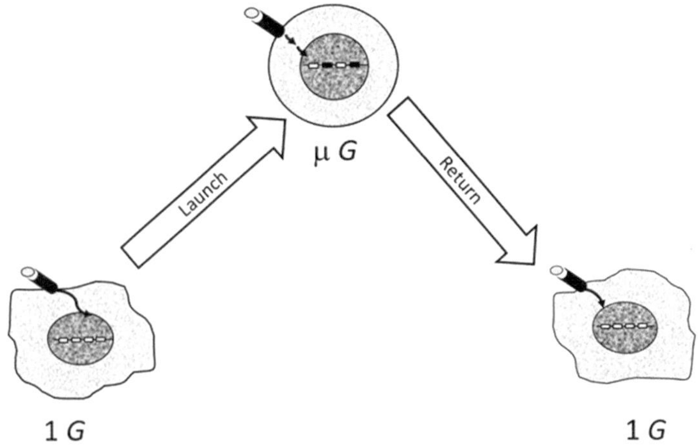

Fig. 1.4 Responses of mammalian cells to the microgravity environment. The transition from 1 × *g* to µg is attended by changes that occur within seconds and those which occur over the ensuing hours and days. The majority of these changes resolve upon return to 1 × *g* over a similar timescale to the microgravity adaptation phase. The most notable phenotypic change is in the shape of the cell which initiates a signaling cascade resulting in downstream changes

their surface area-to-volume ratio and upon return to unit gravity resume the normal morphology as it was prior to flight (Fig. 1.4). As we observe cells adapting to the new environment, we have to question how many of the responses that we see in the adaptation profile are actually a direct response to microgravity or are they due to the forced shape change, as suggested in the tensegrity models [69]. The following have been reported over the years: (1) the shape change and some redistribution of cytoskeleton and subcellular organelles; (2) noted changes in cytoskeletal deployment; (3) fluid flows and distributions; (4) changes in gene expression; (5) transmembrane signaling; (6) differentiation; and (7) metabolic changes. In addition to the aforementioned, the changes are seen in membrane alterations in structure, composition, bi-leaflet organization, lipid rafts, and association with the cytoskeleton. At present, we presume that many of these changes are the consequence of the initial response of the cell minimizing its surface area-to-volume ratio. Others may be direct responses to microgravity mediated by mechanosensitive constituents of the cell.

Later chapters will detail the kinds of changes that occur in microbial populations following replication for multiple generations in microgravity and/or microgravity analogue culture (see Chaps. 10–14). In the context of our discussion here, we will introduce the observation that microorganisms bounded by cell walls probably do not undergo the same kind of morphological changes observed in animal cells. Yet in response to both microgravity and microgravity analogue culture, bacteria and fungi exhibit dramatic and global changes in their molecular genetic and phenotypic characteristics, some of which are similar to animal cells (Fig. 1.5) and will be reviewed in the subsequent chapters. It is tempting to speculate that forces are exerted on the cell wall following the diminution of gravity and that the changes are sensed by mechanosensitive transducers that signal the cell to respond

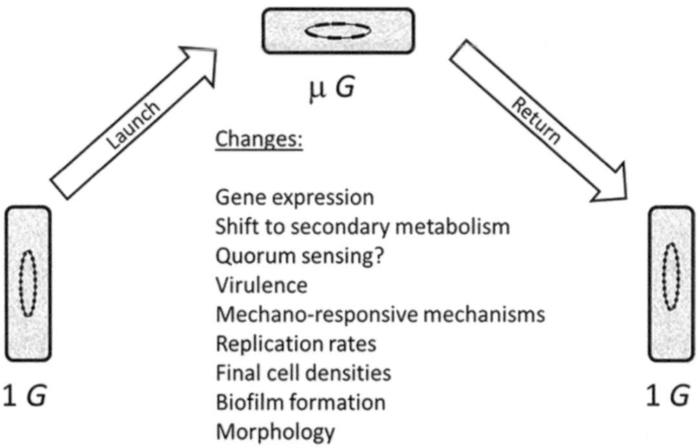

Fig. 1.5 Response of bacterial cells to the microgravity environment. Unlike mammalian cells, bacteria that are bound by a rigid cell wall may not undergo an immediate shape change in microgravity. However there are profound changes in metabolism, virulence, signal transduction, and colonial morphology

to the new environment. Some of the changes observed are in gene (mRNA) and protein expression, a shift to secondary metabolism, alterations in virulence, quorum sensing, biofilm formation, differences in antibiotic resistance, morphological changes, and identification of an evolutionarily conserved master response regulator that acts across bacterial species during culture in the microgravity and microgravity analogue environments [3, 4, 70, 71]. A significant finding for long-term space travel is the increase in virulence and/or pathogenesis-related characteristics observed in some of the microbes flown in space (see Chaps. 10–14) [7].

Our discussion has highlighted the ability of cells to respond to changes in the physical force of gravity, however, whether these changes are the result of a direct or indirect effect remains to be elucidated. There are a number of possibilities as to how the cascade of events begins. Historically, we have known about mechanosensitive channels and their ability to inform cells of the environment with regard to hydrodynamic/fluid shear, physical impact, stretching, and perhaps even microgravity itself. These channels are responsible for modulation of ion flow in and out of the cell and, as a consequence, can modulate a number of cascading activities beyond that. In addition, more recently described structures in mammalian cells, known as the primary cilium, may be an important transducer capable of responding to very subtle force changes, such as gravitational-induced changes in subcellular levels. The primary cilium is a non-motile cilium characterized by nine pairs of microtubules about the periphery of the cilium shaft [26, 27, 72, 73]. It is devoid of the central shaft microtubules and therefore non-motile. Its primary purpose is sensing small changes in the force environment. The changes most noted today involved fluid flow [72, 74]. The primary cilia are critically important during embryogenesis and defects in their structure manifest in significant morphological change [18]. The primary cilia are sensitive to the very subtle forces at play in organs, tissues, and subcellular environments [18, 75].

1.10 Translational Strategies

Cells in microgravity and in microgravity analogue culture behave quite differently than when in the normal Earth gravitational field. These differences are well-suited for and already have been the basis for a number of translational strategies for applied science. Recent meetings (International Space Station (ISS) Research and Development Conference 2012) featured approaches to the use of microgravity and microgravity technology for advancing research beneficial to understanding and treating human diseases. Below is a sampling of some of the experiments.

1. ***Drug Therapy of Duchenne Muscular Dystrophy with Inhibitors of Hematopoietic Prostaglandin D Synthase***—*Yoshihiro Urade, Osaka Bioscience Institute*
2. ***The Microgravity Research Platform: Novel Insights into the Mechanisms and Treatment of Infectious Disease***—*Cheryl Nickerson, Arizona State University*
3. ***NASA's Current Evidence and Hypotheses for the Visual Impairment Intracranial Pressure Risk***—*Christian Otto, Universities Space Research Association and Bill Tarver, NASA Human Health and Performance Directorate*

In microgravity, cells have a tendency to remain suspended for a period of time without sedimenting to a surface. While in suspension, cells like any inert particles have a tendency to aggregate. It is unclear which physicochemical forces are at play to promote this spontaneous association. Nevertheless, for many types of cells, a nidus forms from which tissue begins to develop. Thus, the first stage of bioengineering a piece of tissue from individual cells is satisfied by merely placing them in microgravity or in a microgravity culture analogue. Thereafter not sedimenting, the cells propagate from this assembly with three-dimensional (3-D) freedom. Following a period of replication, cellular differentiation commences, and tissue morphogenesis proceeds. In the ensuing processes, native ECM is synthesized and deployed within the intercellular spaces. This results in 3-D organotypic constructs that exhibit key structural and functional properties akin to the parental tissue and which provide human surrogate models for application in a wide range of experiments with biomedical and biotechnological applications. These include studies of tissue homeostasis and organogenesis, disease progression, and regenerative medicine [8, 76]. This aspect of space cell biology is covered in (see Chaps. 4–9).

1.10.1 Flight and Analogue Research

NASA research in the Space Bioreactors resulted in a number of patents in the early years. The bioreactors are currently produced (Synthecon, Inc., Houston, TX) and used worldwide as a microgravity analogue and as an enabling tool for diverse studies including tissue morphogenesis, host-pathogen interactions, and disease mechanisms. Other actual spaceflight research successfully used the bioreactor to study breast cancer [77], colon cancer [44], and prostate cancer [78]. NASA licensed the patents to private companies, including Synthecon, Inc. (http://www.synthecon.com), for commercial manufacturing of the equipment.

Ground-based cellular research with this technology covers a wide range of areas, including cancer, stem cells, diabetes, cartilage, nerve, skin, kidney, liver, heart, lung, blood vessels, infectious disease, and immunology. Thus, translation of this research into patents (Table 1.1) and novel findings that could be converted into patents, has gone much farther than early concepts of engineering implantable tissue, to include living ex vivo organic life support systems, such as artificial livers, as well as ex vivo generation of functional lung tissue that could ultimately be used for transplantation by using cadaveric decellularized lung scaffolds that have been recellularized with stem cells [79, 80].

Table 1.1 Patents related to analogue microgravity technologies

Match	Document	Document title
1	5638303	*Non-contacting isolated stabilized microgravity platform system* Microgravity sensitive payloads placed on a platform are isolated from external acceleration forces over extended periods of time in the weightless environment inside an orbiting space vehicle by…
2	US2008/075818	*Methods and compositions based on culturing microorganisms in low sedimental fluid shear conditions*
3	5844815	*Umbilical and follower assembly utilized in microgravity platform system* An umbilical and follower assembly for preventing unwanted movements of a frame from being transmitted by an umbilical cord to a microgravity platform as may be utilized on an orbiting space…
4	US20080138828	*Microgravity bioreactor systems for production of bioactive compounds and biological macromolecules* The present invention relates to compositions and methods of plant, fungal, and bacterial cell culture that can be effectively utilized for three-dimensional plant, fungal, and bacterial cell growth…
5	5243544	*Microgravity accelerometer data acquisition system* Accelerometer apparatus for a space vehicle includes sensors for sensing accelerations on board the vehicle which provide accelerometer input signals. Interface circuits are coupled to receive the…
6	EP1988984B1	*PROCESS AND APPARATUS FOR INTRODUCTION OF GAS INTO AND DEGASSING OF LIQUIDS, IN PARTICULAR IN BIOTECHNOLOGY AND ESPECIALLY OF CELL CULTURES*
7	7182810	*Protein temperature evaporation-controlled crystallization device and method thereof* A temperature- and evaporation-controlled device for the crystallization of proteins from a protein-containing solution. The device comprises a compartment, such as a microcapillary tube, for…
8	WO/2009/117098A2	*METHODS FOR STEM CELL PRODUCTION AND THERAPY* The present invention relates to methods for rapidly expanding a stem cell population with or without culture supplements in simulated microgravity conditions. The present invention relates to…

(continued)

Table 1.1 (continued)

Match	Document	Document title
9	6409832	*Protein crystallization in microfluidic structures* A device for promoting protein crystal growth (PCG) using microfluidic channels. A protein sample and a solvent solution are combined within a microfluidic channel having laminar flow…
10	6596081	*Dynamically controlled crystallization method and apparatus and crystals obtained thereby* A method and apparatus for dynamically controlling the crystallization of molecules including a crystallization chamber (14) or chambers for holding molecules in a precipitant solution, one or more…
11	US20100062435	*Methods for Stem Cell Production and Therapy* The present invention relates to methods for rapidly expanding a stem cell population with or without culture supplements in simulated microgravity conditions. The present invention relates to…
12	6730498	*Production of functional proteins: balance of shear stress and gravity* The present invention provides a method for production of functional proteins including hormones by renal cells in a three-dimensional co-culture process responsive to shear stress using a rotating…
13	6099864	*In situ activation of microcapsules* Disclosed are microcapsules comprising a polymer shell enclosing two or more immiscible liquid phases in which a drug, or a prodrug and a drug activator are partitioned into separate phases, or…
14	7723085	*Analyzing cells immobilized in block copolymers* The present invention provides the use of a composition comprising a block polymer as a support matrix in the manipulation, processing, or analysis of particles, such as cells and fluorescent beads....
15	WO/2012/092382A2	*ROLE OF MICRORNA IN T CELL IMMUNE RESPONSE* The present disclosure provides methods for increasing or decreasing T cell activation. The methods include contacting a T cell with an miRNA nucleic acid or an inhibitor of an miRNA. In some…
16	7837040	*Acoustic concentration of particles in fluid flow* An apparatus for acoustic concentration of particles in a fluid flow includes a substantially acoustically transparent membrane and a vibration generator that define a fluid flow path there between....
17	WO/1999/042191A1	*DYNAMICALLY CONTROLLED CRYSTALLIZATION METHOD AND APPARATUS AND CRYSTALS OBTAINED THEREBY* A method and apparatus for dynamically controlling the crystallization of molecules including a crystallization chamber (14) or chambers for holding molecules in a precipitant solution, one or more…

(continued)

Table 1.1 (continued)

Match	Document	Document title
18	8266950	*Particle analysis in an acoustic cytometer* The present invention is a method and apparatus for acoustically manipulating one or more particles. Acoustically manipulated particles may be separated by size. The particles maybe flowed in a…
19	6472175	*Sialylation of N-linked glycoproteins in the baculovirus expression vector system* The present disclosure utilizes a novel approach to protein preparation in the baculovirus expression vector system (BEVS). Specifically, the present invention analyzes the effects of supplementing…
20	7700363	*Method for screening crystallization conditions in solution crystal growth* A method of screening protein crystal growth conditions with picogram to microgram amounts of protein in picoliter or nanoliter volumes is provided. A preferred method comprises a microarray with a…
21	WO/2009/087448A3	*DYNAMIC SYSTEMS FOR CULTURING CELLS IN 3D SUPPORTS* The present invention refers to the conception process of new dynamic systems for cell culture in 3D supports appropriate to its growing. These are constituted by: – cylindrical container (1)…
22	WO/2009/087448A2	*DYNAMIC SYSTEMS FOR CULTURING CELLS IN 3D SUPPORTS* The present invention refers to the conception process of new dynamic systems for cell culture in 3D supports appropriate to its growing. These are constituted by: – cylindrical container (1)…
23	7361493	*Production of urokinase in a three-dimensional cell culture* Embodiments of a method for the production of human urokinase are disclosed. Also disclosed are embodiments of a cell culture well-suited for use with the disclosed method. The method involves…
24	6819420	*Fiber optic apparatus and use thereof in combinatorial material science* Methods, systems, and devices are described for rapid characterization and screening of liquid samples to determine properties (e.g., particle size, particle size distribution, molar mass, and/or…
25	6519032	*Fiber optic apparatus and use thereof in combinatorial material science* Methods, systems, and devices are described for rapid characterization and screening of liquid samples to determine properties (e.g., particle size, particle size distribution, molar mass, and/or…
26	7214540	*Method for screening crystallization conditions in solution crystal growth* A method of screening protein crystal growth conditions on a nano or mesoscale comprises providing a micro-electromechanical chip with a plurality of micro-chambers in the micro-electromechanical…

(continued)

Table 1.1 (continued)

Match	Document	Document title
27	7658829	*Integrated microfluidic transport and sorting system* The invention integrates a two-stage dielectrophoretic (DEP) droplet dispensing and distribution system with particulate DEP to create a novel LOC platform capable of manipulating biological cells…
28	8137964	*Method of producing three-dimensional tissue and method of producing extracellular matrix used in the same* The present invention provides a novel method of producing a three-dimensional tissue by which cell lamination can be carried out easily. According to the method, a three-dimensional tissue in…
29	WO/1999/059554A1	*$i(IN SITU) ACTIVATION OF MICROCAPSULES* Disclosed are microcapsules comprising a polymer shell enclosing two or more immiscible liquid phases in which a drug, or a prodrug and a drug activator are partitioned into separate phases, or…
30	US20070020751	*Bioreactor, in particular for NMR spectroscopy* A description is given of a bioreactor (1), in particular for NMR spectroscopy, comprising a container (7) capable of containing a cell culture, a first inlet line (6) for the inward flow of a…
31	US20100203591	*Bioreactor, in particular for NMR spectroscopy* A description is given of a bioreactor (1), in particular for NMR spectroscopy, comprising a container (7) capable of containing a cell culture, a first inlet line (6) for the inward flow of a…
32	WO/2001/005956A2	*SIALYLATION OF N-LINKED GLYCOPROTEINS IN THE BACULOVIRUS EXPRESSION VECTOR SYSTEM* The present disclosure utilizes a novel approach to protein preparation in the baculovirus expression vector system (BEVS). Specifically, the present invention analyzes the effects of supplementing…
33	US20100273253	*DYNAMIC SYSTEMS FOR CULTURING CELLS IN 3D SUPPORTS* The present invention refers to the conception process of new dynamic systems for cell culture in 3D supports appropriate to its growing. These are constituted by:—cylindrical container (1) e…
34	8202701	*Microencapsulation of cells in hydrogels using electrostatic potentials* Compositions and methods for producing encapsulated cells having an average diameter of less than about 200 μm are provided. Methods for using the disclosed encapsulated cells are also provided.
35	WO/2003/094598A1	*PHOTOBIOREACTOR AND PROCESS FOR BIOMASS PRODUCTION AND MITIGATION OF POLLUTANTS IN FLUE GASES* Certain embodiments and aspects of the present invention relate to photobioreactor apparatus (100) designed to contain a liquid medium (108) comprising at least one species of photosynthetic…

(continued)

Table 1.1 (continued)

Match	Document	Document title
36	8293479	*Use of parathyroid hormone-related protein (PTHRP) in the diagnosis and treatment of chronic lung disease and other pathologies* This invention pertains to the discovery that Parathyroid Hormone-related Protein (PTHRP) can be detect and/or stage, and/or treat chronic lung diseases. In particular, it was discovered that PTHRP…
37	7835000	*System and method for measuring particles in a sample stream of a flow cytometer or the like* A system and method for analyzing a particle in a sample stream of a flow cytometer or the like. The system has a light source, such as a laser pointer module, for generating a low powered light…
38	6778724	*Optical switching and sorting of biological samples and microparticles transported in a micro-fluidic device, including integrated bio-chip devices* Small particles, for example 5 μm diameter microspheres or cells, within, and moving with, a fluid, normally water, that is flowing within microfluidic channels within a radiation-transparent …
39	7297494	*Activatable probes and methods for in vivo gene detection* Probes for detecting a target polynucleotide are provided. One aspect provides a molecular beacon probe set wherein the donor molecular beacon comprises a quantum dot and an acceptor molecular…
40	WO/2005/071115A1	*ACTIVATABLE PROBES AND METHODS FOR IN VIVO GENE DETECTION* Probes for detecting a target polynucleotide are provided. One aspect provides a molecular beacon probe set wherein the donor molecular beacon comprises a quantum dot and an acceptor molecular…
41	8008021	*N-terminal truncation of cardiac troponin subunits and their roles in cardiovascular disease* Methods for diagnosing, determining the likelihood of developing cardiac disease by measuring the level of a truncated form of cardiac Troponin T are provided. Also provided are methods for…
42	US20060226012	*INTEGRATED MICROFLUIDIC TRANSPORT AND SORTING SYSTEM* The invention integrates a two-stage dielectrophoretic (DEP) droplet dispensing and distribution system with particulate DEP to create a novel LOC platform capable of manipulating biological cells…
43	US20080032929	*N-terminal truncation of cardiac troponin subunits and their roles in cardiovascular disease* The invention provides methods for diagnosing, preventing, treating, determining the likelihood of developing, or ameliorating a symptom associated with, cardiac disease by administering a…

(continued)

Table 1.1 (continued)

Match	Document	Document title
44	WO/2011/000572A1	*METHOD AND DEVICE FOR DETECTING LONG-TERM BIOLOGICAL EFFECTS IN CELLS* The invention relates to a method for monitoring and regulating cell-based biological systems for detecting long-term biological effects. Among other things, this method allows for examining stress…
45	US20070207540	*Method of producing three-dimensional tissue and method of producing extracellular matrix used in the same* The present invention provides a novel method of producing a three-dimensional tissue by which cell lamination can be carried out easily. According to the method, a three-dimensional tissue in…
46	US20050260553	*Photobioreactor and process for biomass production and mitigation of pollutants in flue gases* Certain embodiments and aspects of the present invention relate to photobioreactor apparatus (100) designed to contain a liquid medium (108) comprising at least one species of photosynthetic…
47	US20060089388	*Use of parathyroid hormone-related protein (PTHRP) in the diagnosis and treatment of chronic lung disease and other pathologies* This invention pertains to the discovery that Parathyroid Hormone-related Protein (PTHRP) can be detect and/or stage, and/or treat chronic lung diseases. In particular, it was discovered that PTHRP…
48	5155034	*Three-dimensional cell to tissue assembly process* The present invention relates to a 3-dimensional cell to tissue and maintenance process, more particularly to methods of culturing cells in a culture environment, either in space or in a gravity…
49	US20040072875	*Use of parathyroid hormone-related protein (PTHRP) in the diagnosis and treatment of chronic lung disease and other pathologies* This invention pertains to the discovery that Parathyroid Hormone-related Protein (PTHRP) can be detect and/or stage, and/or treat chronic lung diseases. In particular, it was discovered that PTHRP…
50	US20050215606	*Use of parathyroid hormone-related protein(PTHRP) in the diagnosis and treatment of chronic lung disease and other pathologies* This invention pertains to the discovery that Parathyroid Hormone-related Protein (PTHRP) can be detect and/or stage, and/or treat chronic lung diseases. In particular, it was discovered that PTHRP…
51	US20090328242	*Replication of Undifferentiated Cells in a Weightless Environment, Uses Thereof and a Facility for Such Replication and the Acceleration of the Evolution of Plants and Animals* The present invention provides manufacturing processes for biological replication of undifferentiated plant and animal cells and tissue in a weightless condition, including those systems used in…

(continued)

Table 1.1 (continued)

Match	Document	Document title
52	US20050239182	*Synthetic and biologically-derived products produced using biomass produced by photobioreactors configured for mitigation of pollutants in flue gases* Certain embodiments and aspects of the present invention relate to photobioreactor apparatus designed to contain a liquid medium comprising at least one species of photosynthetic organisms therein,…
53	US20070021929	*COMPUTING METHODS FOR CONTROL OF HIGH-THROUGHPUT EXPERIMENTAL PROCESSING, DIGITAL ANALYSIS, AND RE-ARRAYING COMPARATIVE SAMPLES IN COMPUTER-DESIGNED ARRAYS* Computer-controlled automated high-throughput systems can be used to design, prepare, process, screen, and analyze a large number of samples in removable sample vials each containing a compound of…
54	US20050287548	*Activatable probes and methods for in vivo gene detection* Probes for detecting a target polynucleotide are provided. One aspect provides a molecular beacon probe set wherein the donor molecular beacon comprises a quantum dot and an acceptor molecular…
55	US20120115224	*CULTURE METHODS OF BONE MARROW STROMAL CELLS AND MESENCHYMAL STEM CELLS, AND MANUFACTURE METHOD OF GRAFT CELLS FOR CENTRAL NERVE SYSTEM DISEASES THERAPY* In a culture method of the present invention, by culturing bone marrow stromal cells or mesenchymal stem cells under a pseudo micro-gravity environment generated by multi-axis rotation, bone marrow…
56	EP2420569A1	*METHOD FOR CULTURING BONE MARROW STROMAL CELLS AND MESENCHYMAL STEM CELLS, AND METHOD FOR PRODUCING GRAFT CELLS FOR TREATING CENTRAL NEUROLOGICAL DISEASE* In a culture method of the present invention, by culturing bone marrow stromal cells or mesenchymal stem cells under a pseudo micro-gravity environment generated by multi-axis rotation, bone marrow…
57	WO/2012/175357A1	*MODULATION OF MICRORNA-138 FOR THE TREATMENT OF BONE LOSS* There is provided nucleic acids (mir-138Antimirs) for use in treating or preventing bone loss in a patient. Also there is provided a method for reducing the levels of endogenous mir-138 in a cell.
58	WO/2003/028308A1	*METHOD FOR CREATING A DYNAMIC ADDRESS TABLE FOR A SWITCHING NODE IN A DATA NETWORK AND A METHOD FOR TRANSMITTING A DATA MESSAGE* The invention relates to a switching node (2, 3) comprising a dynamic address table (11) in a data network (1), in addition to a method for creating a dynamic address table (11) of this type for…

(continued)

Table 1.1 (continued)

Match	Document	Document title
59	EP2130905A1	*Method for culturing eukaryotic cells* The invention is in the field of biotechnology and production of useful biomolecules. It relates to a method for culturing eukaryotic cells, more in particular adherent cells. The method is useful…
59	WO/1998/045468A1	*PRODUCTION OF FUNCTIONAL PROTEINS: BALANCE OF SHEAR STRESS AND GRAVITY* The present invention provides a method for production of functional proteins including hormones by renal cells in a three dimensional co-culture process responsive to shear stress using a rotating…
60	5740082	*Collocated sensor actuator* A position sensor is structurally integrated within a wide-gap magnetic actuator. Being an integral structure, the sensor provides voltages representative of a relative shift in position of the…
61	7351584	*Hepatocyte bioreactor system for long term culture of functional hepatocyte spheroids* A rotating wall vessel is used as a culture vessel and bioreactor for the cultivation of hepatocytes in the form of spheroids to generate a culture with many properties of the intact liver. These…
62	7575859	*Hepatocyte bioreactor system for long term culture of functional hepatocyte spheroids* A rotating wall vessel is used as a culture vessel and bioreactor for the cultivation of hepatocytes in the form of spheroids to generate a culture with many properties of the intact liver. These…
63	5961934	*Dynamically controlled crystallization method and apparatus and crystals obtained thereby* A method and apparatus for dynamically controlling the crystallization of proteins including a crystallization chamber or chambers for holding a protein in a salt solution, one or more salt…
64	EP1077686B1	*EXTERNALLY TRIGGERED MICROCAPSULES*
65	7939248	*Hepatocyte bioreactor system for long term culture of functional hepatocyte spheroids* A rotating wall vessel is used as a culture vessel and bioreactor for the cultivation of hepatocytes in the form of spheroids to generate a culture with many properties of the intact liver. These…
66	7575911	*Hepatocyte bioreactor system for long term culture of functional hepatocyte spheroids* A rotating wall vessel is used as a culture vessel and bioreactor for the cultivation of hepatocytes in the form of spheroids to generate a culture with many properties of the intact liver. These…
67	US20120196275	*MULTICELLULAR ORGANOTYPIC MODEL OF HUMAN INTESTINAL MUCOSA* Disclosed are methods of preparing multi-cellular three-dimensional tissue constructs, that include fibroblasts, endothelial cells, lymphocytes and epithelial cells. The present methods may include…

(continued)

Table 1.1 (continued)

Match	Document	Document title
68	7198947	*Production of functional proteins: balance of shear stress and gravity* The present invention provides a method for production of functional proteins including hormones by renal cells in a three dimensional co-culture process responsive to shear stress using a rotating…
69	6902909	*Methods for efficient production of mammalian recombinant proteins* A process has been developed for the production of human recombinant polypeptides using transformed mammalian cells cultured in a horizontally rotating culture vessel modulated to create low shear,…
70	WO/2004/033627A2	*METHODS FOR EFFICIENT PRODUCTION OF MAMMALIAN RECOMBINANT PROTEINS* A process has been developed for the production of human recombinant polypeptides using transformed mammalian cells cultured in a horizontally rotating culture vessel modulated to create low shear,…
71	7972821	*Production of functional proteins: balance of shear stress and gravity* A method for the production of functional proteins including hormones by renal cells in a three dimensional culturing process responsive to shear stress uses a rotating wall vessel. Natural mixture…
72	US20030077288	*Compositions and methods for treatment of muscle wasting* The present invention relates to the isolation of cell- or tissue-specific F-box proteins which are involved in ubiquitin-mediated protein degradation in a specific cell- or tissue-type....
73	6001643	*Controlled hydrodynamic cell culture environment for three dimensional tissue growth* A novel hydrodynamic cell culture environment is disclosed for a two-chamber roller bottle. The unstable hydrodynamics of a gas headspace media chamber is coupled through a center opening to a…
74	6946246	*Production of functional proteins: balance of shear stress and gravity* The present invention provides for a method of culturing cells and inducing the expression of at least one gene in the cell culture. The method provides for contacting the cell with a transcription…
75	7122371	*Modular cell culture bioreactor* An apparatus and method for a modular cell culture bioreactor comprises a plurality of chambers for cell culture; at least one reservoir containing a cell support medium; a plurality of conduits…
76	EP2272864A2	*Binding agents which inhibit myostatin* The present invention provides binding agents comprising peptides capable of binding myostatin and inhibiting its activity. In one embodiment the binding agent comprises at least one…

(continued)

Table 1.1 (continued)

Match	Document	Document title
77	5637477	*Recombinant protein production and insect cell culture and process* A process has been developed for recombinant production of selected polypeptides using transformed insect cells cultured in a horizontally rotating culture vessel modulated to create low shear…
78	WO/2001/023595A2	*REDUCED GRAVITY TRANSFORMATION PROCESS AND PRODUCT* A transformation process is carried out in which the competence of a plant or animal cell is increased and/or induced by placing the cell in reduced gravity, and then the cell is transformed in…
79	6875605	*Modular cell culture bioreactor and associated methods* An apparatus and method for a modular cell culture bioreactor comprises a plurality of chambers for cell culture; at least one reservoir containing a cell support medium; a plurality of conduits…
80	WO/2002/102997A2	*ISOLATED HOMOZYGOUS STEM CELLS DIFFERENTIATED CELLS DERIVED THEREFROM AND MATERIALS AND METHODS FOR MAKING AND USING SAME* The present invention discloses and describes pluripotent homozygous stem (HS) cells, and methods and materials for making same. The present invention also provides methods for differentiation of…
81	7288405	*Devices and methods for pharmacokinetic-based cell culture system* Devices, in vitro cell cultures, systems, and methods are provided for microscale cell culture analogous (CCA) device.
82	8030061	*Devices and methods for pharmacokinetic-based cell culture system* Devices, in vitro cell cultures, systems, and methods are provided for microscale cell culture analogous (CCA) device.
83	8168568	*Combinatorial therapy for protein signaling diseases* A method for selecting combinations of drugs for treatment of diseases that arise from deranged signaling pathways is disclosed. The method involves measuring the activity states for signaling…
84	6592623	*Engineered muscle* A muscle implant includes an extracellular matrix, tendon and muscle cells. The extracellular matrix is made of a matrix of electrospun polymer fibers. The tendon is made of extruded collagen…
85	7244402	*Microfluidic protein crystallography* The use of microfluidic structures enables high throughput screening of protein crystallization. In one embodiment, an integrated combinatoric mixing chip allows for precise metering of reagents to…
86	7144727	*Interlinked culture chamber for biologicals* The bioreactor system of the present invention has two fluid-filled culture compartments in which cells, tissues and other biologicals are cultured. The two culture compartments are in fluid…

1.10.2 *Tissue Engineering and Regenerative Medicine*

Microgravity provides an advantageous environment in tissue engineering. Cells are not driven to sediment to a surface and remain suspended in fluid during microgravity culture. For reasons yet unknown, cells spontaneously aggregate in microgravity as they proximate and attraction forces on the cell surface interact. The more interaction, the stronger the binding, and the cells form the rudiments of a stable tissue construct. Three-dimensional growth is achieved because the cells are not driven against the surface. They do not grow at a solid–liquid interface, which is what occurs at the bottom of a standard culture vessel on Earth. Following aggregation and stabilization, the tissue assemblies commence to produce ECM material. This stage is critical to structural integrity and to provision of essential signals for differentiation and maturation.

A filtered search was set in the FPO (Free Patents Online) database for tissue engineering in microgravity. Table 1.1 profiles patents in this field. Inventions range in variety from compositions and methods of plant, fungal, and bacterial cell culture for expression of biological macromolecules and biopharmaceuticals to actual three-dimensional cartilage tissue growth for wound healing applications [81]. Eighty-six relevant patents in tissue engineering and regenerative medicine are listed in Table 1.2.

Table 1.2 Relevant patents in tissue engineering and regenerative medicine

Match	Document	Document title
1	US20070116676	Method for three-dimensional cartilage tissue engineering using bone marrow cells in tissue engineering bone marrow cells in simulated microgravity environment
2	US20100221835	Method for cartilage tissue regeneration via simulated microgravity culture using scaffolds
3	US2002/011088	Methods for modeling infectious disease and chemosensitivity in cultured cells and tissues
4	US20080138828	Microgravity bioreactor systems for production of bioactive compounds and biological macromolecules
5	7615373	Electroprocessed collagen and tissue engineering
6	6787357	Plasma-derived fibrin-based matrices and tissue
7	8338114	Engineered human broncho-epithelial tissue-like assemblies
8	US20040096966	Replication of biological tissue
9	US20060030043	Neural stem cell-collagen-bioreactor system to construct a functional embryonic brain-like tissue…
10	EP2450707A1	Lung tissue model
11	US20090227025	Ex vivo human lung/immune system model using tissue engineering for studying microbial pathogens with lung tropism
12	WO/2007/076865A1	BIOREACTOR FOR CELL AND TISSUE CULTURE
13	7657292	Method for evaluating extracellular water concentration in tissue
14	6756194	Control samples for use as standards for evaluating apoptosis in a selected tissue

(continued)

Table 1.2 (continued)

Match	Document	Document title
15	8137972	Biocompatible, biodegradable polymer-based, lighter than or light as water scaffolds for tissue engineering and methods for preparation and use thereof
16	8293532	Method and apparatus for tissue transfer
17	US20040037813	Electroprocessed collagen and tissue engineering
18	8197553	Composite scaffolds and methods using same for generating complex tissue grafts
19	US20080262744	Systems and Methods for Tissue Engineering Tubular Biological Structures
20	5851816	Cultured high-fidelity three-dimensional human urogenital tract carcinomas and process
21	US20100120136	BIOREACTOR FOR CELL AND TISSUE CULTURE
22	US20050059043	Control samples for use as standards for evaluating apoptosis in a selected tissue
23	7759082	Electroprocessed fibrin-based matrices and tissues
24	5858783	Production of normal mammalian organ culture using a medium containing mem-alpha, leibovitz L-15, glucose galactose fructose…
25	8137964	Method of producing three-dimensional tissue and method of producing extracellular matrix used in the same
26	WO/2001/068800A1	BIOREACTOR FOR GENERATING FUNCTIONAL CARTILAGINOUS TISSUE
27	US20060026694	Method for generating immune-compatible cells and tissues using nuclear transfer techniques
28	US20020106625	Bioreactor for generating functional cartilaginous tissue
29	US20120246746	Method for Generating Immune-Compatible Cells and Tissues Using Nuclear Transfer Techniques
30	US20090035349	Composite scaffolds and methods using same for generating complex tissue grafts
31	6808704	Method for generating immune-compatible cells and tissues using nuclear transfer techniques *This invention relates to methods for making immune-compatible tissues and cells for the purpose of transplantation and tissue engineering, using the techniques of nuclear transfer and cloning....*
32	US20020046410	Method for generating immune-compatible cells and tissues using nuclear transfer techniques
33	US20040203146	Composite scaffolds and methods using same for generating complex tissue grafts
34	5780299	Method of altering blood sugar levels using non-transformed human pancreatic cells that have been expanded in culture
35	WO/2012/068071A2	IN VITRO UROGENITAL CO-CULTURE MODELS
36	6875176	Systems and methods for making noninvasive physiological assessments
37	EP1300435A1	ULTRAFINE METAL PARTICLE/POLYMER HYBRID MATERIAL

(continued)

Table 1.2 (continued)

Match	Document	Document title
38	7239902	Device and method for monitoring body fluid and electrolyte disorders
39	US20030114568	Ultrafine metal particle/polymer hybrid material
40	EP1300435A4	ULTRAFINE METAL PARTICLE/POLYMER HYBRID MATERIAL
41	7361493	Production of urokinase in a three-dimensional cell culture
42	6099864	In situ activation of microcapsules
43	7840272	Methods for modulating osteochondral development using bioelectrical stimulation
44	US20120156670	IN VITRO UROGENITAL CO-CULTURE MODELS
45	5947334	Post-mix beverage system for use in extra-terrestrial space
46	7723085	Analyzing cells immobilized in block copolymers
47	6472175	Sialylation of N-linked glycoproteins in the baculovirus expression vector system
48	7579189	Aligned scaffolds for improved myocardial regeneration
49	8195263	Pulse oximetry motion artifact rejection using near infrared absorption by water
50	WO/2009/087448A3	DYNAMIC SYSTEMS FOR CULTURING CELLS IN 3D SUPPORTS
51	5798261	Distributed pore chemistry in porous organic polymers
52	7384786	Aligned scaffolds for improved myocardial regeneration
53	6569466	Conditioning of bioactive glass surfaces in protein containing solutions
54	WO/2006/132855A3	METHODS FOR MODULATING OSTEOCHONDRAL DEVELOPMENT USING PULSED ELECTROMAGNETIC FIELD THERAPY

1.10.3 Cellular Locomotion

An important component in lymphocyte function is the locomotion through the interstitium. The movement of lymphocytes through the intercellular matrix is critical to the inflammatory process and to the renewal process in lymphoid organs. Immune manifestations such as resistance to infection, delayed hypersensitivity, allograft rejection, tumor rejection, and autoimmunity all rely on the movement of lymphocytes through matrix to accumulate at sites of antigen recognition. Studies were conducted in ground-based modeled microgravity and on spaceflight experiments (Space Shuttle Missions STS-54 and STS-56). The studies demonstrated that microgravity inhibits lymphocyte locomotion through type I collagen [61]. Signaling experiments suggest that the loss of locomotory function and T cell activation is probably due to lesion(s) in transmembrane signaling [2, 51, 52, 82]. This relates the depressed immunity observed in long-term space missions, at least in part, to the suppression of lymphocyte locomotion and activation as a consequence of a failed signal transduction process. Gene expression analysis on normal and activated peripheral blood lymphocytes cultured in ground-based modeled microgravity indicate that selective genes involved in inflammatory processes are affected by changes in gravity [82].

1.10.4 *Investigating Inflammation*

Up-regulation of cardiovascular atherogens (agents that induce atherosclerosis) such as placental growth factor (PlGf) coupled with the down-regulation of heat shock protein 90 (HSP 90) and pro-inflammatory cytokines such as Interleukin 4 (IL-4) suggest an overall increase in inflammation and further in vivo studies are essential to implicate nitric oxide and increased oxidative stress. This could lead to a signal blueprint stress response suite in microgravity. These could also be used as biomarkers to assess health of a particular tissue and overall health of crew members. Gene expression in microgravity is significantly different in many cell types as opposed to 1 *g* conditions. Microgravity reorders forces exerted on cells. The cell's response to this will provide novel insights into cellular and functional mechanisms [82, 83].

1.11 Intellectual Property: Microgravity Biotechnology

The top seven categories for intellectual property development include: Drugs and Formulations, Biomarkers, Biorepair and Bioreplicators, Biological Compounds, Gene Displays, Biochambers and Micromechanical Devices, and Bioreactive Surfaces (Table 1.1). Among the top seven categories, biotechnologies capture nearly 36 % of the total field (277 patents granted). It is nearly triple that of the next most productive category—analytical instrumentation. While this may suggest that biotechnologies represent greater potential for useful applications in microgravity conditions, it may be due to the somewhat lower resource demand of biotechnology experiments; therefore, enjoying greater access to space platforms. Equipment requirements to conduct basic biological studies at the molecular, cellular, and microbe levels are often less complex and more economical than those required for animal experiments and for processing of toxic substances at high temperatures and pressures. Biotechnologies include 36 % out of a total of 277 [84]. Key scientific findings with potential for future translational capabilities include:

- *Biomarkers*—Key cytokine regulation was observed in the formation of 3D thyroid carcinoma cell spheroid formation in the random positioning machine (spaceflight analogue). Cytokines such as IL-6, IL-8, CD 44, were significantly up-regulated in microgravity models [85].
- *Biochambers and micromechanical devices*—Sertoli cell biochambers (spaceflight analogue) were successfully generated as a start towards long-term treatment for diabetic rats and Parkinson's disease rats [86].
- *Bioreactive surfaces*—Hollow bioceramic microspheres and degradable composite microspheres were successfully developed to use as microcarriers for 3-D bone tissue engineering (spaceflight analogue) [87]. ECM formation and mineralization were documented in the osteoblast constructs generated.
- *Colloidal structures* [Natural matrix (type 1 collagen) and surrogate matrices (PGA ceramics)]—Gravity influences the velocity of propagating waves in excitable

media of neuronal tissue [88]. Living skin equivalent (LSE) constructs of normal human epidermal keratinocytes (NHEK) were cultured in submerged microgravity environments (spaceflight analogue) using a submerged fibroblast and type 1 collage matrix. Elimination of the air–liquid interface in this manner enhanced skin cell proliferation in the presence of microgravity [89]. Modeled microgravity was also used as a tool to improve the cell cluster morphology to increase quality of conformal polymer coatings for fabrication of coherent microcapsules in neonatal porcine cell models for pancreatic islet transplantation enhancement. This approach might promote good encapsulated cell engraftment [90].

1.12 Future Directions

In biotechnology and other cellular-based systems, microgravity is pursued by nations with space programs. Early findings show opportunities in disease modeling, tissue engineering for research and transplantation, new biopharmaceuticals, vaccine development, propagation of stem cells, and drug testing.

Microgravity provides environmental characteristics that are difficult to obtain on Earth. Using extreme conditions that modify the physical forces in the environment to investigate life processes affords opportunities for discovery and development of applications that significantly enhance research. For example, the simple modification of elevation of cell culture temperatures brought new concepts to cell biology in the discovery of heat shock proteins and the refined stress response suites observed in gene expression [91]. Indeed, *Thermophilus aquaticus* [92] isolated from aquatic thermal vents provided the Taq polymerase [93] that is so critical in molecular genetics advances, as it revolutionized the applications of the "Polymerase Chain Reaction" (PCR) technology. Indeed, microgravity is a provocative environment for terrestrial life systems and as highlighted in subsequent chapters, has unveiled novel insight into how human and microbial cells behave normally or transition to disease-causing phenotypes. Moreover, both prokaryotic and eukaryotic cell culture in Earth-based spaceflight analogues are easily accommodated and offer promising early applications [94, 95].

In summary, parallel development of space cell culture capabilities and of a ground-based analogues of some microgravity conditions, provided the folio of initial observations to spur the interest in the potential for applied research in microgravity. The ground-based RWV is an analogue culture system which serves as a platform for hypotheses development and refining protocols for flight experiments.

The experiment showed the capability of the technology and the ability to promote differentiation and engineer tissue from individual cells. Microgravity has provided avenues in the study of physiologically appropriate 3D environments of difficult cancers, such as pancreatic carcinoma [96], immune suppression studies in normal human lymphoid tissue [9], and lymphocyte trafficking [96]. The analogue system affords insight into the changes in cell products that space may bring to biotechnology.

1.13 Significance

The value of cell-based research (using both human and microbial cells) is well established in the biomedical community. Likewise, this line of research affords a similar value as we embark on an intensified era of space exploration and research. As such, investigations into the response of cells during microgravity and ground-based microgravity analogue culture have and will continue to unveil novel biomedically relevant responses that cannot be observed using conventional experimental approaches. Indeed, many breakthroughs in biological research and translational advancements to the healthcare setting have been achieved through studying the response of biological systems to extreme environments. The spaceflight and ground-based spaceflight analogue environments are no exception to this rule. Our intent in this book is to emphasize the immeasurable opportunity of these innovative research platforms to fundamentally advance our understanding of how human and microbial cells behave normally or transition to disease-causing phenotypes and the potential for translation of these findings into novel strategies to advance human health and quality of life.

As we proceed into the intensive investigation phase of cell biology and microbiology spaceflight research, and as spaceflight biological hardware continues to be advanced in terms of its capabilities, modularity, and experimental flexibility, it is essential that we keep the following questions as a backdrop for our plans. Indeed evolution has not designed a proscribed format for the adaptation of terrestrial life to microgravity. As we observe this transition over the next century, we will gain useful insight into the role of gravitational as well as other forces in life processes that will afford new opportunities to understand how cells operate in these conditions, and give new fundamental mechanistic insight into disease processes that may lead to new countermeasures for the general public.

Fundamental Questions

- What is the basis of the response to microgravity—direct, indirect, or both?
  - Intrinsic response element in the cell—gravisensor/force transducer? (direct)
  - Response to environmental changes induced by gravity. (indirect)

    With the diminution of gravity, other forces, such as fluid shear and hydrostatic pressure, take on a new dominance.

    Compensatory shape change in mammalian cells initiates a cascade of molecular and cellular responses. Microorganisms with cell walls also initiate global changes in molecular genetic and phenotypic responses, but may be independent of changes in cell shape.
- What is the threshold "G" necessary for normal function? This needs to be addressed with partial gravity studies.
- How is response different in microbial cells vs. mammalian cells? Does this impact microbial responses in both pathogens and commensals?

- How does microgravity change cell response thresholds to other stimuli (radiation, magnetic fields, fluid shear, toxins, pathogens, other chemicals)?
- How do the changes in individual cells relate to tissues, organs, organisms, and microbial populations?

Acknowledgments The author acknowledges the steadfast dedication and expertise of Universities Space Research Association employee Mildred D. Young in the preparation of the manuscript.

References

1. Cogoli, A. (1996). Gravitational physiology of human immune cells: A review of in vivo, ex vivo and in vitro studies. *Journal of Gravitational Physiology, 3*(1), 1–9.
2. Sundaresan, A., Risin, D., & Pellis, N. R. (2004). Cell growth in microgravity. In R. A. Meyers, A. Sendtko, & P. Henheik (Eds.), *Encyclopedia of molecular cell biology and molecular medicine* (Vol. 2, 2nd ed., pp. 303–321). Weinheim, Germany: Wiley-VCH.
3. Nickerson, C. A., Ott, C. M., Wilson, J. W., Ramamurthy, R., & Pierson, D. L. (2004). Microbial responses to microgravity and other low-shear environments. *Microbiology and Molecular Biology Reviews, 68*(2), 345–361.
4. Horneck, G., Klaus, D. M., & Mancinelli, R. L. (2010). Space microbiology. *Microbiology and Molecular Biology Reviews, 74*(1), 121–156.
5. Kim, W., Tengra, F. K., Shong, J., Marchand, N., Chan, H. K., Young, Z., et al. (2013). Effect of spaceflight on *Pseudomonas aeruginosa* final cell density is modulated by nutrient and oxygen availability. *BMC Microbiology, 13*, 241.
6. Kim, W., Tengra, F. K., Young, Z., Shong, J., Marchand, N., Chan, H. K., et al. (2013). Spaceflight promotes biofilm formation by *Pseudomonas aeruginosa*. *PloS One, 8*(4), e62437.
7. Ott, C. M., Crabbe, A., Wilson, J. W., Barrila, J. B., & Nickerson, C. A. (2011). Effect of spaceflight on microbial gene expression and virulence. In A. Chouker (Ed.), *Stress challenges and immunity in space* (pp. 203–225). Heidelberg, Germany: Springer.
8. Vunjak-Novakovic, G., Searby, N., De Luis, J., & Freed, L. E. (2002). Microgravity studies of cells and tissues. *Annals of the New York Academy of Sciences, 974*, 504–517.
9. Fitzgerald, W., Chen, S., Walz, C., Zimmerberg, J., Margolis, L., & Grivel, J. C. (2009). Immune suppression of human lymphoid tissues and cells in rotating suspension culture and onboard the international space station. *In Vitro Cellular & Developmental Biology Animal, 45*(10), 622–632.
10. Wilson, J. W., Ott, C. M., Honer zu Bentrup, K., Ramamurthy, R., Quick, L., Porwollik, S., et al. (2007). Space flight alters bacterial gene expression and virulence and reveals a role for global regulator Hfq. *Proceedings of the National Academy of Sciences of the United States of America, 104*(41), 16299–16304.
11. Wilson, J. W., Ott, C. M., Quick, L., Davis, R., Honer zu Bentrup, K., Crabbe, A., et al. (2008). Media ion composition controls regulatory and virulence response of Salmonella in spaceflight. *PloS One*, *3*(12), e3923.
12. Crabbe, A., Schurr, M. J., Monsieurs, P., Morici, L., Schurr, J., Wilson, J. W., et al. (2011). Transcriptional and proteomic responses of *Pseudomonas aeruginosa* PAO1 to spaceflight conditions involve Hfq regulation and reveal a role for oxygen. *Applied and Environmental Microbiology, 77*(4), 1221–1230.
13. Crabbe, A., Nielsen-Preiss, S. M., Woolley, C. M., Barrila, J., Buchanan, K., McCracken, J., et al. (2013). Spaceflight enhances cell aggregation and random budding in Candida albicans. *PloS One, 8*(12), e80677.

14. Ma, X., Wehland, M., Schulz, H., Saar, K., Hubner, N., Infanger, M., et al. (2013). Genomic approach to identify factors that drive the formation of three-dimensional structures by EA. hy926 endothelial cells. *PloS One, 8*(5), e64402.
15. Leach, J. E., Ryba-White, M., Sun, Q., Wu, C. J., Hilaire, E., Gartner, C., et al. (2001). Plants, plant pathogens, and microgravity--a deadly trio. *Gravitational and Space Biology Bulletin, 14*(2), 15–23.
16. Fengler, S., Spirer, I., Neef, M., Ecke, M., Nieselt, K., & Hampp, R. (2015). A whole-genome microarray study of *Arabidopsis thaliana* semisolid callus cultures exposed to microgravity and nonmicrogravity related spaceflight conditions for 5 days on board of Shenzhou 8. *BioMed Research International, 2015*, 547495.
17. Ferl, R. J., Koh, J., Denison, F., & Paul, A. L. (2015). Spaceflight induces specific alterations in the proteomes of *Arabidopsis*. *Astrobiology, 15*(1), 32–56.
18. Moorman, S. J., & Shorr, A. Z. (2008). The primary cilium as a gravitational force transducer and a regulator of transcriptional noise. *Developmental Dynamics, 237*(8), 1955–1959.
19. Ingber, D. E., Wang, N., & Stamenovic, D. (2014). Tensegrity, cellular biophysics, and the mechanics of living systems. *Reports on Progress in Physics, 77*(4), 046603.
20. Paul, A. L., Wheeler, R. M., Levine, H. G., & Ferl, R. J. (2013). Fundamental plant biology enabled by the space shuttle. *American Journal of Botany, 100*(1), 226–234.
21. Ingber, D. (1999). How cells (might) sense microgravity. *The FASEB Journal, 13*(Suppl), S3–15.
22. Albrecht-Buehler, G. (1991). Possible mechanisms of indirect gravity sensing by cells. *ASGSB Bulletin, 4*(2), 25–34.
23. Iskratsch, T., Wolfenson, H., & Sheetz, M. P. (2014). Appreciating force and shape-the rise of mechanotransduction in cell biology. *Nature Reviews Molecular Cell Biology, 15*(12), 825–833.
24. Orr, A. W., Helmke, B. P., Blackman, B. R., & Schwartz, M. A. (2006). Mechanisms of mechanotransduction. *Developmental Cell, 10*(1), 11–20.
25. Vogel, V., & Sheetz, M. P. (2009). Cell fate regulation by coupling mechanical cycles to biochemical signaling pathways. *Current Opinion in Cell Biology, 21*(1), 38–46.
26. Singla, V., & Reiter, J. F. (2006). The primary cilium as the cell's antenna: Signaling at a sensory organelle. *Science, 313*(5787), 629–633.
27. Satir, P., Pedersen, L. B., & Christensen, S. T. (2010). The primary cilium at a glance. *Journal of Cell Science, 123*(Pt 4), 499–503.
28. Bhat, R., & Bissell, M. J. (2014). Of plasticity and specificity: Dialectics of the microenvironment and macroenvironment and the organ phenotype. *Wiley Interdisciplinary Reviews: Developmental Biology, 3*(2), 147–163.
29. Polacheck, W. J., German, A. E., Mammoto, A., Ingber, D. E., & Kamm, R. D. (2014). Mechanotransduction of fluid stresses governs 3D cell migration. *Proceedings of the National Academy of Sciences of the United States of America, 111*(7), 2447–2452.
30. Bissell, M. J., & Barcellos-Hoff, M. H. (1987). The influence of extracellular matrix on gene expression: Is structure the message? *Journal of Cell Science Supplement, 8*, 327–343.
31. Nelson, C. M., & Bissell, M. J. (2005). Modeling dynamic reciprocity: Engineering three-dimensional culture models of breast architecture, function, and neoplastic transformation. *Seminars in Cancer Biology, 15*(5), 342–352.
32. Roskelley, C. D., & Bissell, M. J. (1995). Dynamic reciprocity revisited: A continuous, bidirectional flow of information between cells and the extracellular matrix regulates mammary epithelial cell function. *Biochemical Cell Biology, 73*(7–8), 391–397.
33. Xu, R., Boudreau, A., & Bissell, M. J. (2009). Tissue architecture and function: Dynamic reciprocity via extra- and intra-cellular matrices. *Cancer and Metastasis Reviews, 28*(1–2), 167–176.
34. Wang, N., Butler, J. P., & Ingber, D. E. (1993). Mechanotransduction across the cell surface and through the cytoskeleton. *Science, 260*(5111), 1124–1127.
35. Ciobanasu, C., Faivre, B., & Le Clainche, C. (2013). Integrating actin dynamics, mechanotransduction and integrin activation: The multiple functions of actin binding proteins in focal adhesions. *European Journal of Cell Biology, 92*(10–11), 339–348.

36. Vogel, V., & Sheetz, M. (2006). Local force and geometry sensing regulate cell functions. *Nature Reviews Molecular Cell Biology, 7*(4), 265–275.
37. Humphrey, J. D., Dufresne, E. R., & Schwartz, M. A. (2014). Mechanotransduction and extracellular matrix homeostasis. *Nature Reviews Molecular Cell Biology, 15*(12), 802–812.
38. Olesen, S. P., Clapham, D. E., & Davies, P. F. (1988). Haemodynamic shear stress activates a K+ current in vascular endothelial cells. *Nature, 331*(6152), 168–170.
39. Engler, A. J., Sen, S., Sweeney, H. L., & Discher, D. E. (2006). Matrix elasticity directs stem cell lineage specification. *Cell, 126*(4), 677–689.
40. Reed, A., Kohl, P., & Peyronnet, R. (2014). Molecular candidates for cardiac stretch-activated ion channels. *Global Cardiology Science & Practice, 2014*(2), 9–25.
41. Ingber, D. E. (1997). Integrins, tensegrity, and mechanotransduction. *Gravitational and Space Biology Bulletin, 10*(2), 49–55.
42. Ingber, D. (1998). In search of cellular control: Signal transduction in context. *Journal of Cellular Biochemistry Supplement, 30–31*, 232–237.
43. Zahir, N., & Weaver, V. M. (2004). Death in the third dimension: Apoptosis regulation and tissue architecture. *Current Opinion in Genetics & Development, 14*(1), 71–80.
44. Jessup, J. M., Frantz, M., Sonmez-Alpan, E., Locker, J., Skena, K., Waller, H., et al. (2000). Microgravity culture reduces apoptosis and increases the differentiation of a human colorectal carcinoma cell line. *In Vitro Cellular & Developmental Biology Animal, 36*(6), 367–373.
45. Schatten, H., Lewis, M. L., & Chakrabarti, A. (2001). Spaceflight and clinorotation cause cytoskeleton and mitochondria changes and increases in apoptosis in cultured cells. *Acta Astronautica, 49*(3–10), 399–418.
46. Gaboyard, S., Blanchard, M. P., Travo, C., Viso, M., Sans, A., & Lehouelleur, J. (2002). Weightlessness affects cytoskeleton of rat utricular hair cells during maturation in vitro. *Neuroreport, 13*(16), 2139–2142.
47. Sytkowski, A. J., & Davis, K. L. (2001). Erythroid cell growth and differentiation in vitro in the simulated microgravity environment of the NASA rotating wall vessel bioreactor. *In Vitro Cellular & Developmental Biology Animal, 37*(2), 79–83.
48. Ingber, D. E. (2003). Mechanobiology and diseases of mechanotransduction. *Annals of Medicine, 35*(8), 564–577.
49. Hashemi, B. B., Penkala, J. E., Vens, C., Huls, H., Cubbaqge, M., & Sams, C. F. (1999). T cell activation responses are differentially regulated during clinorotation and in spaceflight. *The FASEB Journal, 13*(14), 2071–2082.
50. Pellis, N. R. (1998, November/December). *Microgravity: A medical research and treatment tool* (Vol. 6, Issue 6, pp. 3–6). NASA Office of Aero-Space Technology.
51. Cooper, D., & Pellis, N. R. (1998). Suppressed PHA activation of T lymphocytes in simulated microgravity is restored by direct activation of protein kinase C. *Journal of Leukocyte Biology, 63*(5), 550–562.
52. Sundaresan, A., Risin, D., & Pellis, N. R. (2002). Loss of signal transduction and inhibition of lymphocyte locomotion in a ground-based model of microgravity. *In Vitro Cellular & Developmental Biology Animal, 38*(2), 118–122.
53. Uva, B. M., Masini, M. A., Sturla, M., Prato, P., Passalacqua, M., Giuliani, M., et al. (2002). Clinorotation-induced weightlessness influences the cytoskeleton of glial cells in culture. *Brain Research, 934*(2), 132–139.
54. Uva, B. M., Masini, M. A., Sturla, M., Tagliafierro, G., & Strollo, F. (2002). Microgravity-induced programmed cell death in astrocytes. *Journal of Gravitational Physiology, 9*(1), 275–276.
55. Moorman, S. J., Shimada, N., Sokunbi, G., & Pfirrm, C. (2007). Simulated-microgravity induced changes in gene expression in zebrafish embryos suggest that the primary cilium is involved in gravity transduction. *Gravitational and Space Biology Bulletin, 20*(2), 79–86.
56. Altshuler, B. (1981). Modeling of dose–response relationships. *Environmental Health Perspectives, 42*, 23–27.

57. DeLean, A., Munson, P. J., & Rodbard, D. (1978). Simultaneous analysis of families of sigmoidal curves: Application to bioassay, radioligand assay, and physiological dose–response curves. *The American Journal of Physiology, 235*(2), E97–102.
58. Gargaud, M. (2011). Encyclopedia of Astrobiology. In M. Gargaud, R. Amils, J. Cernicharo Quintanilla, H. J. Cleaves, W. M. Irvine, D. Pinti, & M. Viso (Eds.), *Encyclopedia of astrobiology*. New York, NY: Springer.
59. Dickson, K. J. (1991). Summary of biological spaceflight experiments with cells. *ASGSB Bulletin, 4*(2), 151–260.
60. Johnston, R. S., & Dietlein, L. F. (1977). *Biomedical results from Skylab* (Vol. SP-377). Washington, DC: NASA Government Printing Office.
61. Pellis, N. R., Goodwin, T. J., Risin, D., McIntyre, B. W., Pizzini, R. P., Cooper, D., et al. (1997). Changes in gravity inhibit lymphocyte locomotion through type I collagen. *In Vitro Cellular & Developmental Biology Animal, 33*(5), 398–405.
62. Cogoli, A. (1997). Signal transduction in T lymphocytes in microgravity. *Gravitational and Space Biology Bulletin, 10*(2), 5–16.
63. Cogoli, A., & Cogoli-Greuter, M. (1997). Activation and proliferation of lymphocytes and other mammalian cells in microgravity. *Advances in Space Biology and Medicine, 6*, 33–79.
64. Manchester, J. K., Chi, M. M., Norris, B., Ferrier, B., Krasnov, I., Nemeth, P. M., et al. (1990). Effect of microgravity on metabolic enzymes of individual muscle fibers. *The FASEB Journal, 4*(1), 55–63.
65. Gao, H., Liu, Z., & Zhang, L. (2011). Secondary metabolism in simulated microgravity and space flight. *Protein & Cell, 2*(11), 858–861.
66. Huang, B., Liu, N., Rong, X., Ruan, J., & Huang, Y. (2015). Effects of simulated microgravity and spaceflight on morphological differentiation and secondary metabolism of *Streptomyces* coelicolor A3(2). *Applied Microbiology and Biotechnology, 99*(10), 4409–4422.
67. Unsworth, B. R., & Lelkes, P. I. (1998). Growing tissues in microgravity. *Nature Medicine, 4*(8), 901–907.
68. Freed, L. E., Langer, R., Martin, I., Pellis, N. R., & Vunjak-Novakovic, G. (1997). Tissue engineering of cartilage in space. *Proceedings of the National Academy of Sciences of the United States of America, 94*(25), 13885–13890.
69. Sultan, C., Stamenovic, D., & Ingber, D. E. (2004). A computational tensegrity model predicts dynamic rheological behaviors in living cells. *Annals of Biomedical Engineering, 32*(4), 520–530.
70. Tixador, R., Richoilley, G., Gasset, G., Templier, J., Bes, J. C., Moatti, N., et al. (1985). Study of minimal inhibitory concentration of antibiotics on bacteria cultivated in vitro in space (Cytos 2 experiment). *Aviation, Space, and Environmental Medicine, 56*(8), 748–751.
71. Kacena, M. A., & Todd, P. (1999). Gentamicin: Effect on *E. coli* in space. *Microgravity Science and Technology, 12*(3–4), 135–137.
72. Malicki, J., & Avidor-Reiss, T. (2014). From the cytoplasm into the cilium: Bon voyage. *Organogenesis, 10*(1), 138–157.
73. Berndt, J. D. (2013). The cilium gets excited. *Science Signaling, 6*(306), ec312.
74. Louvi, A., & Grove, E. A. (2011). Cilia in the CNS: The quiet organelle claims center stage. *Neuron, 69*(6), 1046–1060.
75. Resnick, A., & Hopfer, U. (2007). Force-response considerations in ciliary mechanosensation. *Biophysical Journal, 93*(4), 1380–1390.
76. Navran, S. (2008). The application of low shear modeled microgravity to 3-D cell biology and tissue engineering. *Biotechnology Annual Review, 14*, 275–296.
77. Becker, J. L., & Blanchard, D. K. (2007). Characterization of primary breast carcinomas grown in three-dimensional cultures. *The Journal of Surgical Research, 142*(2), 256–262.
78. Zhau, H. E., Goodwin, T. J., Chang, S. M., Baker, T. L., & Chung, L. W. (1997). Establishment of a three-dimensional human prostate organoid coculture under microgravity-simulated conditions: Evaluation of androgen-induced growth and PSA expression. *In Vitro Cellular & Developmental Biology Animal, 33*(5), 375–380.

79. Cortiella, J., Niles, J., Cantu, A., Brettler, A., Pham, A., Vargas, G., et al. (2010). Influence of acellular natural lung matrix on murine embryonic stem cell differentiation and tissue formation. *Tissue Engineering Part A, 16*(8), 2565–2580.
80. Crabbe, A., Liu, Y., Sarker, S. F., Bonenfant, N. R., Barrila, J., Borg, Z. D., et al. (2015). Recellularization of decellularized lung scaffolds is enhanced by dynamic suspension culture. *PloS One, 10*(5), e0126846.
81. Kida, N., Mochizuki, Y., & Taguchi, F. (2007). Sporicidal activity of the KMT reagent in its vapor phase against *Geobacillus stearothermophilus* spores. *Microbiology and Immunology, 51*(1), 99–103.
82. Sundaresan, A., & Pellis, N. R. (2009). Cellular and genetic adaptation in low-gravity environments. *The Annals of the New York Academy of Sciences, 1161*, 135–146.
83. Sundaresan, A. (2011). A possible cardiovascular predictor of susceptibility to microgravity. *International Journal of Transport Phenomena, 12*(1–2), 93–100.
84. Uhran, M. (2012, June 26–29). Microgravity related patent history. In *First Annual International Space Station (ISS) Research and Development Conference*. Denver, CO.
85. Grosse, J., Wehland, M., Pietsch, J., Schulz, H., Saar, K., Hubner, N., et al. (2012). Gravity-sensitive signaling drives 3-dimensional formation of multicellular thyroid cancer spheroids. *The FASEB Journal, 26*(12), 5124–5140.
86. Cameron, D. F., Hushen, J. J., Nazian, S. J., Willing, A., Saporta, S., & Sanberg, P. R. (2001). Formation of Sertoli cell-enriched tissue constructs utilizing simulated microgravity technology. *The Annals of the New York Academy of Sciences, 944*, 420–428.
87. Qiu, Q. Q., Ducheyne, P., & Ayyaswamy, P. S. (2001). 3D bone tissue engineered with bioactive microspheres in simulated microgravity. *In Vitro Cellular & Developmental Biology Animal, 37*(3), 157–165.
88. Hanke, W., Fernandes de Lima, V. M., Wiedemann, M., & Meissner, K. (2006). Microgravity dependence of excitable biological and physicochemical media. *Protoplasma, 229*(2–4), 235–242.
89. Doolin, E. J., Geldziler, B., Strande, L., Kain, M., & Hewitt, C. (1999). Effects of microgravity on growing cultured skin constructs. *Tissue Engineering, 5*(6), 573–582.
90. Blasi, P., Luca, G., Mancuso, F., Schoubben, A., Calvitti, M., Giovagnoli, S., et al. (2013). Conformal polymer coatings for pancreatic islets transplantation. *International Journal of Pharmaceutics, 440*(2), 141–147.
91. Holland, P. M., Abramson, R. D., Watson, R., & Gelfand, D. H. (1991). Detection of specific polymerase chain reaction product by utilizing the 5′–3′ exonuclease activity of thermus aquaticus DNA polymerase. *Proceedings of the National Academy of Sciences of the United States of America, 88*(16), 7276–7280.
92. Brock, T. D. (1967). Life at high temperatures. Evolutionary, ecological, and biochemical significance of organisms living in hot springs is discussed. *Science, 158*(3804), 1012–1019.
93. Saiki, R. K., Gelfand, D. H., Stoffel, S., Scharf, S. J., Higuchi, R., Horn, G. T., et al. (1988). Primer-directed enzymatic amplification of DNA with a thermostable DNA polymerase. *Science, 239*(4839), 487–491.
94. Vastag, B. (2001). Cell biology update: A decade of simulating space on earth. *Journal of the American Medical Association, 285*(17), 2181–2182.
95. Nakamura, K., Kuga, H., Morisaki, T., Baba, E., Sato, N., Mizumoto, K., et al. (2002). Simulated microgravity culture system for a 3-D carcinoma tissue model. *Biotechniques, 33*(5), 1068–1070, 1072, 1074–1076.
96. Margolis, L. B., Fitzgerald, W., Glushakova, S., Hatfill, S., Amichay, N., Baibakov, B., et al. (1997). Lymphocyte trafficking and HIV infection of human lymphoid tissue in a rotating wall vessel bioreactor. *AIDS Research and Human Retroviruses, 13*(16), 1411–1420.
97. Hatton, J. P., Gaubert, F., Cazenave, J. P., & Schmitt, D. (2002). Microgravity modifies protein kinase C isoform translocation in the human monocytic cell line U937 and human peripheral blood T-cells. *Journal of Cellular Biochemistry, 87*(1), 39–50.

Chapter 2
Principles of Analogue and True Microgravity Bioreactors to Tissue Engineering

David A. Wolf and Stanley J. Kleis

2.1 Introduction

2.1.1 The Early Days of Cell Culture in Microgravity

NASA, in the 1980s, embarked upon a program to enable pilot scale cell culture under microgravity conditions utilizing the Space Shuttle as a laboratory. The initial program goal was to enable culture of mammalian cells as a space-based bioprocess. It was ultimately discovered that both ground-based analogue and true microgravity offered powerful advantages for the in vitro reproduction of natural tissue cytoarchitecture. Initial work on ground-based analogues centered on development of continuous perfusion bioreactors and prototype microgravity compatible culture vessels with microcarrier beads as the cell attachment substrate. Fluid dynamic studies addressed optimizing internal vessel designs to simultaneously achieve, on Earth, low hydrodynamic stress, 3-dimensional spatial freedom, and co-location of particles with differing sedimentation rates while sustaining optimal mass transport and compatibility with microgravity operations [1, 2, 18]. It is these three simultaneous conditions that achieve the simulated microgravity conditions when operated on Earth in unit gravity.

D.A. Wolf (✉)
School of Aeronautics and Astronautics, Purdue University, West Lafayette, IN, USA

The Children's Museum of Indianapolis, Indianapolis, IN, USA

EarthTomorrow, Inc., Houston, TX, USA

SpektronSystems, Jonesborrow, AR, USA
e-mail: davidwolf@earthtomorrow.net; http://www.earthtomorrow.net

S.J. Kleis
Department of Mechanical Engineering, University of Houston, Houston, TX, USA
e-mail: kleis@central.uh.edu

C.A. Nickerson et al. (eds.), *Effect of Spaceflight and Spaceflight Analogue Culture on Human and Microbial Cells*, DOI 10.1007/978-1-4939-3277-1_2

Initial designs led to a horizontal orientation of a cylindrical zero headspace culture vessel with an internal vane to achieve suspension of culture materials. This early design, while compatible with microgravity operations and enabling suspension of the culture, introduced damaging hydrodynamic forces similar to conventional vessels also utilizing internal lift mechanisms. Laser doppler measurements by Stan Kleis' team at the University of Houston verified the expected source of the damaging (to the cultured cells) hydrodynamic forces to be from high velocity gradients, and associated shear stress, through the boundary layer associated with fluid moving inside the vessel relative to the internal vessel walls.

Ray Schwarz, Tihn Trihn, and David Wolf, of the NASA design team, recognized that wall rotation alone, without any internal lift mechanism, such as a vane, was sufficient to achieve suspension. The internal vane was removed and the rotating fluid itself served as the lift mechanism. This wall rotation effectively negated the boundary layer velocity gradient near the vessel walls resulting in greatly reduced fluid shear stress in this region. These Rotating Wall Vessels (RWV's) were designed for continuous perfusion with controlled gas partial pressures and dissolved chemical species concentrations and also a simplified batch fed version was designed for use in an incubator.

The operational rotation rate typically was 10–30 rotations per minute (rpm) to avoid excessive sedimentation, minimize centrifugation, and achieved a homogenous distribution of culture materials including attachment substrates and cells. This new configuration, in 1986, demonstrated a dramatic improvement in cell viability for the cultured baby hamster kidney cell line (BHK-21) and primary human embryonic kidney cells (HEK). The very low hydrodynamic shear stress near the walls of the rotating zero headspace vessels produced an excellent microgravity analogue enabling projection of expected results in actual microgravity as well as enabling many of the benefits of space based microgravity culture on Earth.

Experience with microcarrier-based suspension mammalian cultures, using the rotational "low shear modelled microgravity" or as it is sometimes called "simulated microgravity" technique, demonstrated the formation of large viable three-dimensional (3-D) tissue-like aggregates bridging multiple beads. We recognized that the quiescent conditions within the RWV produced a uniformly distributed suspension of 3-D tissue-like aggregates similar to the anticipated behavior in microgravity. Thus, 3-D tissue modeling in analogue microgravity established an early form of 3-dimensional "Tissue Engineering."

2.1.2 *Benefits of Ground-Based Analogue and True Spaceflight Bioreactor Culture*

Earth-based analogue and actual Space-based microgravity Bioreactors, both based on the RWV technology, offer unique physical conditions that enable high fidelity in vitro culture of 3-D organotypic tissue. RWV technology is applicable over a wide range of normal and transformed mammalian cell types (including cell lines and primary cells) to establish organotypic/organoid models that reproduce key aspects of the 3-D cytoarchitecture and function found in the parental tissue. Accurate models of tissues, which behave closely to the corresponding in vivo tissue, enable basic biological research to address understanding human disease mechanisms and treatments. As discussed in Chaps. 4–9, the application of these in vivo-like 3-D tissue aggregates to study the transition between normal cellular homeostasis and disease has unveiled novel mechanisms not possible using conventional flat 2-D monolayers grown on plastic.

Typically, delicate anchorage dependent mammalian cells are seeded onto suspended microcarrier beads or other substrates in a media within a bioreactor vessel volume. Rotating bioreactors provide the physical, biochemical, and oxygen requirements to support 3-D volumetric culture [3–6]. Common to these are (1) a means to support a delicate culture against the force of gravity in 3 dimensions, and (2) minimize the introduction of undesired hydrodynamic forces, and (3) allow co-spatial location over extended periods of time of culture components with differing sedimentation rates. Conventional monolayers grown on flat flasks and plates inherently drive tissue growth to a surface not supportive of 3-D tissue growth. In contrast, spinner flask systems incorporate internal stirring mechanisms that introduce disruptive hydrodynamic forces. Hollow fiber systems introduce limitations on the free mobility, homogeneity, and sampling of the culture. Fluidized beds act to separate culture assembles by sedimentation properties and introduce fluid velocity gradients, shear stress, and instability. Gels and encapsulation techniques limit diffusion and mobility of culture constituents.

RWV bioreactors are fundamentally compatible with microgravity in that they are completely fluid filled, without gas headspace. When operated under gravity conditions they are rotated at an appropriate rate to achieve suspension of the culture constituents by viscous fluid coupling without internal mechanical lift mechanisms or associated fluid velocity gradients.

True microgravity bioreactor culture allows for optimal realization of the 3 simultaneous physical conditions in a free aqueous environment: (1) three-dimensional freedom for cell and substrate interaction, (2) low fluid mechanical stress, and (3) co-localization of particles with differing sedimentation rates. However, due to the absence of gravity driven culture component drift through the media (in actual microgravity) one must be careful to actively introduce a method to facilitate adequate and homogeneous mass transfer by induced fluid motion. This avoids a diffusion limited culture condition. The quiescent environment enabled by the 3 simultaneous conditions permit organized self-directed tissue assembly with minimal disruption or influence from the culture system per se. Counter to these extreme quiescent conditions is the necessity to introduce sufficient perfusion and mixing to meet the culture metabolic demand, support a "healthy" cell and tissue micro-environment, provide adequate mass transport and sustain a reasonably homogeneous overall vessel macro-environment minimizing deleterious dead spaces or shunts from the vessel fluid inlets to outlets. Essentially, the absence of dead spaces and shunts may be verified by how close to optimal is the residence distribution curve as measured at the vessel inlet and outlet under the specific operating conditions and with the culture materials selected. In order to achieve these conditions, the Space Bioreactor baseline design incorporates independently rotatable inner and outer cylindrical walls enclosing a 125 ml culture volume in a concentric "couette" configuration. The inner cylinder is covered with a porous membrane providing the vessel media outlet. Differential rotation of the cylindrical walls sets up a forced convective flow pattern, particularly important in microgravity conditions, to effect mass transport of dissolved species including respiratory gases, nutrients, and waste products. Fixed to one end of the inner cylinder, adjacent to the media inlet, is a disk shaped "viscous pump," introduced by Dr. Kleis' laboratory, which introduces momentum asymmetry to incoming media to assure a single cell toroidal circulation pattern and reduce shunting of media directly from the inlet to the outlet. In the absence of this viscous pump the couette configuration results in a two cell flow pattern [10] with a rather astonishingly low crossover of dissolved species between cells and in a sense results in two separate volumes operating nearly independently. Numerical calculations over a range of operating conditions show that the fluid shear stress, when operating under actual microgravity conditions is quite low, typically on the order of 0.002 dynes/cm^2. This "residual" fluid shear stress, present even when operating under actual microgravity conditions, largely results from intentionally introduced forced perfusion flow and may be reduced to nearly arbitrarily low levels for significant periods of time by reducing the perfusion flow rate and "suffering" the resulting period of reduced mass transport. In unit gravity ($1 \times g$ on Earth), much of the fluid shear results from gravity driven continuous settling of tissue aggregates at very nearly terminal velocity through the fluid media and is much higher, on the order of 1 dynes/cm^2 for typical tissue aggregates [7, 8]. Ground-based RWV bioreactors utilize this gravity-induced sedimentation to effect mass transport often alleviating the need for additional forced convection by differential wall rotation or perfusion. This enables simplified batch fed reactors to operate, without differential wall rotation (as compared to a central annular cylinder – typically a gas exchange membrane) in an incubator environment. In contrast, the continuously perfused Space Bioreactor version does not require daily media change and can be perfused by a reservoir that is replenished at intervals, often weekly, and therefore is very advantageous during spaceflight operations. However, in this actual microgravity condition, means to actively induce mass transport by forced fluid flow are necessary to obtain a homogenous and well controlled microenvironment.

Accordingly, modeled (or simulated) microgravity in the ground-based RWV bioreactors is achieved in a calculable, deterministic, and defined fashion which enables methods for testable hypothesis driven experiments designed to causally link the output biological results to the input fluid dynamic conditions which are coupled to gravity in a defined way [1, 2, 5, 9–13].

True microgravity and analogue (Earth based RWV-generated) organotypic tissue constructs demonstrate the principles of gravitationally controlled culture techniques, and include 3-D organotypic models of human liver, intestine, lung, neuronal, cardiac muscle, prostate, cartilage, and lymphoid tissue, among numerous other organ sources [5, 14–16]. These 3-D tissue constructs have been applied in interdisciplinary studies over the past several decades to advance our understanding of how cells behave normally or transition to disease (see Chaps. 4–9). The ability to culture primary and transformed human cells in 3-D reproduces key aspects of the morphology and function of the corresponding in vivo tissue and is a significant advancement to tissue engineering [3, 5, 14–17]. Importantly, mechanical as well as biochemical conditions imposed upon the tissue microenvironment are key factors in determining functionality and cytoarchitecture. These microenvironmental conditions can be different for each particular cell/tissue system.

NASA's development of RWV bioreactors provides a simultaneous combination of three-dimensionality, gentle and physiological fluid dynamics, and spatial-co location of mixed cell populations (and substrates) in free suspension sustaining adequate aeration and mass transport, [18]. Interestingly, the RWV provides conditions that are consistent with high mass transport analogues. Still, comparisons of data obtained in these bioreactors across unmatched platforms (i.e., on Earth versus in space) may introduce uncontrolled effects that weaken attribution of the RWV results to gravity. The NASA Space Bioreactor is designed to minimize confounding variables by providing a high degree of biochemical and fluid dynamic control, to be compatible with both Earth based control and spaceflight operations on the International Space Station (ISS) National Laboratory, as well as provide high performance analogue microgravity culture conditions on Earth, i.e. one is able to utilize the same bioreactor instrumentation both on Earth and in Space while also achieving simulated microgravity for the Earth based controls. The Space Bioreactor, and its Earth-based simulated microgravity RWV versions, have demonstrated high performance with respect to the ability to propagate and support functional, cytoarchitecturally correct, 3-D mammalian tissue constructs derived from a variety of tissue types [1, 3–6, 15, 17, 19] that have demonstrated their ability to serve as predictive human surrogate platforms (see Chaps. 4–9).

2.2 Operation and Examples of Ground-Based RWV Cultures and Comparison to Actual Spaceflight Cultures

To facilitate cell growth, microcarrier or other biocompatible anchorage substrates/scaffolds are added prior to cell culture. The vessel is then seeded with one or more cell types at the appropriate density, which then rapidly attach, spread, and multiply on the microcarrier or substrate. (Note: suspension cell culture in the RWV can also be performed without scaffolding, depending on the experimental objectives). This laboratory initially utilized baby hamster kidney (BHK21) cells and Cytodex-3 microcarrier beads as the anchorage dependent substrate to demonstrate the concept and feasibility of modeled (or equivalently stated, simulated) microgravity tissue culture in the RWV bioreactor. Initial observations revealed a high order of cellular assembly into a viable tissue-like cytoarchitecture [18] characterized by ordered and regular intercellular membrane close associations. Rapid growth, high metabolic activity, and uniform macroscopic tissue-like formations

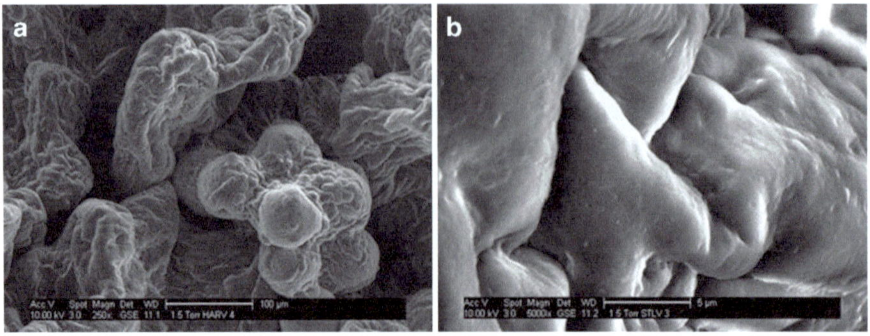

Fig. 2.1 Baby Hamster Kidney cell line, BHK21, cultured in the RWV in modeled microgravity on Cytodex-3 microcarrier beads. Rapidly proliferating cells on confluent 3-D cell-bead aggregates show tissue-like bridging across beads ((**a**) 250×; (**b**) 5000×) demonstrating the ability to form high order 3-D tissue-like assemblies; photo courtesy of NASA

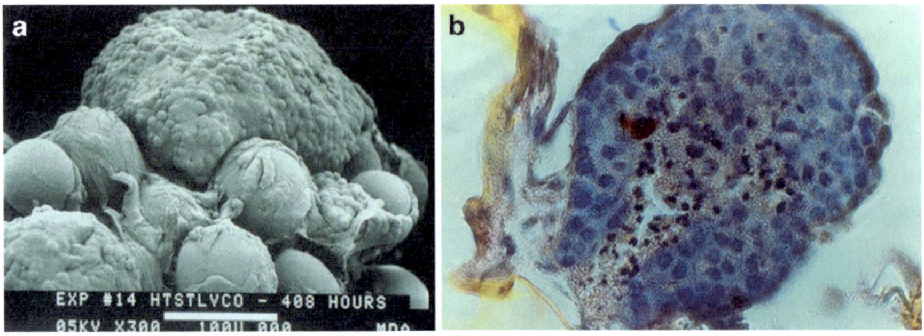

Fig. 2.2 Modeled microgravity co-culture of HT-29 colon carcinoma epithelial cell line with normal human colon fibroblasts. (**a**) SEM (200×), demonstrates assembly of macroscopic "polypoid" structures. The fibroblasts form a confluent feeder layer on Cytodex 3 microcarrier beads upon which the colon epithelial cells proliferate and differentiate. (**b**) Mucicarmine stain of a 10 μm section (200×) demonstrates fibroblast infiltration (*yellow color*) and sinusoid formation extending from the base feeder layer into the epithelial polypoid tumor [20]

were observed (Fig. 2.1a, b). Successful demonstration with BHK21 cells led to subsequent investigation of difficult to grow primary human embryonic kidney cells where the RWV growth conditions enabled rapid growth and high viability of 3-D tissue constructs (unpublished data).

This early work was extended to small and large intestine co-culture models, normal and neoplastic, respectively [20, 21]. Specifically, similar high order 3-D structure was achieved for colon carcinoma epithelial cell lines, including those derived from HT-29 cells both in monotypic culture (comprised of a single cell type, here HT-29 cells alone) and in co-culture (Fig. 2.2a, b) with normal human colon fibroblasts as a mesenchymal feeder layer [20, 21]. This study demonstrates the typical pattern of high order self-organizing 3-D cytoarchitecture obtained for different cell types cultured in the RWV, and has since been found by many investigators to also occur in numerous other neoplastic, transformed cell lines, primary cells, and normal cells derived from cardiac, liver, prostate, skin, cornea, bone, cartilage, nerve, intestinal, lung, placental, vaginal, bladder, and lymphoid tissues [14, 17, 22–44]. While this chapter provides a general overview of the basic principles and applications of the RWV bioreactor in early 3-D cell culture studies, Chaps. 4–9 describe in detail (1) the specific 3-D cell culture models advanced using this technology (including immunocompetent co-culture models), (2) their molecular genetic and phenotypic validation demonstrating in vivo-like characteristics of the parental tissue, and (3) their application to serve as predictive platforms to study the transition between normal tissue organogenesis and disease phenotypes. This includes high fidelity models of infectious disease and cancer, the latter of which includes the establishment of models of both malignancy and phenotypic reversion to investigate mechanisms of tumorigenesis and cancer dormancy, respectively. In addition, RWV-derived 3-D cell cultures have shown promise for future use in translational studies in regenerative medicine and pharmaceutical development.

Initial true microgravity spaceflight testing of mammalian cell cultures conducted on the Space Shuttle and the MIR Space Station are largely consistent with those projected by use of RWV culture on Earth, particularly with respect to achieving a large and well-organized 3-D architecture and molecular and phenotypic indicators of proliferation, differentiation, and function [45–48]. For example, cultured lymphocytes showed a similar set of responses to both modeled and actual microgravity, characterized by reduced locomotion, blunted chemotaxis, and reduced mitogen response [47, 49]. Interestingly, these characteristics are similar to the responses of lymphocytes drawn from astronauts during actual spaceflight that are hypothesized as a potential risk of spaceflight induced immune dysfunction [50]. Conversely, some findings from actual microgravity cultures are not expressed under modeled microgravity conditions in the RWV, such as glycosaminoglycan distribution of highly mechanoreceptive cartilage tissue [6, 19], reflecting the differences between RWV and actual spaceflight growth conditions and presumably attributed to the more aggressive (but still very mild) hydrodynamic conditions introduced by RWV modeled microgravity as compared to actual microgravity. This "imperfect" modeling is rather dramatic to the naked eye when one observes the far more quiescent conditions of actual microgravity culture of cells as compared to the RWV (operated on Earth), as larger tissue aggregates form and begin to increasingly respond to the gravity vector (as their size and sedimentation rate increases). This loss of microgravity simulation becomes progressively more severe as tissue aggregates increase in size in the RWV and their Stokes terminal velocity of sedimentation attains an increased magnitude [1, 5]. This increased magnitude further degrades the 3-D tissue aggregates by damaging effects, such as vessel wall impacts and non-uniform shear stresses resulting from directional entrainment of the growing tissue aggregates and direct surface shear effects as they are pulled through the rotating culture media by gravity at very near terminal velocity [12].

2.3 Ground Based and Spaceflight Bioreactor Design

In addition to deterministic gravity coupling, a defined, spatially homogeneous and time invariant mass transport condition in a cell culture vessel was determined to be beneficial in order to conduct meaningful experiments with results causally attributable to the unique "microgravity" fluid mechanics. These conditions drove the fundamental bioreactor design requirements to be a continuous flow perfused zero headspace cylindrical rotating wall reactor vessel, 250 ml in volume, composed of independently rotating concentric couette configured inner and outer walls designated as the "Rotating Wall Perfused Vessel" (RWPV). The perfused culture system provided for respiratory gas exchange, external to the culture vessel, by means of a high surface area gas exchanger in an external media flow loop and offered reduced temporal and spatial variability and improved control for dissolved gas species, as well as the opportunity for introducing the benefits of direct organoid perfusion [29]. A simplified batch fed version, for incubator use, enabled low overhead operation for developmental testing and use by conventional laboratories without extensive engineering staff. This batch system was designated the "Slow Turning Lateral Vessel" (STLV). This simplified instrument, and it's related "High Aspect Rotating Vessel" (HARV) which enables greater diffusional respiratory gas exchange, provides less defined and less controllable fluid mechanical properties but functions well for initial Earth-based studies while still providing basic modeled microgravity conditions. These incubator batch fed systems provide respiratory gas exchange with incubator atmosphere by means of a membrane within the vessels themselves. One approach for development and optimization of a spaceflight mammalian cell culture experiment is to begin with the ground-based STLV and HARV systems and then progress to cultures in the RWPV to optimize for flight (Fig. 2.3). Detailed methods for conducting such work may be referenced in Freed

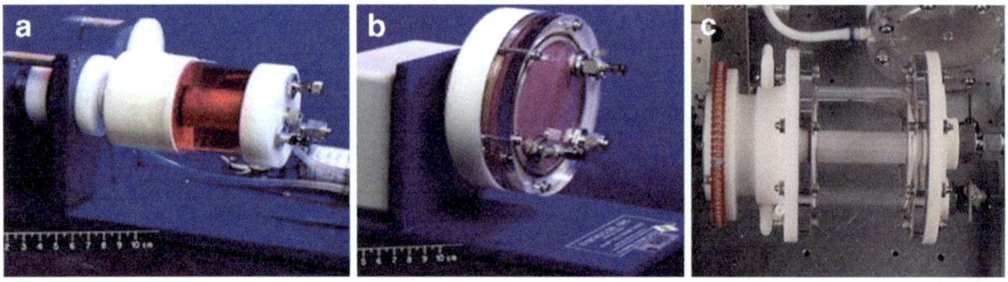

Fig. 2.3 The three primary RWV bioreactors which produce modeled microgravity conditions by zero head space rotation. The Slow Turning Lateral Vessel (STLV) (**a**) with a center core cylindrical gas exchange membrane, and the High Aspect Rotating Vessel (HARV) (**b**) with a high surface-to-volume ratio gas exchange membrane, are batch fed and used in a conventional incubator. The Rotating Wall Perfused Vessel (RWPV) (**c**) supports an external media perfusion loop with associated external oxygenator (not shown). Actual microgravity bioreactors incorporate the RWPV that allows independently adjustable rotation of the inner wall (the spin filter media exit) and outer walls in order to set up controlled fluid dynamic conditions including mass transport and fluid shear

Fig. 2.4 The Biospecimen Temperature Controller (BSTC) provides a CO_2 controlled environment to hold up to 32 Tissue Culture Modules (TCMs), of 25 ml volume each, for static mammalian tissue culture in space. Media may be manually exchanged and cell samples extracted for fixation with syringes from a support kit (not shown). High order tissue structure is demonstrated by the neuroendocrine pheochromocytoma derived PC12 cell line. This structure is estimated at 3 mm in length and was obtained by placing a drop of culture directly on a zero focus distance video camera lens to obtain macro-photography. These frond-like structures do not appear in Earth-based simulated microgravity or conventional culture. Photos courtesy of NASA

et al. [19], describing both ground-based controls and inflight cultures of bovine cartilage tissue on polyglycolic acid matrix during the MIR 3 mission. An excellent fluid dynamic numerical analysis of this experiment is provided by Lappa [12] in which the physical conditions are related to the tissue morphological development. In this case, several parameters, such as glycosaminoglycan content, were found superior in the Earth-based simulated microgravity controls, as compared to actual microgravity for this mechanosensitive tissue. While both culture in the RWPV and actual microgravity provided models that were superior to conventional culture techniques, this experiment highlighted the necessity to modulate the extreme quiescence of actual microgravity to the desired mass transfer and mechanical stress operating points to optimize for the particular tissue system. This greatly extended operating range enabled for actual microgravity tissue culture is an important use of the Space-based laboratory to conduct studies not possible on Earth.

A spaceflight certified companion static tissue culture device, the Biotechnology Specimen Temperature Controller (BSTC, Fig. 2.4) system, was also developed and operationally deployed. This instrument sustains a 5 % CO_2 gas mix in a 37° chamber around a maximum of 32 gas permeable Tissue Culture Modules (TCM's, Fig. 2.4) each of 25 ml volume. Although the fluid dynamics are not specifically controlled or adjustable, nor is modeled microgravity achieved for Earth based controls, these modules have the advantage over the Space Bioreactor of multiple replicates. The media may be exchanged manually via syringes and the cells may be similarly sampled for preservation. The neuroendocrine PC12 cell line derived from human pheochromocytoma cells (Fig. 2.4) was successfully cultured on the MIR Space Station during the NASA 6 mission demonstrating the capacity for delicate high order structure of cell cultures even under less than fully controlled conditions.

2.4 Fluid Dynamics of RWV in Unit Gravity (1 × *g* Ground Based) and Microgravity (Spaceflight)

Fluid dynamics is an important aspect of the mechanical interaction between the biological cell/tissue culture and its environment. When designing a vessel for the culture of anchorage dependent cells in microgravity, special care must be paid to avoid conditions leading to inadequate mass transport. The convective transport conditions nearly always present in Earth's gravity (1 × *g*) due to slight density differences between culture media and cells/microcarriers or cell aggregates are reduced by several orders of magnitude in microgravity. As such, it is important to partially replace this enhanced mass transport by forced or even natural convection present in 1 *g* by a controlled three-dimensional flow field in microgravity to provide adequate transport of scalar quantities such as dissolved oxygen, CO_2, nutrients, and cellular waste products.

The dominant feature in the presence of gravity is the distortion of the paths of suspended particles (including cells/aggregates) from the background circular fluid streamlines within an RWV, as these particles track the gravity vector (Fig. 2.5a). Analysis and experimental verification of this first order effect is provided by NASA Technical Papers [1, 2] that explored various rotational rates and particle terminal velocities under unit gravity and during low gravity aircraft parabolic flight. Cell/tissue aggregates and their associated substrates/scaffolds accelerate to terminal velocity very quickly, on the order of milliseconds. The RWV rotates through a negligible angle during the time required for these constituents to reach 98 % of terminal velocity. Essentially, the 3-D tissue aggregates are always settling through the media at terminal velocity (in a circular motion unless there is contact with structure) at the same speed as if they were settling linearly downwards in a non-

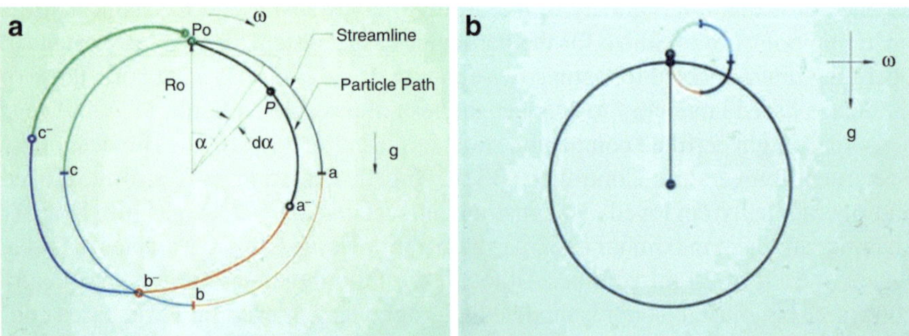

Fig. 2.5 Observation of a particle within the RWV bioreactor. (**a**) As viewed from the external (non-rotating) reference frame, one observes a rotating culture vessel with the path of particles suspended inside deviating from the background circular streamline. In single revolution, the particles do not quite return to their original location due to centrifugal and Coriolis pseudo forces. They slowly drift outwards and lag in phase, cumulatively over time. (**b**) The same physical situation as viewed from the rotating reference frame (i.e., spinning at the same rate as the RWV). In this case, one observes particles tracking the rotating gravity vector at very nearly terminal velocity [11, 13]

rotating system. Several investigators have verified this gravity-induced motion by various analytical, numerical, and empirical imaging techniques [11, 13]. This "path distortion" may be visualized more clearly as observed from the rotating reference frame of the RWV vessel (Fig. 2.5b). In this reference frame, a rotating gravity vector, g, is observed from the point of view of an apparently non-rotating vessel, i.e. as one would observe if one was inside the vessel rotating with it. The 3-D tissue constructs are observed to follow a circular path with a tangential speed very close to terminal velocity. The circle diameter increases with increasing terminal velocity and with decreasing rotation rate. Note that all the particles in the RWV move in phase with each other. In this regard, a smaller slower settling particle and a larger aggregate which settles more rapidly remain co-located in the same spatial region, co-orbiting for extended periods of time. This allows cells, substrates, and growing tissue aggregates to interact without encountering disrupting regions of high fluid shear stress. However, particles near the vessel wall and those that sediment rapidly will be exposed to wall impacts at speeds as high as their terminal sedimentation rates. Residual fluid shear stress in unit gravity is the result of particles settling through the culture media at terminal velocity. Under the true microgravity conditions of spaceflight, sedimentation is negligible and a much lower residual fluid shear stress is governed by the intentionally introduced background forced convective flow field as introduced to achieve necessary mass transport. Centrifugal and Coriolis pseudo-forces, apparent in the circularly accelerated reference frame, are much lower than forces resulting from gravity, but these do have a cumulative effect over time causing particles denser than the media to migrate outwards, or those less dense than media to migrate inwards [11]. In practice, as 3-D tissue aggregates progressively increase in size and density, the microgravity simulation is progressively lost, and the advantages of RWV cell culture are progressively diminished. However, this gravity-induced motion through the culture media does serve to increase mass transport and to reduce the size of the local region around a metabolizing tissue aggregate through which diffusion must occur. For example, mechanically sensitive tissue such as cartilage will respond to these forces [12, 19].

In Fig. 2.6, Lappa [12] shows flow and relative fluid shear calculations superimposed with cartilage tissue, approximately 1 cm long, that were originally grown by Martin et al. [51] under RWV conditions. Hypertrophic growth response of the tissue is in agreement with local regions of higher shear stress at the tissue surface. In Fig. 2.7, Nauman et al., demonstrated the local disturbance to circular flow and resulting fluid shear stress introduced by beads of differing sizes in bacterial cells cultured in the RWV [52]. The authors made use of this property to intentionally introduce known regions of fluid shear to measure bacterial molecular genetic and pathogenesis-related stress responses to that stimulus.

The gravitationally induced particle motions, as described above, contribute to mass transport of the scalar dissolved concentrations under unit gravity conditions. The motion may be considered to be convection. Modeling of the scalar transport may be accomplished for microgravity conditions. Mass transport in microgravity, as in 1 g, can be thought of as a diffusion process due to concentration gradients and a convection process due to relative motions between the fluid and particles.

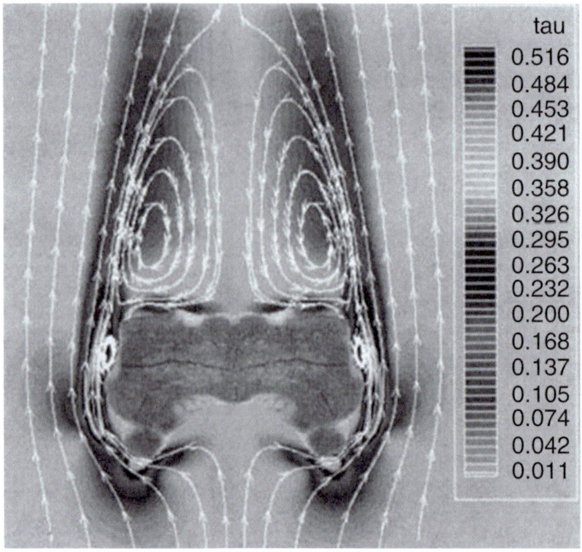

Fig. 2.6 Lappa [12] mathematically evaluated the expected tissue shape, fluid shear stress distribution, and flow velocity field and compared these calculations to the histologic cross section of cartilage grown for 60 days in RWV conditions (in unit gravity) by Martin et al. [51]

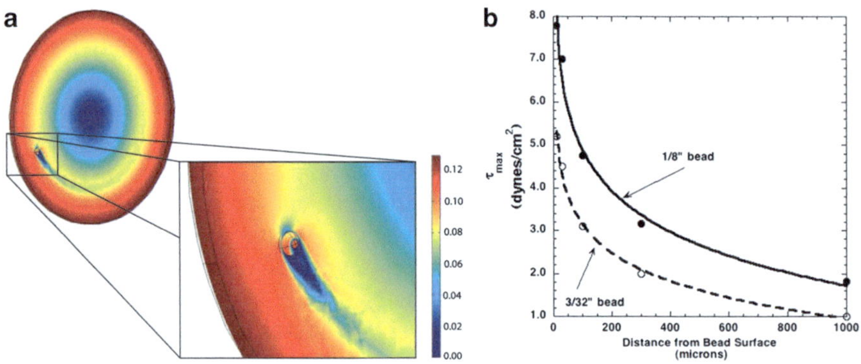

Fig. 2.7 Fluid velocity distribution and maximum shear stress in the RWV bioreactor. (**a**) Nauman et al. [52] model the distribution of fluid velocity in the RWV when a spherical bead is added to disrupt the velocity field, thus inducing elevated shear stress. (**b**) The induced fluid shear stress surrounding beads of two sizes (1/8 and 3/32 in. diameter) when added to the RWV. Copyright © American Society for Microbiology, [*Applied and Environmental Microbiology*, 73, 699–705, 2011]

However, in microgravity, convection must be intentionally introduced. Dependence on diffusion for mass transport quickly leads to limiting the culture to smaller quantities of cells and influences the cell microenvironment and health of the cultures. In microgravity, the relative motion to achieve sufficient mass transport is due only to the nature of the flow field, since the dominant effect of gravity is absent. Introduction of convective fluid flow also serves to reduce the size of the local viscous layer at the tissue surface through which diffusion must ultimately occur to effect transport at the cell surface.

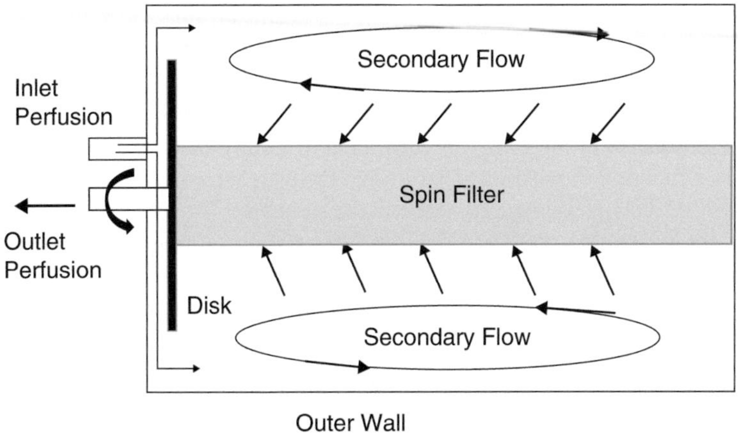

Fig. 2.8 Rotating Wall Perfused Vessel as used in true microgravity and on Earth in unit gravity. The disk on the left side of the spin filter acts as a "viscous pump" and promotes a single cell secondary flow. In true microgravity, the spin filter is typically operated from 5 to 10 rpm and the outer vessel wall at 1 rpm. Some outer wall rotation is required to stabilize the secondary flow cell. On Earth, to sustain tissue suspension, these rates are biased upwards by 10–20 rpm

In space, in the absence of a dominant gravitational acceleration and resulting particle tracking of the gravity vector, simple RWV bioreactors such as the High Aspect Ratio Vessel (HARV) would move the cells on microcarriers to the outside wall of the vessel. To avoid this situation, a three-dimensional flow field is required. A controllable three-dimensional flow field is obtained by differential rotation of concentric cylinders similar to the Taylor/Couette geometry. However, the symmetry of this geometry leads to two or more counter rotating fluid zones, with very little scalar transport between these "fluid cells." To avoid this situation, the RWPS was designed with a solid disk, termed a "viscous pump" attached to one end of the inner cylinder, Fig. 2.8. Specifically, the inner cylinder (spin filter) with disk (viscous pump) are rotated independently from the outer cylinder vessel wall to create the axial/radial secondary flow, which when added to the primary azimuthal flow, completes the required three-dimensional flow field. The spin filter is covered with a fine filter material to retain the cells on microcarriers within the RWPS, while the media is withdrawn for external conditioning before returning to the vessel at the perfusion inlet. Fresh media enters the vessel from behind the rotating disk near the disk tip and the vessel outer wall.

To achieve the desired large cell number per volume for the RWPS culture, the anchorage dependent cells are grown on porous microcarriers, approximately 170 μm in diameter and coated with extracellular matrix (ECM) proteins. As the cells grow, they spread over the microcarrier surface, forming a confluent unilayer. The individual microcarriers coated with cells then come in contact, allowing cell–cell-junctions to form under low fluid shear conditions, resulting in formation of 3-D tissue-like aggregates over time. Low fluid shear effectively stabilizes the delicate intermediate state required to achieve the more stable and larger 3-D tissue aggregates.

To model the transport of scalars from the perfusion inlet of the RWPS to the surface of the cells, the process was separated into multiple steps as shown schematically in Fig. 2.9. First, the scalar concentration field in the fluid media was computed by numerically solving the transport equations, subject to the assumption that the only effect of the cells was equivalent to a uniform consumption rate of the scalar as a function of position. That is, that the cells were uniformly distributed in space and all cells consumed the scalar at the same rate. The effects of the presence of the cells on the flow field and fluid properties were assumed negligible. Inflight video imaging shows this to be a reasonable assumption.

After the scalar concentration field was computed, the transport of the scalar from the fluid to the surface of the 3-D aggregates was needed. At the very low relative velocities typical of operating conditions in microgravity (low Reynolds number flow), there is a relatively large viscous region of fluid that moves with the cell aggregates (local domain). The scalar transport must traverse this region to reach the aggregate surface. Figure 2.10 shows the model for the solution of mass transport through the local domain. The boundary conditions for the local solution are obtained by solving for the trajectory of the cell aggregate. The relative velocity and concentration value of the scalar at the local domain outer boundary are taken to be the aggregate/fluid relative velocity and the scalar concentration at the aggregate location in the scalar concentration field, respectively. For details of the mass transport calculations in the RWPS geometry, see Rivera-Solorio [53], Kleis and Rivera-Solorio [7], and Rivera-Solorio and Kleis [8].

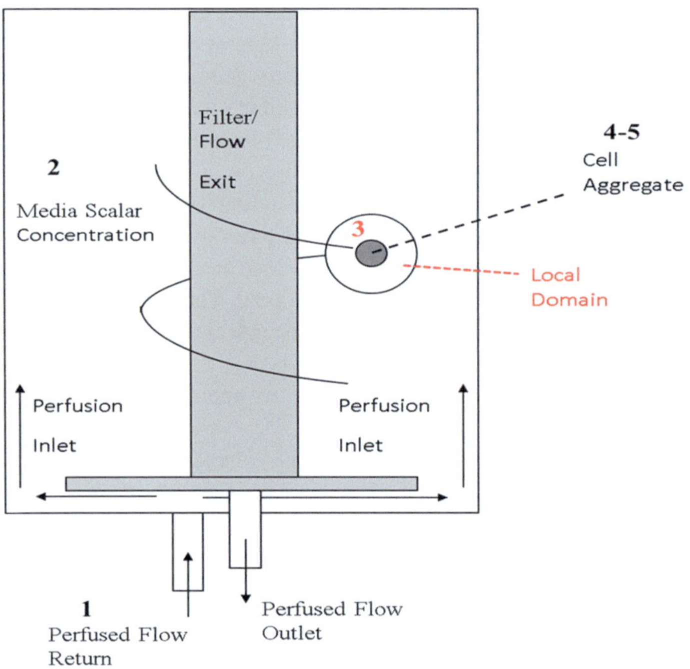

Fig. 2.9 Schematic of mass transport model. Scalar transport from inlet, *1*, through medium, *2*, to the moving local domain, *3*, to the aggregate surface, *4*, to the cells within the aggregate, *5*

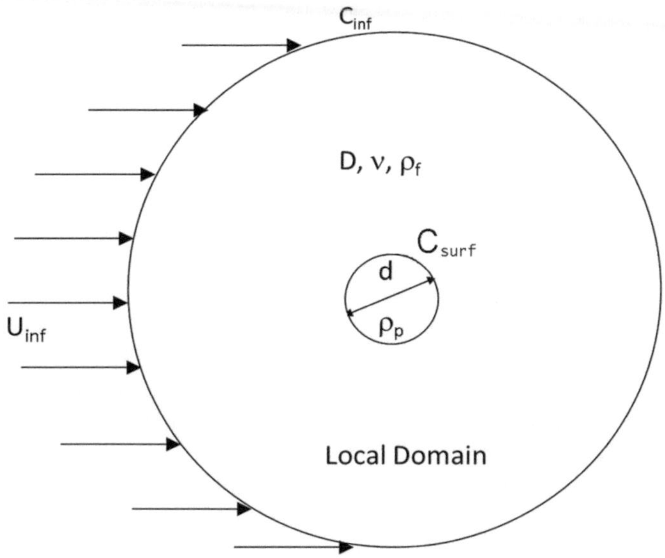

Fig. 2.10 Schematic for local domain scalar concentration solution. Domain size, D, with conditions at outer boundary, U_{inf}, medium viscosity, ν, and density, ρ_f. Inner surface of particle size, d, density, ρ_p, and surface concentration, C_{surf}

In calculation of the 3-D aggregate trajectory, when the acceleration of gravity is removed, the dominant forces acting on the aggregate are the Coriolis and centrifugal forces. Since these forces are much smaller than gravity acting on the density differences, the time constant for the response of the aggregate is much longer than the 1 *g* situation. This means that the implicit assumption of equilibrium conditions at all times generally made in analyses in 1 *g* may no longer be valid. Specifically, the assumption is often made that steady state drag coefficients based upon the instantaneous velocity difference can be used to compute force balances. When the time constant becomes comparable with the time for one axial orbit of the aggregate, this assumption is no longer valid. The drag and lift forces on the aggregate are then functions of the instantaneous relative velocity and the history of the relative velocity. Also, the large viscous domain moving with the aggregate at these low Reynolds numbers cause a significant increase in the mass being accelerated by the forces. This "added mass" must be accounted for when computing the 3-D aggregate trajectories.

2.5 Fluid Dynamics Results and Discussion

To demonstrate the importance of using a more complete model for the calculations of 3-D tissue aggregate trajectories and local mass transport, a few examples are given with operating conditions typical of cultures of anchorage dependent cells in the RWPS operating in microgravity. Because the basic flow field is axisymmetric, a stream function can be defined. The secondary flow is then shown by the intersection of the stream function constant surfaces as they intersect a radial/axial plane. Figure 2.11 shows the streamlines for three rotational conditions. The streamline pattern changes with rotation conditions show similar and controllable secondary flows as the inner vessel wall rotation rate is changed. Also, as expected, a strong axial flow is induced by adding a small amount of outer wall rotation. This is a well-known effect in rotating fluid systems often referred to as Taylor columns, after the pioneering work of G. I. Taylor [54]. This relatively strong axial flow is ideal for the RWPS to help distribute inlet fluid throughout the vessel.

Figure 2.12 shows what can happen when the outer vessel wall is held stationary. Three counter rotating fluid cells are formed along the axis. Exchange of scalar concentrations between fluid cells would be limited to diffusion. This situation should be avoided for cell culture in the RWPS. Once the basic flow pattern has been computed, the 3-D aggregate trajectories can be computed from the equations of motion. Figure 2.13 shows a typical single axial orbit and a surface of many axial orbits traversed by a single 3-D cell "aggregate," affected very little by centrifugal

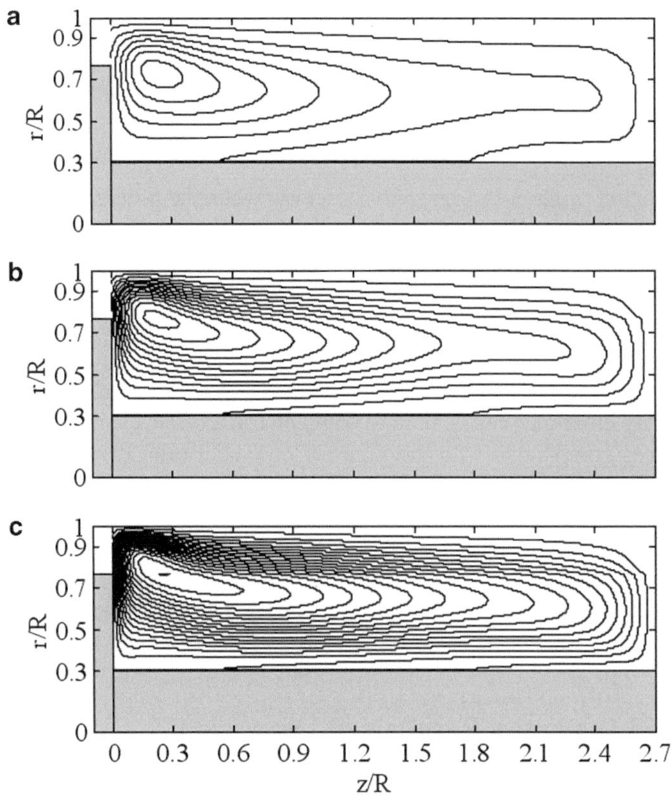

Fig. 2.11 Streamlines in radial/axial planes for 10 cc/min perfusion and outer vessel wall at 1 rpm. The inner wall is varied from (**a**) 5 rpm, (**b**) 10 rpm, and (**c**) 15 rpm. Mass transport is effectively increased by increasing the inner wall rotation rate

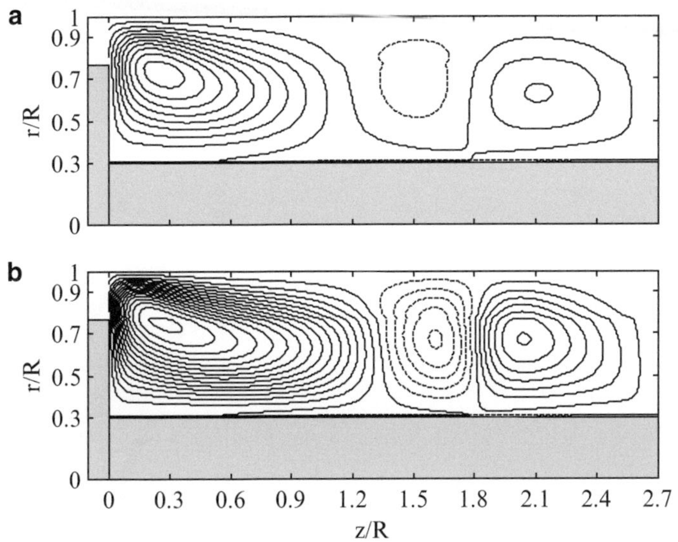

Fig. 2.12 Streamlines in the RWPS for zero outer wall rate and 10 cc/min perfusion. The inner wall is set to (**a**) 6 rpm and (**b**) 11 rpm. *Solid lines* represent clockwise cell rotation and *dashed lines*, counterclockwise. One observes unstable fluid flow breaking up into multiple cells in the absence of outer wall rotation

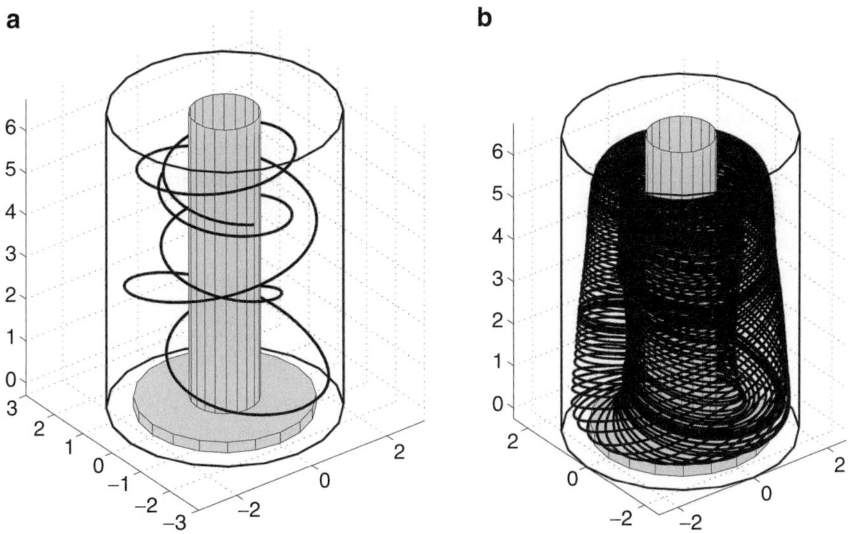

Fig. 2.13 Trajectories for (**a**) one axial orbit and (**b**) multiple axial orbits

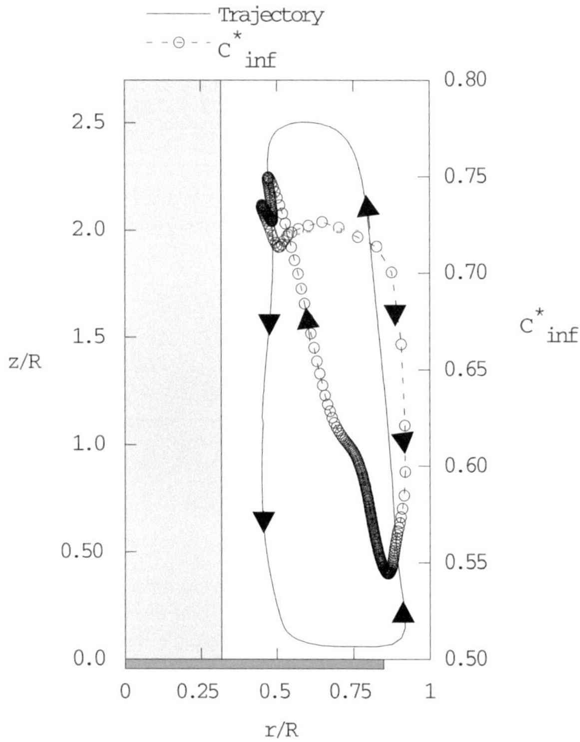

Fig. 2.14 Trajectory and concentration history for a 3 mm aggregate

effects. This allows one to visualize the underlying fluid flow pattern. The basic motion is for the 3-D aggregate to move axially toward the disk at a radial position closer to the axis, followed by (1) a rapid increase in radial position as the aggregate approaches the disk, (2) axial motion away from the disk at the larger radial position, and finally, (3) a decrease in radial position at the axial position farthest from the disk. The aggregate roughly follows the streamline in the radial/axial plane, deviating somewhat due to the centrifugal force.

Once the basic flow field and 3-D tissue aggregate trajectories are computed, the boundary conditions for the local domain mass transport problem are known. Figure 2.14 shows a plot of the aggregate trajectory and resulting normalized oxygen concentration distribution for a 3 mm diameter aggregate. Note that the trajectory and concentration curves are traversed in opposite directions. The concentration numbers are for dissolved oxygen, normalized by the saturated condition at the perfusion inlet. The values are those from the concentration distribution in the media at the center of the aggregate at each time point. This is assumed to be the outer boundary condition for the local domain, which follows the 3-D aggregate position in time.

The results of numerical fluid dynamics analysis highlight that care must be exercised when operating bioreactors in microgravity to ensure sufficient mass transport for the cell culture. Operating conditions must be selected that provide sufficient transport even at the expense of operating at the minimum fluid shear levels possible. A complete fluid dynamics analysis should be conducted for any bioreactor system to be operated in microgravity in order to properly select the desired conditions and interpret results in a causal manner. RWV culture on Earth, in unit gravity, serves as an analogue to true microgravity but is limited in operating range due to gravity driven sedimentation of the culture components. As microgravity analogues, they are useful for experiment development, selecting high value spaceflight research, anticipating results, and as controls for spaceflight tissue culture.

Gravity introduces fundamental limitations to the adjustable parameters as well as co-mingling critical variables such as fluid mixing required to suspend a 3-D culture versus that required to effect mass transport. Analogue microgravity culture achieves partial relief from such gravity-induced limitations, often achieving state of the art results in terms of establishing 3-D organotypic cell culture models. In particular, the simultaneous achievement of (1) 3-D spatial freedom, (2) low hydrodynamic stress, and (3) co-spatial localization of cells/tissues and their associated scaffolds with differing sedimentation properties is considered to be the primary mechanism of gravity coupling which enables enhanced formation of functional high fidelity 3-D tissue. Analytical and numerical methods, verified by experimental measurements, have been conducted to quantify these simultaneous conditions for true and analogue microgravity RWV culture. The high fidelity 3-D tissue models achieved are central to the emerging field of tissue engineering as it is applied to regenerative medicine, ex-vivo organ development, infectious disease, cancer, pharmaceutical development, and tissue biology basic research.

2.6 Summary

Understanding the effects of microgravity per se on cultured cells is greatly advanced by the use of fluid physics and computational fluid dynamics. It is essential that the design and operation of cell culture systems for application in microgravity and microgravity analogues are accompanied by thorough analysis of the physical environment in the context of gravity and fluid properties. This becomes critically important when assessing whether the cells are responding directly to microgravity as opposed to the indirect cell response to environmental (fluid) conditions created by microgravity. The development of the bioreactor systems described herein was an orderly process of developing and optimizing analytical and numerical models and validation with live cultures and inert measurements. The approach was to deterministically relate gravitational effects to the culture environment and resulting in-vitro tissues. Without this deliberate approach, it would be difficult to advance our understanding of low gravity and low fluid shear environments and their impact on cellular biology and to exploit the advantage of controlled gravitational tissue culture.

In addition to culture of mammalian cells under low fluid shear stress, the RWV bioreactor has also been used to culture microbial pathogens under conditions of physiological low fluid shear that are relevant to those encountered in the infected host during the natural course of infection. These studies have revealed novel insight into pathogenic mechanisms used by microbes during the infection process, which are not observed during conventional culture, and along with spaceflight studies, are the topic of Chaps. 10–14.

References

1. Wolf, D.A., & Schwarz, R.P. (1991). *Analysis of gravity-induced particle motion and fluid perfusion flow in the NASA-designed rotating zero-head-space tissue culture vessel.* Washington, DC: National Aviation and Space Administration, NASA/TP-3143.
2. Wolf, D.A., & Schwarz, R.P. (1992). *Experimental measurement of the orbital paths of particles sedimenting within a rotating viscous fluid as influenced by gravity.* Washington, DC: National Aviation and Space Administration, NASA/TP-3200.
3. Martin, I., Wendt, D., & Heberer, M. (2004). The role of bioreactors in tissue engineering. *Trends in Biotechnology, 22*(2), 80–86.
4. Lyons, E., & Pandit, A. (2005). Design of bioreactors for cardiovascular applications. In N. Ashammakhi & R. L. Reis (Eds.), *Tissue Engineering, 446*(4), 408–420.
5. Unsworth, B. R., & Lelkes, P. I. (1998). Growing tissues in microgravity. *Nature Medicine, 4*(8), 901–907.
6. Freed, L. E., & Vunjak-Novakovic, G. (1997). Microgravity tissue engineering. *In Vitro Cellular & Developmental Biology. Animal, 33*(5), 381–385.
7. Kleis, S. J., & Rivera-Solorio, I. (2003). Time scales for unsteady mass transfer from a sphere at low-finite Reynolds numbers. *Journal of Heat Transfer, 125*(4), 716–723.
8. Rivera-Solorio, I., & Kleis, S. J. (2006). Model of the mass transport to the surface of animal cells cultured in a rotating bioreactor operated in micro gravity. *Biotechnology and Bioengineering, 94*(3), 495–504.
9. Ju, Z. H., et al. (2006). Numerical simulation of microcarrier motion in a rotating wall vessel bioreactor. *Biomedical and Environmental Sciences, 19*(3), 163–168.
10. Tsao, Y.-M., et al. (1994). Fluid dynamics within a rotating bioreactor in space and Earth environments. *Journal of Spacecraft and Rockets, 31*(6), 937–943.
11. Pollack, S. R., et al. (2000). Numerical model and experimental validation of microcarrier motion in a rotating bioreactor. *Tissue Engineering, 6*(5), 519–530.
12. Lappa, M. (2003). Organic tissues in rotating bioreactors: Fluid-mechanical aspects, dynamic growth models, and morphological evolution. *Biotechnology and Bioengineering, 84*(5), 518–532.
13. Botta, G. P., et al. (2006). Real-time assessment of three-dimensional cell aggregation in rotating wall vessel bioreactors in vitro. *Nature Protocols, 1*(4), 2116–2127.
14. Barrila, J., et al. (2010). Organotypic 3D cell culture models: Using the rotating wall vessel to study host-pathogen interactions. *Nature Reviews Microbiology, 8*(11), 791–801.
15. Duray, P. H., Hatfill, S. J., & Pellis, N. R. (1997). Tissue culture in microgravity. *Science and Medicine, 4*(3), 46–55.
16. Vunjak-Novakovic, G., et al. (2002). Microgravity studies of cells and tissues. *Annals of the New York Academy of Sciences, 974*, 504–517.
17. Margolis, L. B., et al. (1997). Lymphocyte trafficking and HIV infection of human lymphoid tissue in a rotating wall vessel bioreactor. *AIDS Research and Human Retroviruses, 13*(16), 1411–1420.
18. Schwarz, R. P., Goodwin, T. J., & Wolf, D. A. (1992). Cell culture for three-dimensional modeling in rotating-wall vessels: An application of simulated microgravity. *Journal of Tissue Culture Methods, 14*(2), 51–57.

19. Freed, L. E., et al. (1997). Tissue engineering of cartilage in space. *Proceedings of the National Academy of Sciences of the United States of America, 94*(25), 13885–13890.
20. Goodwin, T. J., Jessup, J. M., & Wolf, D. A. (1992). Morphologic differentiation of colon carcinoma cell lines HT-29 and HT-29KM in rotating-wall vessels. *In Vitro Cellular & Developmental Biology, 28A*(1), 47–60.
21. Jessup, J. M., et al. (1997). Induction of carcinoembryonic antigen expression in a three-dimensional culture system. *In Vitro Cellular & Developmental Biology. Animal, 33*(5), 352–357.
22. Zhau, H. E., et al. (1997). Establishment of a three-dimensional human prostate organoid coculture under microgravity-simulated conditions: Evaluation of androgen-induced growth and PSA expression. *In Vitro Cellular & Developmental Biology. Animal, 33*(5), 375–380.
23. Wang, R., et al. (2005). Three-dimensional co-culture models to study prostate cancer growth, progression, and metastasis to bone. *Seminars in Cancer Biology, 15*(5), 353–364.
24. Lelkes, P. I., et al. (2001). Neuroendocrine tissue engineering in rotating wall vessel bioreactors under simulated microgravity conditions. In: *23rd Annual International Conference of the IEEE Engineering in Medicine and Biology Society*. Istanbul, Turkey: Engineering in Medicine and Biology Society.
25. Akins, R. E., et al. (1999). Cardiac organogenesis in vitro: Reestablishment of three-dimensional tissue architecture by dissociated neonatal rat ventricular cells. *Tissue Engineering, 5*(2), 103–118.
26. Khaoustov, V. I., et al. (1999). Induction of three-dimensional assembly of human liver cells by simulated microgravity. *In Vitro Cellular & Developmental Biology. Animal, 35*(9), 501–509.
27. Doolin, E. J., et al. (1999). Effects of microgravity on growing cultured skin constructs. *Tissue Engineering, 5*(6), 573–582.
28. Chen, J., Chen, R., & Gao, S. (2007). Morphological characteristics and proliferation of keratocytes cultured under simulated microgravity. *Artificial Organs, 31*(9), 722–731.
29. Rauh, J., et al. (2011). Bioreactor systems for bone tissue engineering. *Tissue Engineering. Part B, Reviews, 17*(4), 263–280.
30. Hidaka, M., et al. (2007). Transplantation of engineered bone tissue using a rotary three-dimensional culture system. *In Vitro Cellular & Developmental Biology. Animal, 43*(2), 49–58.
31. Low, H. P., Savarese, T. M., & Schwartz, W. J. (2001). Neural precursor cells form rudimentary tissue-like structures in a rotating-wall vessel bioreactor. *In Vitro Cellular & Developmental Biology. Animal, 37*(3), 141–147.
32. Duke, J., et al. (1996). Chondrogenesis in aggregates of embryonic limb cells grown in a rotating wall vessel. *Advances in Space Research, 17*(6–7), 289–293.
33. Baker, T. L., & Goodwin, T. J. (1997). Three-dimensional culture of bovine chondrocytes in rotating-wall vessels. *In Vitro Cellular & Developmental Biology. Animal, 33*(5), 358–365.
34. Freed, L. E., et al. (1998). Chondrogenesis in a cell-polymer-bioreactor system. *Experimental Cell Research, 240*(1), 58–65.
35. Goodwin, T. J., et al. (1993). Rotating-wall vessel coculture of small intestine as a prelude to tissue modeling: Aspects of simulated microgravity. *Proceedings of the Society for Experimental Biology and Medicine, 202*(2), 181–192.
36. Radtke, A. L., et al. (2010). Analysis of interactions of Salmonella type three secretion mutants with 3-D intestinal epithelial cells. *PLoS One, 5*(12), e15750.
37. Suderman, M. T., et al. (2006). *Three-dimensional human bronchial-tracheal epithelial tissue-like assemblies as hosts for severe acute respiratory syndrome-COV infection.* Houston, TX: National Aviation and Space Administration.
38. Crabbe, A., et al. (2011). Alveolar epithelium protects macrophages from quorum sensing-induced cytotoxicity in a three-dimensional co-culture model. *Cellular Microbiology, 13*(3), 469–481.
39. Crabbe, A., et al. (2015). Recellularization of decellularized lung scaffolds is enhanced by dynamic suspension culture. *PLoS One, 10*(5), e0126846.
40. LaMarca, H. L., et al. (2005). Three-dimensional growth of extravillous cytotrophoblasts promotes differentiation and invasion. *Placenta, 26*(10), 709–720.

41. Hjelm, B. E., et al. (2010). Development and characterization of a three-dimensional organotypic human vaginal epithelial cell model. *Biology of Reproduction, 82*(3), 617–627.
42. Swan, K. F., et al. (2015). [317-POS]: In vitro model of placental trophoblast differentiation and cytomegalovirus infection. *Pregnancy Hypertens, 5*(1), 155–156.
43. Smith, Y. C., et al. (2006). Novel three-dimensional organoid model for evaluation of the interaction of uropathogenic Escherichia coli with terminally differentiated human urothelial cells. *Infection and Immunity, 74*(1), 750–757.
44. Cortiella, J., et al. (2010). Influence of acellular natural lung matrix on murine embryonic stem cell differentiation and tissue formation. *Tissue Engineering Part A, 16*(8), 2565–2580.
45. Jessup, J. M., et al. (2000). Microgravity culture reduces apoptosis and increases the differentiation of a human colorectal carcinoma cell line. *In Vitro Cellular & Developmental Biology. Animal, 36*(6), 367–373.
46. Fitzgerald, W., et al. (2009). Immune suppression of human lymphoid tissues and cells in rotating suspension culture and onboard the International Space Station. *In Vitro Cellular & Developmental Biology. Animal, 45*(10), 622–632.
47. Pellis, N. R., et al. (1997). Changes in gravity inhibit lymphocyte locomotion through type I collagen. *In Vitro Cellular & Developmental Biology. Animal, 33*(5), 398–405.
48. Chen, J., et al. (2011). The simulated microgravity enhances the differentiation of mesenchymal stem cells into neurons. *Neuroscience Letters, 505*(2), 171–175.
49. Cooper, D., & Pellis, N. R. (1998). Suppressed PHA activation of T lymphocytes in simulated microgravity is restored by direct activation of protein kinase C. *Journal of Leukocyte Biology, 63*(5), 550–562.
50. Taylor, G. R. (1993). Immune changes in humans concomitant with space flights of up to 10 days duration. *Physiologist, 36*(1 Suppl), S71–S74.
51. Martin, I., et al. (2000). Modulation of the mechanical properties of tissue engineered cartilage. *Biorheology, 37*(1–2), 141–147.
52. Nauman, E. A., et al. (2007). Novel quantitative biosystem for modeling physiological fluid shear stress on cells. *Applied and Environmental Microbiology, 73*(3), 699–705.
53. Rivera-Solorio, I. (2001). *Local mass transfer for bioreactors in microgravity, in mechanical engineering*. Houston, TX: University of Houston.
54. Velasco Fuentes, O. U. (2009). Kelvin's discovery of Taylor columns. *European Journal of Biomechanics - B/Fluids, 28*(3), 469–472.

Chapter 3
Immune Dysfunction in Spaceflight: An Integrative View

Buqing Yi, Brian Crucian, Svantje Tauber, Oliver Ullrich, and Alexander Choukèr

Historical Landmarks

1986—Taylor et al. show changes in leukocyte distribution patterns in space shuttle crew members at landing [1].

1995—Muller et al. report cell-mediated immune dysregulation in Antarctic wintering personnel [2].

2000—Crucian et al. identify altered cytokine production by specific human peripheral blood cell subsets following short-duration spaceflight [3].

2002—Shearer et al. report elevation of proinflammatory cytokine IFN-gamma and suppression of anti-inflammatory plasma cytokines IL-10 and IL-1RA during isolation of the Antarctic winter, suggesting isolation related T-cell activation [4].

2005—Pierson et al. demonstrate that Epstein-Barr virus shedding increases in astronauts during space shuttle missions [5].

2008—Cohrs et al. show asymptomatic reactivation and shed of infectious varicella zoster virus (VZV) in astronauts during space shuttle missions [6].

2013—Crucian et al. demonstrate that immune system dysregulation occurs during short-duration spaceflight on board the space shuttle [7].

(continued)

B. Yi • A. Choukèr (✉)
Department of Anesthesiology, Stress and Immunology Lab, Hospital of the University of Munich, Marchioninstraße 15, 81377 Munich, Germany
e-mail: achouker@med.uni-muenchen.de

B. Crucian
Biomedical Research and Environmental Sciences Division, NASA Johnson Space Center, Houston, TX, USA

S. Tauber • O. Ullrich
Division of Cellbiology, Institute of Anatomy, University of Zurich, Winterthurerstr. 190, CH-8057 Zurich, Switzerland

C.A. Nickerson et al. (eds.), *Effect of Spaceflight and Spaceflight Analogue Culture on Human and Microbial Cells*, DOI 10.1007/978-1-4939-3277-1_3

(continued)

2014—Yi et al. report elevated cortisol levels, increased lymphocyte amount, and heightened immune responses in crew members during a simulated Mars mission (Mars520 study), indicating that chronic stress imposed by prolonged isolation and confinement is able to trigger leukocyte phenotype changes and poorly controlled immune responses [8].

2014—Crucian et al. report cytokine dysfunction patterns in crew members on ISS during long-duration spaceflight, suggesting multiple physiological adaptations persist during spaceflight, including inflammation, leukocyte recruitment, angiogenesis, and thrombocyte regulation [9].

3.1 Introduction

It is well established that spaceflight conditions have a fundamental impact on a wide variety of physiological functions in humans. This includes spaceflight-induced alterations in immune responses and associated immune dysfunction [10, 11]. Changes reported include alterations in the number, proliferation and function of immune cell populations, cytokine production, and neuroendocrine responses [10, 11]. In alignment with these findings, increased susceptibility to infection in astronauts can be dated back to the Apollo era. After the first Apollo missions, more than half of the astronauts were afflicted with bacterial or viral infections [12]. Even after initiation of the preflight health stabilization program by NASA in 1971 to limit crew interactions with the public prior to spaceflight in order to mitigate infectious disease risks, combined with control of food quality, use of ground-based clean rooms to prepare materials destined for delivery to crew, and stringent selection of healthy individuals as astronauts, there were still a surprising number of reported infectious disease incidences on board several Shuttle missions [13]. Furthermore, reactivation of latent viruses has been repeatedly reported in the crew during short-duration spaceflight [5, 6, 14, 15]. Recent investigations on crew members of long-duration space missions have revealed the potential of immune dysfunction in two directions: immune hyperactivity (which may result in risks such as hypersensitivities or autoimmunity) and immune hypoactivity (which means increased risks for infectious diseases and viral reactivation, as well as other disorders) [9].

Consistent with the above findings, exposure of subjects to ground-based models of spaceflight analogue conditions, such as chronic bed-rest, Arctic and Antarctic winter-over analogue, and prolonged isolation-and-confinement such as simulated Mars mission (Mars520), has also caused changes in the immune system [2, 8, 16, 17]. However, the specific cause(s) and potential clinical relevance of spaceflight-induced immune dysregulation is not currently understood. Several factors make an accurate assessment of clinical incidence during spaceflight challenging. All events are bound by the constraints ascribed to private medical information. Also, although crew surgeons maintain electronic medical records, the recording practice may vary from crew to crew, or surgeon to surgeon. Similar variance may occur among incidence data collected for specific research studies. Therefore, those observations recorded during spaceflight may only reflect a tabulation of observed symptomology, since diagnoses may not be confirmed by serology during spaceflight.

Although the correlation of these changes to immune problems in spaceflight has not been fully established, an improved understanding of the stressors associated with the spaceflight environment and their effects on the immune response and development of crew countermeasures could translate toward treatments for immune problems on Earth. This includes an understanding of both the physiological and psychological stressors, and the interconnections between them, that underlie spaceflight-induced alterations in immune regulation.

During spaceflight there are several environmental risk factors, such as stressors caused by prolonged isolation, ecologically and environmentally closed systems, microgravity, radiation, circadian misalignment, which altogether can provoke an allostatic overload (an imbalance of cellular and organ functions, and a disruption of the activities of the central nervous system (CNS)) [18, 19]. If this allostatic load is maintained and becomes an overload, a variety of immune related disorders and health problems can occur (Fig. 3.1). Among these factors, an altered gravitational field is a unique environmental condition of spaceflight. Exposure to extremely reduced levels of gravity (microgravity) in low Earth orbit or in deep space is a critical stressor that could alter immune responses in space, either directly or indirectly. And for future interplanetary space missions, such as to Mars, the journey will take a minimum of 520 days. During the mission, the crewmembers have to live in an extremely isolated condition owing to highly limited communication with Earth. The effects of prolonged isolation-and-confinement are a serious concern for the health of the space travelers. Due to the difficulty of isolating all the variable factors that might affect immune system function in spaceflight, any effect observed on the immune system in astronauts should be assumed as the combined effect of all environmental risk factors found in space.

Human immunity works through several interconnected approaches: innate immunity, humoral and cell-mediated immunity, and mucosal immunity. Host defenses against pathogens are achieved by physical barriers (such as skin, mucous membranes, peristalsis, and fluid flow) and the coordinated, amplified and redundant actions of innate, humoral, cell-mediated, and mucosal immunity. Innate immunity is the primary line of host defense and yields an immediate nonspecific response, which is mediated by neutrophils, monocytes, macrophages, dendritic cells, and natural killer cells, together with cytokines, defensins, complement, and acute phase reactants such as C-reactive protein [20, 21]. Humoral immunity is antibody-mediated and dependent upon B lymphocytes which express antigen-specific surface receptors. Cellular immunity is coordinated by antigen-specific T lymphocytes. Both humoral and cell-mediated adaptive immune responses are dependent upon helper T lymphocytes, therefore T-lymphocyte proliferation is a fundamental requirement for a functional adaptive immune response. Mucosal immunity is mediated by single-layer epithelium that is covered by mucus and antimicrobial products and fortified by both innate and adaptive components of host defense [22].

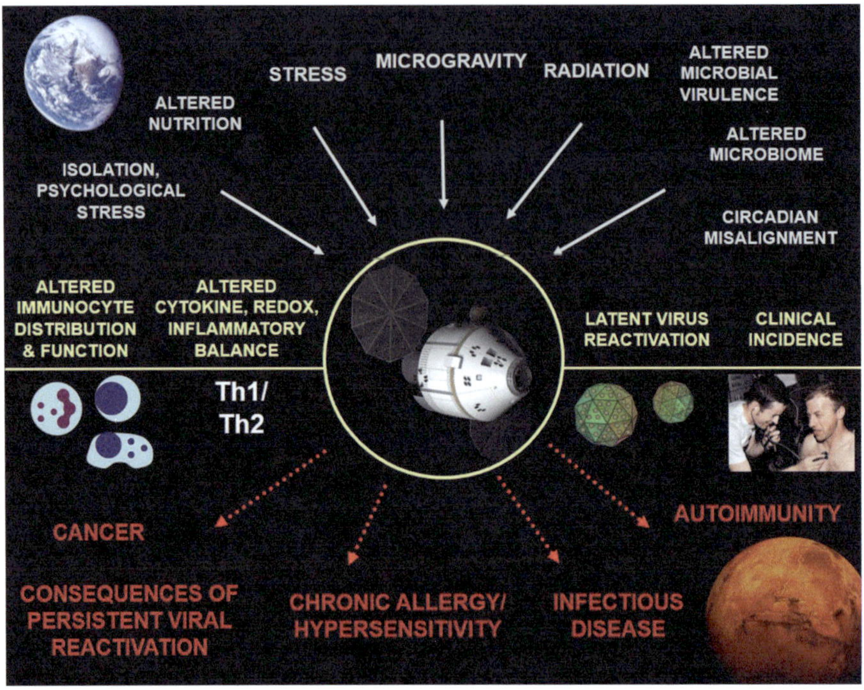

Fig. 3.1 The primary spaceflight-associated factors that influence the immune system during spaceflight (*grey*), which lead to observable changes in the immune system of crewmembers immune systems during spaceflight (*yellow*). Of concern is the elevation of specific clinical risks (*red*) during prolonged deep-space missions, characterized by potentially increased risks of autoimmunity, hypersensitivity, cancer, allergy, viral reactivation, and infectious disease

In addition to animal models and humans covered in this chapter, spaceflight and spaceflight analogue studies on immune function have also been carried out using in vitro mammalian cell cultures (covered in Chap. 6).

3.2 Immune Responses of Mammalian and Other Organisms to Spaceflight and Spaceflight Analogue Environments

3.2.1 Humans in Space

Immune studies directly carried out on crew members who had flown to space are summarized in Table 3.1. An inhibition of mitogen-induced proliferation of lymphocytes obtained from astronauts and cosmonauts was observed in a wide variety of spaceflight missions [10]. A decrease in production of interferon (IFN)-α/β has been reported when the blood of cosmonauts was sampled and tested

Table 3.1 Publications on the immune effects of spaceflight in humans

Author	Type of sample	Flight	Assay	Observations
Talas et al. [23]	Blood	Post	Cytokine production; natural killer cell activity	IFN-a/β reduced, natural killer cell activity reduced
Taylor et al. [1]	Blood	Post	Leukocyte subset distribution	B lymphocyte, monocyte, and lymphocyte numbers decreased
Meehan et al. [27]	Blood	Post	Leukocyte subset distribution; lymphocyte subset analysis	Peripheral blood leucocytes, granulocytes, and monocytes increased, lymphocytes decreased; CD4+ T-inducer and CD8+ T-cytotoxic decreased, monocytes increased
Taylor and Janney [35]	Skin test	In	Delayed-type hypersensitivity	Immune response decreased
Gmunder et al. [36]	Skin test	In	Delayed-type hypersensitivity	Immune response decreased
Stowe et al. [30]	Blood	Post	Leukocyte subset distribution	Neutrophil increased, lymphocyte slightly decreased, monocytes variable
Payne et al., [50]	Saliva	In	Viral shedding	EBA reactivation increased
Mehta et al. [41]	Urine & blood	In&Post	PCR analysis of CMV shedding; CMV antibody titer	CMV reactivation occurred before spaceflight and may continue during spaceflight
Crucian et al. [3]	Blood	Post	Leukocyte subset distribution; cytokine production; lymphocyte subset analysis	Granulocytes increased, lymphocytes, and monocytes unchanged; T cell cytokine secretion profiles changed; T cells percentage decreased, percentages of B cells and natural killer (NK) cells unchanged, CD14+ CD16+ monocytes decreased, CD4/CD8 T cell ratio increased
Mills et al. [28]	Blood & urine	Post	Leukocyte subset distribution; catecholamine	Neutrophils, monocytes, T-helper cells, and B cells increased, natural killer cells decreased; plasma norepinephrine increased
Stowe et al. [18]	Blood & urine	In&Post	Leukocyte and lymphocyte subsets; neuroendocrine hormones; Herpes-virus reactivation	During flight, cytomegalovirus (CMV) shedding in urine increased, and CMV antibody increased too; Cortisol increased in-flight

(continued)

Table 3.1 (continued)

Author	Type of sample	Flight	Assay	Observations
Stowe et al. [29]	Blood & urine	Post	Leukocyte and lymphocyte subsets; cortisol, urinary catecholamines	White blood cells, polymorphonuclear leukocytes, and CD4+ T cells increased; monocyte varied after short or long flight; cortisol and urinary Varied after short- or long-flight catecholamines
Mehta et al. [38]	Saliva	In&Post	PCR analysis of VZV DNA	Increased VZV DNA copies during and shortly after spaceflight
Kaur et al. [53]	Blood	Post	Neutrophil phagocytosis, oxidative burst, and degranulation	Neutrophils increased at landing compared to preflight levels; Before and after 9- to 11-day missions, phagocytosis and oxidative burst capacities were lower than control values; No consistent changes in degranulation or expression of surface markers
Pierson et al. [5]	Saliva & blood	In&Post	PCR analysis of EBV DNA; titers of antibody to EBV viral capsid antigen	Increased EBV DNA copies in-flight compared with pre- and post-flight; higher levels of titers at landing compared with baseline
Kaur et al. [25]	Blood & urine	Post	Antimicrobial functions of monocytes	No significant changes of monocyte amount; Following 5–11 days of spaceflight, the phagocytic index of monocytes significantly reduced; The expression of two surface markers CD32 and CD64 altered; No changes of urine cortisol, epinephrine, and norepinephrine levels after spaceflight
Rykova et al. [34]	Blood	Post	Immunoglobulins (IgG, IgA, IgM, IgE), specific IgE-antibodies, and interleukin-4 (IL-4) were measured	No significant changes of immunoglobulins, IgE-antibodies or IL-4
Cohrs et al. [6]	Saliva & blood	In&Post	Saliva VZV DNA and serum antibody titer; Virus isolation and culture	Increase VZV DNA copies during flight and at landing; the samples obtained on day 2 post-flight from two subjects yielded infectious VZV
Crucian et al. [31]	Blood	Post	Peripheral leukocyte subset analysis, early T cell activation potential, and intracellular/secreted cytokine profiles	Peripheral leukocyte distribution altered after landing; The ISS crewmembers demonstrated a reduction in early T cell activation potential immediately post-flight; The percentage of T cells capable of producing IL-2 was reduced, but IFNγ percentages were unchanged
Kaur et al. [24]	Blood	Post	Monocytes challenged with gram-negative endotoxin	IL-6&IL-1β lower level, IL-1ra and IL-8 higher level; Toll-like receptor 4 increased, CD14 decreased

Stowe et al. [26]	Blood & urine	Post	Leukocyte subset distribution; Cytokine production; adrenocortisol measurement	Stress hormone levels increased and leukocyte subsets altered following both short and long-duration; following long-duration, plasma and urinary cortisol increased more significantly as compared to that following short-during
Crucian et al. [33]	Blood	Post	Cytometric bead array analysis following activation with LPS or PMA+ ionomycin	Following LPS stimulation of monocytes, IL-6, IL-10 & TNFα reduced, IL-1b increased, IL-8 varied; following PMA + ionomycin stimulation of all leukocyte populations, only IL-6 reduced
Rykova [32]	Blood	Post	Expression of Toll-like receptors on the surface of white blood cells	After orbital spaceflight, a decrease in the content of circulating monocytes and granulocytes expressing TLR2, TLR4, and TLR6 was detected
Crucian et al. [7]	Blood	In&Post	Leukocyte distribution, T cell blastogenesis and cytokine production profiles	WBC, differential, lymphocyte subsets were unaltered during spaceflight, but were altered following landing. CD8+ T cell subsets were altered during spaceflight. T cell early blastogenesis varied by culture mitogen. Response to anti-CD3/28 antibodies was elevated post-flight. Plasma levels of IFNα, IFNγ, IL-1β, IL-4, IL-10, IL-12, and TNF-α were elevated in-flight, while IL-6 was significantly elevated at R + 0. Cytokine production profiles following mitogenic stimulation were significantly altered both during, and following spaceflight with reduced IFNγ, IL-17 and IL-10 and elevated TNFα and IL-8 during spaceflight
Crucian et al. [9]	Blood	In&Post	Plasma cytokines	An increase in the plasma cytokines: TNF-α, IL-8, IL-1ra, Tpo, VEGF, CCL2, CCL4, and CXCL5, while no significant alterations of the inflammatory or adaptive/T-regulatory cytokines: IL-1α, IL-1β, IL-2, IFN-γ, IL-17, IL-4, IL-5, IL-10, G-CSF, GM-CSF, FGF basic, CCL3, or CCL5

immediately after return from flight (post-flight) [23]. Moreover, the response of monocytes isolated from the crew immediately post-flight showed significant dysregulation, including altered response to Gram negative endotoxin and reduced ability to phagocytose *Escherichia coli* and induce an oxidative burst [24, 25]. Cell population analysis of blood samples obtained post-flight have shown alterations in the distribution of peripheral blood leukocytes, reductions in the function of specific immune cell subpopulations, and alterations in stress hormone levels [3, 26–32]. One recent study of post-flight assessment indicated that short-duration spaceflight could influence both monocyte phenotype and cytokine production profiles [33]. Many studies showed consistent similar results indicating an increase of granulocytes and reduction of lymphocytes and monocytes. It also appears that T cell and natural killer (NK) cell percentages are commonly reduced, whereas B cell percentages are commonly elevated, and the T cell CD4+:CD8+ ratio is either unaltered or increased. Specific subset findings may vary among the studies. Furthermore, no significant changes of immunoglobulins and IgE antibodies were detected [34]. Cytokine dysregulation was reported for crew members on short- and long-duration on the ISS [9, 31]. The results from long-duration spaceflights showed an increase in some specific cytokines associated with inflammation and other regulatory processes: tumor necrosis factor-α (TNFα), IL-8, IL-1Ra, thrombopoietin (Tpo), vascular endothelial growth factor (VEGF), C-C motif chemokine ligand 2 (CCL2), chemokine ligand 4/macrophage inhibitory protein 1b (CCL4), and C-X-C motif chemokine 5/epithelial neutrophil-activating protein 78 (CXCL5), while no significant alterations of the other cytokines associated with inflammatory or adaptive/T--regulatory cytokines: IL-1α, IL-1β, IL-2, IFN-γ, IL-17, IL-4, IL-5, IL-10, G-CSF, GM-CSF, FGF basic, CCL3, or CCL5 [9].

The majority of the existing spaceflight immunology database has been established from post-flight assessments. Although post-flight assays may produce valuable information about potential in-flight alterations of immune responses, they do not necessarily reflect the in-flight status of the immune system. The process of re-entry and landing, as well as the re-adaptation to gravity on Earth causes elevated physiological stress to the human body, which may to a large extent influence post-flight assessments. One test reflecting the efficiency of cellular immunity was carried out during spaceflight for a few times, showing changes of cellular immune response [35, 36]. Recently, more comprehensive NASA and ESA-Roscosmos immune studies onboard both the Space Shuttle and ISS have confirmed that dysregulation of some immunological parameters observed post-flight before also occurs during short-duration spaceflight and possibly persists during long-duration spaceflight [7, 9, 15, 32, 37]. In particular, several studies demonstrated the increase of Epstein-Barr virus (EBV) shedding in astronauts during space shuttle mission [5], and asymptomatic reactivation and shed of infectious VZV in astronauts were also confirmed during space shuttle missions [6, 38]. Further spaceflight immune studies are planned for ISS that will assess previously uninvestigated aspects of immunity, as well as translational aspects with other physiological systems.

Spaceflight experiments are highly limited and very expensive. Therefore, many investigations have been done in ground-based facilities that simulate spaceflight conditions, which are so-called analogue studies. The advantage of analogue studies is that they provide scope for the isolation of diverse factors that occur in spaceflight conditions, yielding insight into possible mechanisms. There is also a downside, as the analogue of spaceflight on ground does not fully reflect the influence of real spaceflight. For example, radiation is a serious concern for human health during spaceflight, but due to the variety of radiation types in space and their distinct ways of depositing energy, it is difficult to calculate the absorbed doses and the induced biological effects for their simulation in ground-based analogue studies. Even though no analogue offers a perfect reproduction of the spaceflight environment, their easiness and cost-effectiveness make these platforms very useful for gaining basic scientific information, assessing the individual effects of certain spaceflight variables which may be reproduced to high fidelity, and for planning future spaceflight experiments.

Isolation and confinement are among the major stressors in space. They may potentially induce considerable psychological and physiological modifications. In the past few decades several variables, such as habitat volume, mission duration, crew size, and gender composition during group confinement have been investigated for their effects on stress response through ground-based analogue studies. In 1999, the SFINCSS-99 (Simulation of Flight of International Crew on Space Station) study was performed to examine the effects of confinement on two groups of four healthy men living in a low-volume containment chamber for either 110 or 240 days with limited contact with the outside world. Results from this study indicate that the innate immune system shows a tendency of activation, which is demonstrated by increased neutrophil counts and up-regulated β2-integrin on the surface of neutrophils [39]. The same analogue facility in Moscow was used for the Mars520 study simulating a long-duration space mission to Mars, which is so far the longest interplanetary spaceflight simulation study. During the 520-d isolation and confinement period, stress hormone cortisol was measured as a typical neuroendocrine parameter, and the changes of immune activity were monitored using an ex vivo simulated viral infection assay. The participants of the Mars520 mission displayed persistent high levels of cortisol during the mission period, and the results of immune analysis revealed increased lymphocyte amount and heightened immune responses, suggesting that prolonged isolation and confinement acting as chronic stressors may trigger leukocyte phenotype changes and poorly controlled immune responses, which may result in clinical risks such as hypersensitivities or autoimmunity [8].

During Antarctic winter-over (AWO), subjects experience prolonged periods of social isolation with limited telecommunications, confinement, minimal rescue/abort possibilities, and circadian misalignment, which are similar to those of long-duration spaceflight and have the capacity to influence immune homeostasis. Immune studies performed during winter-over missions have shown decreased cell-mediated immune responses in a few studies [2, 40, 41]. Interestingly, altered cytokine production profiles were detected during the isolation of the Antarctic winter characterized by elevation of proinflammatory cytokine IFN-gamma and suppression of anti-inflammatory plasma cytokines IL-10 and IL-1RA, suggesting isolation related T-cell activation [4, 42]. Latent herpes virus reactivation was also reported in the winter-over personal [41, 42]. A study of antibody production following immunization during AWO found no mid-mission alterations [43], potentially indicating that humoral immunity is unchanged in the presence of altered cellular immunity. Whether these cellular immune changes are aggravated by the conditions of hypoxia (it is considered to be the atmospheric condition in future exploration class mission vehicles and habitats), is under current investigation in Antarctica (at Concordia Station). Initial study results assessing early adaption to deployment at Concordia display significant innate immune changes [44].

In addition, the prolonged exposure to supine (6° tilt) head-down bed-rest (HDBR) model has been developed as a ground-based analogue of spaceflight to mimic the impact of microgravity on humans. This model leads to a loss of load bearing and a shift of fluid to the head, simulating the physiological conditions of space travelers. In addition to microgravity, HDBR also models several other effectors observed during spaceflight: confinement, immobilization, nutritional, and psychological stress. A few studies have been performed with HDBR to investigate the impact of microgravity on human immunity. Immune alterations were often observed during bed-rest periods, such as a significant decrease in the production of interleukin-2 (IL-2), interferon-γ (IFN-γ), and tumor necrosis factor-α (TNF-α) following immunization [45]. A recent bed-rest study demonstrated that regular exercise could accelerate primary antibody production and increase antibody levels to bacteriophage φX-174 [16]. It appears that countermeasures for other health problems during spaceflight, such as bone and muscle loss, could correct immune changes as well. However, intermittent short centrifugation protocols did not affect the bed-rest related granulocyte responses indicating innate immune activity [46] which should be taken into consideration for future development of countermeasures.

Addressing the risks of infectious diseases in the course of spaceflight yielded several effective countermeasures to prevent the damaging effects of spaceflight on human health upon return to Earth. For example, astronauts currently participate in a crew health stabilization program which restricts interactions with the public right before and after flight, respectively. This approach has curbed respiratory tract infections in the crew [13]. However, in-flight bacterial infection still occurred. During Apollo 13, a crew member was infected by a life threatening *Pseudomonas aeruginosa* infection [47, 48]. It has also been reported that spaceflight conditions can increase the virulence of the food borne pathogen *Salmonella enterica* serovar Typhimurium [49]. In addition, studies have been implemented to specify the impacts of spaceflight on herpes viral reactivation. The reactivation of the VZV has been detected in the saliva of astronauts suggesting an asymptomatic reactivation [6]. The results from these studies also indicated that astronauts during short-duration spaceflight had increased reactivation of Epstein-Barr virus (EBV), cytomegalovirus (CMV), and urinary catecholamine excretion [5, 14, 18, 50, 51]. These observations from short-duration flight indicate that exposure to spaceflight may lead to altered immune responses which allow the reactivation of latent viruses. However, it is currently unknown if this reactivation also occurs during long-duration space missions.

3.2.2 *Animals in Space*

Rodents

The rat has been used as a model organism in many spaceflight immunological studies. These in vivo flight studies performed with rat models have shown changes in mass [52] as well as hypoplasia in both the spleen and thymus [35]. Functionally, exposure to the spaceflight environment has resulted in decreases in the proliferative responses of lymphocytes, which were isolated from lymph nodes [53] and the spleen [1], respectively. To determine the effects of spaceflight on the distribution of leukocyte subsets, spleens from rats flown to space were dissociated into individual cells, and individual samples were stained with antibodies directed against different cell surface markers on rat leukocytes and analyzed using a flow cytometer. Alterations were observed in the following cell populations compared with those from ground-based controls: total T lymphocytes, CD8+ T lymphocytes, and interleukin-2 receptor bearing T lymphocytes [54]. The altered ratio of T lymphocyte subsets may at least partially account for the inhibited proliferative post-flight responses. Other studies describe changes in cell-mediated immunity [54], cytokine production [55], signal transduction, and natural killer cell activity [23]. Although not all the results within the different experiments are identical (due in part to a lack of standardization in experimental design), they all indicate that spaceflight consistently results in measurable alterations of the rat immune system.

One of the limitations of spaceflight studies with rat models is that most investigations have been carried out post-flight. To provide insight into true spaceflight-induced alterations in the microgravity environment, one study was carried out with tissues obtained while the rats were still subjected to space. Inhibition of leukocyte proliferation and NK cell activity was observed in samples obtained during spaceflight, which indicates that the actual in-flight conditions indeed affect immunological parameters [56]. The influence of spaceflight on the development of the immune system was also investigated in a study that subjected pregnant rats to spaceflight conditions during most of the gestation period. Blastogenesis of spleen cells in response to mitogen and the response of bone marrow cells to colony-stimulating factor observed in the offspring of pregnant rats showed no obvious difference from ground controls [55].

To date, there have been limited scientific studies in which the impact of spaceflight on mice immunity was investigated. Gridley and Pecaut first performed such studies in space with a mouse model and their results were largely consistent with those observed earlier in rats, including alteration of leukocyte subset populations and cytokine production [57, 58]. In a recent series of experiments with mice flown on the Space Shuttle Endeavour for 13 days, Baqai et al. characterized several parameters of C57BL/6 mice immediately after landing. Data from this study indicated that flight mice showed decreased spleen and thymus mass, as well as lower numbers of splenic leukocytes compared with their ground controls. Furthermore, when the isolated spleen T cells were stimulated with PHA, they secreted lower levels of IFN-γ, IL-1, and IL-4, indicating that to some extent the adaptive immune response was suppressed after spaceflight [59]. Interestingly, gene expression analysis has shown that spaceflight may lead to significant changes in thymus cell gene expression profiles, especially in those genes involved in the regulation of stress responses, glucocorticoid receptor metabolism, and T cell activity [60].

Other Animal Models

In addition to rodents, many other animal models have been used to investigate biological phenomena, such as the fruit fly *Drosophila melanogaster,* the nematode *Caenorhabditis elegans* (*C. elegans*), and the frog *Xenopus laevis*. Studies of these model organisms have greatly contributed to our understanding of human physiology because of the conservation of developmental and metabolic pathways over the course of evolution [61, 62]. Both *Drosophila* and *C. elegans* are well suited to address innate immune mechanisms, as they lack an adaptive immune response and have been used as human surrogates in infection studies [63–66]. The discovery of Toll signaling pathways in the innate immune response in *Drosophila* has paved the way for our understanding of key aspects of mammalian innate immune responses [67, 68].

Over the last few decades, the nematode *C. elegans* and its comparative model *Pristionchus pacificus* (*P. pacificus*) have been extensively used as model organisms in the research of molecular and developmental biology [69]. In recent years, nematodes have also been established as a powerful model for the study of stress responses [70–73]. A great deal of conservation has been confirmed between the signal transduction pathways of nematodes and mammals. Apart from the conservation of a large number of protein targets in the stress response, many of the critical regulatory mechanisms are preserved, and on many occasions differ only in their level of expression [74, 75]. Therefore, the nematode model serves as a human surrogate to provide an opportunity for an integrated understanding of the organismal response to diverse environmental and physiological stress, including spaceflight conditions.

Many spaceflight biology studies have been carried out using these invertebrate model organisms. One recent report using *Drosophila* showed that larvae born in the microgravity environment were affected in both cellular and humoral immune responses, whereas adult organisms appeared less susceptible to spaceflight conditions [76]. The recent Micro-5 spaceflight study on the International Space Station by the Nickerson team is the first real-time monitoring of the infection and virulence process when both the host (*C. elegans*) and pathogen (*Salmonella*) are simultaneously exposed to the spaceflight environment, and will profile host cellular, molecular, and innate immune responses, and will also test the efficacy of a nutritional countermeasure to prevent in-flight infection. Studies using the amphibian *Pleurodeles waltl* indicated that as compared to ground controls, adult organisms showed changes in the expression of immunoglobulin VH genes (encoding antibody heavy-chain), thus suggesting that spaceflight has an influence on antibody production and the humoral immune response in these vertebrate organisms [77]. Considering the operational convenience for the human surrogate models discussed above, and the conservation of many signal transduction pathways between these invertebrate and vertebrate organisms, the application of such models in future space immunity studies can be productive and informative for understanding human immunity in space.

3.2.3 Imbalance of Immune Homeostasis Resulted from Multiple Influential Factors

As discussed above, during spaceflight, multiple factors affect the body's psychological and physiological stress responses, which in turn influence the immune system and immune functions. However, the detailed nature and associated mechanisms of this phenomenon has yet to be established. The interconnection between stressors, the neurological system, the immune system, and other aspects of human physiology requires further investigation.

A few recent space or space analogue studies have revealed elevated cortisol responses during the mission (e.g., [8, 14]). Glucocorticoids (GCs) are a class of steroid hormones (cortisol is the most important human glucocorticoid), which are present in almost every vertebrate animal cell. The effects of GCs on innate immune responses have been studied in detail. GCs cause their effects by binding to the glucocorticoid receptor (GR). The activated GR complex, in turn, regulates target gene transcription in two different ways, by binding to hormone-response elements in the promoters of various genes, or acting through direct interaction with other transcription factors, such as NF-κB or AP1. For example, NF-κB normally occurs in an inactive state bound to its inhibitor (called IκB). Once the cell is activated, NF-κB dissociates from IκB and it moves into the nucleus, where it transactivates proinflammatory cytokines [78].

However, cortisol responses can also be regulated by the immune system. Actually, a bidirectional network exists between the neuroendocrine system and the immune system in which hormones and neurotransmitters are able to affect immune function and, thereby also affect immune mediators that are then again capable of inducing neuroendocrine changes [79]. This two-way communication is possible as both systems share receptors for common ligands and their coupled signalling pathways. The bidirectional "back and forth" flow of information between the neuroendocrine and the immune systems maintains and protects the internal homeostasis of the organism. Therefore, it seems that multidirectional communication networks are in effect that allow the signal transmission between the different organ systems and to hereby maintain homeostasis when the human is subjected to conditions of stress and disease [80].

Based on the results from recent NASA long-duration spaceflight studies [9], it seems the immune system gets "confused" in space with some cell functions reduced while some cell activities increased. Mars520 as the longest ever isolation-and-confinement study revealed that the chronic stress of prolonged isolation can trigger heightened immune responses, indicating an increased risk of chronic inflammation and autoimmune symptoms [8]. However, the microgravity environment has been shown to reduce cellular immune activity [10], suggesting increased susceptibility to infection. In long-duration space mission, astronauts are exposed to multiple environmental risk factors, such as prolonged isolation, ecologically and environmentally closed systems, microgravity, radiation, and circadian misalignment, which may altogether provoke an imbalance of immune homeostasis. All the effects observed on the immune system during spaceflight should be assumed as the combined effect of all environmental risk factors found in space, but it is important to isolate single influential factors to facilitate development of mitigation strategies. On-ground analogue study for single environmental factors has contributed significantly to the understanding of spaceflight related immune problems in the past, and we could assume that future analogue studies will further contribute to the development of effective mitigation strategies.

3.3 Outlook and Conclusion

It has been confirmed by multiple studies using cell culture systems (see Chap. 6), animal models and human subjects, that the myriad of variables associated with spaceflight cause changes to innate immune responses, humoral immune responses, and adaptive cell-mediated immune responses. Such alterations, should they persist during prolonged exploration-class deep-space missions, could lead to compromised defenses against infectious diseases as well as other diseases associated with immune imbalance including chronic inflammation, autoimmune diseases and cancer [9, 11]. The mechanism(s) of spaceflight-induced alterations in immune function remains to be established. It is likely that multiple factors, including microgravity, radiation, and chronic stress imposed by prolonged isolation and confinement, are involved in altering certain aspects of the immune function with interconnected mechanisms.

Given the physiological integration of the human "system," there is clearly a need for the implementation of even more translational and interdisciplinary protocols on the ISS and in analogue studies, such as in Antarctica or in more controlled conditions of bed-rest or isolation studies such as Mars520. The Mars520 study serves as a good example for the application of interdisciplinary protocols. In this study, the integrated information achieved from psychological, cognitive, cardiovascular and immune function evaluation, and analyses of molecular changes, is unique since this was collected from healthy crew members during such prolonged isolation and confinement [8, 81]. It will not only allow us to better judge the possible risks of a real Mars mission and develop appropriate mitigation strategies, but also bring insight into our understanding about immune *im*balance in space.

The ISS and ground-based spaceflight analogues provide excellent, complimentary opportunities to study humans under extreme environmental conditions. Space related health problems are linked to diseases on Earth, and space medical study has become an important contributor in conquering health problems on Earth. Systematic studies of space stress-responses and adaptation, in most cases carried out with spaceflight analogues on Earth, have and will continue to help increase our understanding of the pathophysiology of diseases and their associated mechanisms, and accelerate the development of new non-invasive countermeasures to prevent and treat disease. While a *single* spaceflight experiment will probably not save a life or cure a disease, the cumulative knowledge obtained from iterative discoveries using both spaceflight and spaceflight analogue immunological studies will advance our mechanistic understanding of disease and holds exciting potential for translational health benefits for the general public. Moreover, this knowledge will ultimately determine whether long-term and interplanetary spaceflights are an unacceptable risk or a challenging, but realistic option for mankind.

References

1. Taylor, G. R., Neale, L. S., & Dardano, J. R. (1986). Immunological analyses of U.S. Space Shuttle crewmembers. *Aviation, Space, and Environmental Medicine, 57*(3), 213–217.
2. Muller, H. K., Lugg, D. J., & Quinn, D. (1995). Cell mediated immunity in Antarctic wintering personnel; 1984-1992. *Immunology and Cell Biology, 73*(4), 316–320.
3. Crucian, B. E., Cubbage, M. L., & Sams, C. F. (2000). Altered cytokine production by specific human peripheral blood cell subsets immediately following space flight. *Journal of Interferon and Cytokine Research, 20*(6), 547–556.
4. Shearer, W. T., Lee, B. N., Cron, S. G., Rosenblatt, H. M., Smith, E. O., Lugg, D. J., et al. (2002). Suppression of human anti-inflammatory plasma cytokines IL-10 and IL-1RA with elevation of proinflammatory cytokine IFN-gamma during the isolation of the Antarctic winter. *Journal of Allergy and Clinical Immunology, 109*(5), 854–857.
5. Pierson, D. L., Stowe, R. P., Phillips, T. M., Lugg, D. J., & Mehta, S. K. (2005). Epstein-Barr virus shedding by astronauts during space flight. *Brain, Behavior, and Immunity, 19*(3), 235–242.
6. Cohrs, R. J., Mehta, S. K., Schmid, D. S., Gilden, D. H., & Pierson, D. L. (2008). Asymptomatic reactivation and shed of infectious varicella zoster virus in astronauts. *Journal of Medical Virology, 80*(6), 1116–1122.
7. Crucian, B., Stowe, R., Mehta, S., Uchakin, P., Quiriarte, H., Pierson, D., et al. (2013). Immune system dysregulation occurs during short duration spaceflight on board the space shuttle. *Journal of Clinical Immunology, 33*(2), 456–465.
8. Yi, B., Rykova, M., Feuerecker, M., Jager, B., Ladinig, C., Basner, M., et al. (2014). 520-d Isolation and confinement simulating a flight to Mars reveals heightened immune responses and alterations of leukocyte phenotype. *Brain, Behavior, and Immunity, 40*, 203–210.
9. Crucian, B. E., Zwart, S. R., Mehta, S., Uchakin, P., Quiriarte, H. D., Pierson, D., et al. (2014). Plasma cytokine concentrations indicate that in vivo hormonal regulation of immunity is altered during long-duration spaceflight. *Journal of Interferon and Cytokine Research, 34*(10), 778–786.
10. Gueguinou, N., Huin-Schohn, C., Bascove, M., Bueb, J. L., Tschirhart, E., Legrand-Frossi, C., et al. (2009). Could spaceflight-associated immune system weakening preclude the expansion of human presence beyond Earth's orbit? *Journal of Leukocyte Biology, 86*(5), 1027–1038.
11. Crucian, B., & Sams, C. (2009). Immune system dysregulation during spaceflight: Clinical risk for exploration-class missions. *Journal of Leukocyte Biology, 86*(5), 1017–1018.
12. Hawkins, W. R., & Zieglschmid, J. F. (1975). Clinical aspects of crew health. In R. S. Johnson, L. F. Dietlein, & C. A. Berry (Eds.), *Biomedical results of Apollo* (pp. 43–81). Washington: U.S. Government Printing Office.
13. Mermel, L. A. (2013). Infection prevention and control during prolonged human space travel. *Clinical Infectious Diseases, 56*(1), 123–130.

14. Mehta, S. K., Laudenslager, M. L., Stowe, R. P., Crucian, B. E., Sams, C. F., & Pierson, D. L. (2014). Multiple latent viruses reactivate in astronauts during Space Shuttle missions. *Brain, Behavior, and Immunity, 41*, 210–217.
15. Mehta, S. K., Crucian, B. E., Stowe, R. P., Simpson, R. J., Ott, C. M., Sams, C. F., et al. (2013). Reactivation of latent viruses is associated with increased plasma cytokines in astronauts. *Cytokine, 61*(1), 205–209.
16. Shearer, W. T., Ochs, H. D., Lee, B. N., Cohen, E. N., Reuben, J. M., Cheng, I., et al. (2009). Immune responses in adult female volunteers during the bed-rest model of spaceflight: Antibodies and cytokines. *Journal of Allergy and Clinical Immunology, 123*(4), 900–905.
17. Crucian, B., Lee, P., Stowe, R., Jones, J., Effenhauser, R., Widen, R., et al. (2007). Immune system changes during simulated planetary exploration on Devon Island, high arctic. *BMC Immunology, 8*, 7.
18. Stowe, R. P., Pierson, D. L., & Barrett, A. D. (2001). Elevated stress hormone levels relate to Epstein-Barr virus reactivation in astronauts. *Psychosomatic Medicine, 63*(6), 891–895.
19. McEwen, B., & Karatsoreos, I. (2012). What is stress. In A. Chouker (Ed.), *Stress challenges and immunity in space*. Heidelberg: Springer.
20. Akira, S., Uematsu, S., & Takeuchi, O. (2006). Pathogen recognition and innate immunity. *Cell, 124*(4), 783–801.
21. Medzhitov, R., & Janeway, C. A., Jr. (1997). Innate immunity: The virtues of a nonclonal system of recognition. *Cell, 91*(3), 295–298.
22. Holmgren, J., & Czerkinsky, C. (2005). Mucosal immunity and vaccines. *Nature Medicine, 11*(4 Suppl), S45–S53.
23. Talas, M., Batkai, L., Stoger, I., Nagy, L., Hiros, L., Konstantinova, I., et al. (1983). Results of space experiment program "Interferon". I. Production of interferon in vitro by human lymphocytes aboard space laboratory Solyut-6 ("Interferon I") and influence of space flight on lymphocyte functions of cosmonauts ("Interferon III"). *Acta Microbiologica Hungarica, 30*(1), 53–61.
24. Kaur, I., Simons, E. R., Kapadia, A. S., Ott, C. M., & Pierson, D. L. (2008). Effect of spaceflight on ability of monocytes to respond to endotoxins of gram-negative bacteria. *Clinical and Vaccine Immunology, 15*(10), 1523–1528.
25. Kaur, I., Simons, E. R., Castro, V. A., Ott, C. M., & Pierson, D. L. (2005). Changes in monocyte functions of astronauts. *Brain, Behavior, and Immunity, 19*(6), 547–554.
26. Stowe, R. P., Sams, C. F., & Pierson, D. L. (2011). Adrenocortical and immune responses following short- and long-duration spaceflight. *Aviation, Space, and Environmental Medicine, 82*(6), 627–634.
27. Meehan, R. T., Neale, L. S., Kraus, E. T., Stuart, C. A., Smith, M. L., Cintron, N. M., et al. (1992). Alteration in human mononuclear leucocytes following space flight. *Immunology, 76*(3), 491–497.
28. Mills, P. J., Meck, J. V., Waters, W. W., D'Aunno, D., & Ziegler, M. G. (2001). Peripheral leukocyte subpopulations and catecholamine levels in astronauts as a function of mission duration. *Psychosomatic Medicine, 63*(6), 886–890.
29. Stowe, R. P., Sams, C. F., & Pierson, D. L. (2003). Effects of mission duration on neuroimmune responses in astronauts. *Aviation, Space, and Environmental Medicine, 74*(12), 1281–1284.
30. Stowe, R. P., Sams, C. F., Mehta, S. K., Kaur, I., Jones, M. L., Feeback, D. L., et al. (1999). Leukocyte subsets and neutrophil function after short-term spaceflight. *Journal of Leukocyte Biology, 65*(2), 179–186.
31. Crucian, B. E., Stowe, R. P., Pierson, D. L., & Sams, C. F. (2008). Immune system dysregulation following short- vs long-duration spaceflight. *Aviation, Space, and Environmental Medicine, 79*(9), 835–843.
32. Rykova, M. P. (2013). Immune system in Russian cosmonauts after orbital space flights. *Fiziologiia Cheloveka, 39*(5), 126–136.
33. Crucian, B., Stowe, R., Quiriarte, H., Pierson, D., & Sams, C. (2011). Monocyte phenotype and cytokine production profiles are dysregulated by short-duration spaceflight. *Aviation, Space, and Environmental Medicine, 82*(9), 857–862.

34. Rykova, M. P., Gertsik Iu, G., Antropova, E. N., & Buravkova, L. B. (2006). Immunoglobulin e and allergen-specific IgE antibodies in cosmonauts before and after long-duration missions on the International Space Station. *Aviakosmicheskaia i Ekologicheskaia Meditsina, 40*(2), 19–22.
35. Taylor, G. R., & Janney, R. P. (1992). In vivo testing confirms a blunting of the human cell-mediated immune mechanism during space flight. *Journal of Leukocyte Biology, 51*(2), 129–132.
36. Gmunder, F. K., Konstantinova, I., Cogoli, A., Lesnyak, A., Bogomolov, W., & Grachov, A. W. (1994). Cellular immunity in cosmonauts during long duration spaceflight on board the orbital MIR station. *Aviation, Space, and Environmental Medicine, 65*(5), 419–423.
37. Strewe, C., Feuerecker, M., Nichiporuk, I., Kaufmann, I., Hauer, D., Morukov, B., et al. (2012). Effects of parabolic flight and spaceflight on the endocannabinoid system in humans. *Reviews in the Neurosciences, 23*(5-6), 673–680.
38. Mehta, S. K., Cohrs, R. J., Forghani, B., Zerbe, G., Gilden, D. H., & Pierson, D. L. (2004). Stress-induced subclinical reactivation of varicella zoster virus in astronauts. *Journal of Medical Virology, 72*(1), 174–179.
39. Chouker, A., Smith, L., Christ, F., Larina, I., Nichiporuk, I., Baranov, V., et al. (2002). Effects of confinement (110 and 240 days) on neuroendocrine stress response and changes of immune cells in men. *Journal of Applied Physiology, 92*(4), 1619–1627.
40. Muller, H. K., Lugg, D. J., Ursin, H., Quinn, D., & Donovan, K. (1995). Immune responses during an Antarctic summer. *Pathology, 27*(2), 186–190.
41. Mehta, S. K., Pierson, D. L., Cooley, H., Dubow, R., & Lugg, D. (2000). Epstein-Barr virus reactivation associated with diminished cell-mediated immunity in antarctic expeditioners. *Journal of Medical Virology, 61*(2), 235–240.
42. Tingate, T. R., Lugg, D. J., Muller, H. K., Stowe, R. P., & Pierson, D. L. (1997). Antarctic isolation: Immune and viral studies. *Immunology and Cell Biology, 75*(3), 275–283.
43. Shearer, W. T., Lugg, D. J., Rosenblatt, H. M., Nickolls, P. M., Sharp, R. M., Reuben, J. M., et al. (2001). Antibody responses to bacteriophage phi X-174 in human subjects exposed to the antarctic winter-over model of spaceflight. *Journal of Allergy and Clinical Immunology, 107*(1), 160–164.
44. Feuerecker, M., Crucian, B., Salam, A. P., Rybka, A., Kaufmann, I., Moreels, M., et al. (2014). Early adaption to the antarctic environment at dome C: Consequences on stress-sensitive innate immune functions. *High Altitude Medicine & Biology, 15*(3), 341–348.
45. Kelsen, J., Bartels, L. E., Dige, A., Hvas, C. L., Frings-Meuthen, P., Boehme, G., et al. (2012). 21 Days head-down bed rest induces weakening of cell-mediated immunity—Some spaceflight findings confirmed in a ground-based analog. *Cytokine, 59*(2), 403–409.
46. Feuerecker, M., Feuerecker, B., Matzel, S., Long, M., Strewe, C., Kaufmann, I., et al. (2013). Five days of head-down-tilt bed rest induces noninflammatory shedding of L-selectin. *Journal of Applied Physiology, 115*(2), 235–242.
47. Sonnenfeld, G. (2005). The immune system in space, including Earth-based benefits of space-based research. *Current Pharmaceutical Biotechnology, 6*(4), 343–349.
48. Taylor, G. R. (1974). Recovery of medically important microorganisms from Apollo astronauts. *Aerospace Medicine, 45*(8), 824–828.
49. Wilson, J. W., Ott, C. M., Honer zu Bentrup, K., Ramamurthy, R., Quick, L., Porwollik, S., et al. (2007). Space flight alters bacterial gene expression and virulence and reveals a role for global regulator Hfq. *Proceedings of the National Academy of Sciences of the United States of America, 104*(41), 16299–16304.
50. Payne, D. A., Mehta, S. K., Tyring, S. K., Stowe, R. P., & Pierson, D. L. (1999). Incidence of Epstein-Barr virus in astronaut saliva during spaceflight. *Aviation, Space, and Environmental Medicine, 70*(12), 1211–1213.
51. Stowe, R. P., Mehta, S. K., Ferrando, A. A., Feeback, D. L., & Pierson, D. L. (2001). Immune responses and latent herpesvirus reactivation in spaceflight. *Aviation, Space, and Environmental Medicine, 72*(10), 884–891.

52. Nash, P. V., & Mastro, A. M. (1992). Variable lymphocyte responses in rats after space flight. *Experimental Cell Research, 202*(1), 125–131.
53. Kaur, I., Simons, E. R., Castro, V. A., Mark Ott, C., & Pierson, D. L. (2004). Changes in neutrophil functions in astronauts. *Brain, Behavior, and Immunity, 18*(5), 443–450.
54. Lesnyak, A., Sonnenfeld, G., Avery, L., Konstantinova, I., Rykova, M., Meshkov, D., et al. (1996). Effect of SLS-2 spaceflight on immunologic parameters of rats. *Journal of Applied Physiology, 81*(1), 178–182.
55. Sonnenfeld, G., Foster, M., Morton, D., Bailliard, F., Fowler, N. A., Hakenewerth, A. M., et al. (1998). Spaceflight and development of immune responses. *Journal of Applied Physiology, 85*(4), 1429–1433.
56. Rykova, M. P., Sonnenfeld, G., Lesnyak, A. T., Taylor, G. R., Meshkov, D. O., Mandel, A. D., et al. (1992). Effect of spaceflight on natural killer cell activity. *Journal of Applied Physiology, 73*(2 Suppl), 196S–200S.
57. Gridley, D. S., Nelson, G. A., Peters, L. L., Kostenuik, P. J., Bateman, T. A., Morony, S., et al. (2003). Genetic models in applied physiology: Selected contribution: Effects of spaceflight on immunity in the C57BL/6 mouse. II. Activation, cytokines, erythrocytes, and platelets. *Journal of Applied Physiology, 94*(5), 2095–2103.
58. Pecaut, M. J., Nelson, G. A., Peters, L. L., Kostenuik, P. J., Bateman, T. A., Morony, S., et al. (2003). Genetic models in applied physiology: Selected contribution: Effects of spaceflight on immunity in the C57BL/6 mouse. I. Immune population distributions. *Journal of Applied Physiology, 94*(5), 2085–2094.
59. Baqai, F. P., Gridley, D. S., Slater, J. M., Luo-Owen, X., Stodieck, L. S., Ferguson, V., et al. (2009). Effects of spaceflight on innate immune function and antioxidant gene expression. *Journal of Applied Physiology, 106*(6), 1935–1942.
60. Lebsack, T. W., Fa, V., Woods, C. C., Gruener, R., Manziello, A. M., Pecaut, M. J., et al. (2010). Microarray analysis of spaceflown murine thymus tissue reveals changes in gene expression regulating stress and glucocorticoid receptors. *Journal of Cellular Biochemistry, 110*(2), 372–381.
61. Pires-daSilva, A., & Sommer, R. J. (2003). The evolution of signalling pathways in animal development. *Nature Reviews Genetics, 4*(1), 39–49.
62. Watson, E., MacNeil, L. T., Ritter, A. D., Yilmaz, L. S., Rosebrock, A. P., Caudy, A. A., et al. (2014). Interspecies systems biology uncovers metabolites affecting C. elegans gene expression and life history traits. *Cell, 156*(4), 759–770.
63. Rae, R., Sinha, A., & Sommer, R. J. (2012). Genome-wide analysis of germline signaling genes regulating longevity and innate immunity in the nematode *Pristionchus pacificus. PLoS Pathogens, 8*(8), e1002864.
64. Iatsenko, I., Yim, J. J., Schroeder, F. C., & Sommer, R. J. (2014). B. subtilis GS67 Protects C. elegans from Gram-positive pathogens via fengycin-mediated microbial antagonism. *Current Biology, 24*(22), 2720–2727.
65. Irazoqui, J. E., Urbach, J. M., & Ausubel, F. M. (2010). Evolution of host innate defence: Insights from Caenorhabditis elegans and primitive invertebrates. *Nature Reviews Immunology, 10*(1), 47–58.
66. Pukkila-Worley, R., & Ausubel, F. M. (2012). Immune defense mechanisms in the Caenorhabditis elegans intestinal epithelium. *Current Opinion in Immunology, 24*(1), 3–9.
67. Hoffmann, J. A., & Reichhart, J. M. (2002). Drosophila innate immunity: An evolutionary perspective. *Nature Immunology, 3*(2), 121–126.
68. Taylor, K., Kleinhesselink, K., George, M. D., Morgan, R., Smallwood, T., Hammonds, A. S., et al. (2014). Toll mediated infection response is altered by gravity and spaceflight in Drosophila. *PLoS One, 9*(1), e86485.
69. Sommer, R. J., & Streit, A. (2011). Comparative genetics and genomics of nematodes: Genome structure, development, and lifestyle. *Annual Review of Genetics, 45*, 1–20.
70. Prahlad, V., & Morimoto, R. I. (2009). Integrating the stress response: Lessons for neurodegenerative diseases from C. elegans. *Trends in Cell Biology, 19*(2), 52–61.

71. Burkewitz, K., Choe, K., & Strange, K. (2011). Hypertonic stress induces rapid and widespread protein damage in C. elegans. *American Journal of Physiology. Cell Physiology, 301*(3), C566–C576.
72. McMullen, P. D., Aprison, E. Z., Winter, P. B., Amaral, L. A., Morimoto, R. I., & Ruvinsky, I. (2012). Macro-level modeling of the response of C. elegans reproduction to chronic heat stress. *PLoS Computational Biology, 8*(1), e1002338.
73. Rodriguez, M., Snoek, L. B., De Bono, M., & Kammenga, J. E. (2013). Worms under stress: C. elegans stress response and its relevance to complex human disease and aging. *Trends in Genetics, 29*(6), 367–374.
74. Lant, B., & Storey, K. B. (2010). An overview of stress response and hypometabolic strategies in Caenorhabditis elegans: Conserved and contrasting signals with the mammalian system. *International Journal of Biological Sciences, 6*(1), 9–50.
75. Jones, O. A., Swain, S. C., Svendsen, C., Griffin, J. L., Sturzenbaum, S. R., & Spurgeon, D. J. (2012). Potential new method of mixture effects testing using metabolomics and Caenorhabditis elegans. *Journal of Proteome Research, 11*(2), 1446–1453.
76. Marcu, O., Lera, M. P., Sanchez, M. E., Levic, E., Higgins, L. A., Shmygelska, A., et al. (2011). Innate immune responses of Drosophila melanogaster are altered by spaceflight. *PLoS One, 6*(1), e15361.
77. Bascove, M., Huin-Schohn, C., Gueguinou, N., Tschirhart, E., & Frippiat, J. P. (2009). Spaceflight-associated changes in immunoglobulin VH gene expression in the amphibian *Pleurodeles waltl. FASEB Journal, 23*(5), 1607–1615.
78. Sternberg, E. M. (2006). Neural regulation of innate immunity: A coordinated nonspecific host response to pathogens. *Nature Reviews Immunology, 6*(4), 318–328.
79. Wilder, R. L. (1995). Neuroendocrine-immune system interactions and autoimmunity. *Annual Review of Immunology, 13*, 307–338.
80. Ferone, D., Boschetti, M., Resmini, E., Giusti, M., Albanese, V., Goglia, U., et al. (2006). Neuroendocrine-immune interactions: The role of cortistatin/somatostatin system. *Annals of the New York Academy of Sciences, 1069*, 129–144.
81. Yi, B., Titze, J., Rykova, M., Feuerecker, M., Vassilieva, G., Nichiporuk, I., et al. (2014). Effects of dietary salt levels on monocytic cells and immune responses in healthy human subjects: A longitudinal study. *Translational Research, 166*, 103–110.

Part II
Human Cellular Investigations

Chapter 4
Biomedical Advances in Three Dimensions: An Overview of Human Cellular Studies in Space and Spaceflight Analogues

Cheryl A. Nickerson and C. Mark Ott

Historical Landmarks

1984—Activation of human T lymphocytes is profoundly depressed during in vitro culture in spaceflight microgravity [1].

1995–2002—Ground-based experiments in the RWV bioreactor suggested that microgravity may facilitate engineering advanced three-dimensional (3-D) human surrogate tissue models from individual cells. To test this hypothesis, a spaceflight experiment aboard the Mir space station showed that cells formed cartilage that was significantly more compressible than the control on Earth, due in part to a decrease in production of glycosaminoglycan [2–4].

1997—Functional cardiac tissues engineered in the RWV bioreactor [5].

1997—Primary myoblasts cultured in the RWV demonstrate myotube formation [6].

1997—Pellis et al. use the RWV to analyze microgravity-induced inhibition of lymphocyte locomotion and investigate mechanisms related to blunted lymphocyte movement [7].

1998—Cooper and Pellis use the RWV to characterize suppression of T cell activation observed in microgravity and microgravity analogue culture and demonstrate that signaling pathways upstream of protein kinase C are sensitive to these conditions [8].

(continued)

C.A. Nickerson, PhD (✉)
School of Life Sciences, The Biodesign Institute, Center for Infectious Diseases and Vaccinology, Arizona State University, Tempe, AZ, USA
e-mail: cheryl.nickerson@asu.edu

C. Mark Ott (✉)
Biomedical Research and Environmental Sciences Division, NASA/Johnson Space Center, 2101 NASA Parkway, Mail: SK4, Houston, TX 77058, USA
e-mail: c.m.ott@nasa.gov

C.A. Nickerson et al. (eds.), *Effect of Spaceflight and Spaceflight Analogue Culture on Human and Microbial Cells*, DOI 10.1007/978-1-4939-3277-1_4

(continued)

1999—3-D skeletal muscle organoids constructed from myoblasts demonstrate that microgravity exposure can induce muscle atrophy even without the removal of extrinsic mechanical load [9].

1998–2008—First applications of RWV-derived 3-D cell culture models to study host–pathogen interactions and infectious disease mechanisms using viral, bacterial, and parasitic pathogens [10–12].

2000—Mathematical model of cartilage development in the RWV [13].

2001—Permanent and non-random phenotypic and genotypic changes observed in prostate cancer epithelial cells upon co-culture in 3-D in the RWV with either prostate or bone stromal cells [14].

2001—Use of the RWV in quantitative studies of cartilage healing [15].

2003—Fluid-mechanic analysis of cartilage development in the RWV [16].

2005—3-D skeletal muscle organoids created from myoblasts, fibroblasts, and endothelial cells formed a functional microvascular network as well as contractile myotubes when transplanted into the whole animal [17].

2005—Wang et al. report 3-D co-culture of prostate cancer epithelial cells and bone stromal cells in space [18].

2005—Impaired induction of gene signaling pathways contributes to T cell dysfunction in microgravity [19].

2006—RWV culture of human bone marrow stem cells produced tissue constructs resembling those of trabecular bone [20].

2006–2013—Cartilage formation from bone marrow cells in the RWV [21–23].

2008—Sung et al. use the RWV to demonstrate the co-evolution of cancer and the interacting stromal cells during prostate cancer progression and metastasis [24].

2010 – First study using RWV-derived 3-D intestinal cell culture model that mechanistically confirmed clinical data showing that the Salmonella type three secretion system (T3SS) is not required for enteric infection in humans [35].

2010—First RWV-derived 3-D immunocompetent co-culture model used for studying infectious disease [25].

2013—Brinley et al. reported RWV culture decreased DNA repair in cells infected with Epstein Barr virus (EBV) and caused greater DNA damage from radiation compared to EBV negative cells [26].

2014—First transcriptomic study to reveal molecular genetic differences between epithelial cells grown in 2-D versus RWV-derived 3-D cultures in response to infection [27].

2015—Barrila et al. report first human cell infection conducted in spaceflight aboard Space Shuttle mission STS-131 [28].

4.1 Introduction

Many of the underlying causes of human disease result from the effects of physical/mechanical forces acting on living cells (cellular biomechanics) [29–31]. However, the constant overriding force of gravity precludes our ability to identify the full spectrum of cellular responses to mechanical forces that dictate the transition between homeostasis and disease. One method to elucidate these previously undetected responses is to exploit the quiescent environment of spaceflight as a unique platform to profile cellular responses to microgravity that are directly relevant to normal cellular development and disease progression that cannot otherwise be observed using traditional experimental approaches. Garnering information about cellular structure–function relationships in this environment is not unreasonable, as many breakthroughs in biological research and translational advancements have been achieved through studying the response of biological systems to extreme environments. Accordingly, mammalian cell culture studies using spaceflight platforms like the International Space Station (ISS) are no exception to this rule, and provide a unique opportunity to see life in a new adaptational mode that has not been seen before—adaptation to sustained low gravity.

As discussed in Chap. 1, mammalian cells may not "sense" changes in gravity directly, but rather by sensing indirect mechanical stresses (or the relief from such stresses) associated with the reduced force of gravity during spaceflight via mechanotransductive responses [32]. These responses effectively convert physical force changes at the cell surface (like those induced by mechanical signals exerted on the tension-dependent support of the cytoskeletal architecture) into biochemical responses that orchestrate coordinated cellular behaviors. Accordingly, the response of mammalian cells to culture in the microgravity environment must undergo a careful dissection to understand the direct contribution of microgravity as opposed to the indirect effect of physical changes invoked by cell culture in this unique environment, including alterations in fluid shear, mass transfer, hydrostatic pressure, and lack of sedimentation.

Since access to spaceflight experimental platforms is not yet easily accessible and affordable, NASA engineers designed ground-based spaceflight analogue bioreactor systems, RWVs, to allow scientists on Earth to culture cells in the lab under conditions that simulate aspects of the microgravity environment (Fig. 4.1) (see Chap. 2 for the operational principles of the RWV bioreactor). The RWV has allowed major advances in bioengineering research and our understanding of the mechanisms regulating the structure–function relationship of normal tissues, as well the pathogenesis of a variety of important human diseases, including infectious disease, cancer, immunological disorders, and bone and muscle wasting diseases [3, 14, 15, 18, 20–25, 33–45].

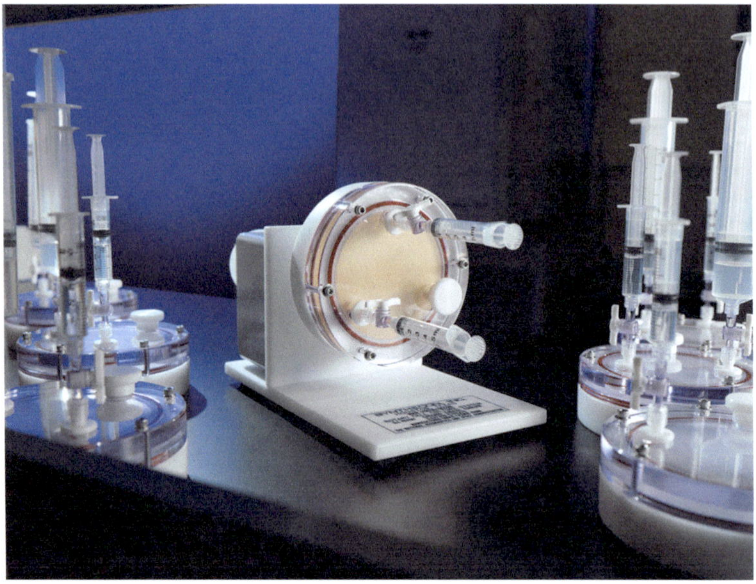

Fig. 4.1 The Rotating Wall Vessel (RWV) bioreactor

4.2 Culture of Three-Dimensional Mammalian Tissue Aggregates during Spaceflight

Mammalian cell culture in microgravity has been shown to promote spontaneous aggregation, which is associated with cellular assembly into high-density 3-D growth of differentiated human cells under these conditions [34, 46–48]. Since cellular assembly and 3-D growth is a prerequisite for functional tissue development, microgravity cell culture has proven beneficial in advancing our understanding of (1) the development and function of tissues and organs, (2) the progressive diversification of cells and tissues during the differentiation process, and (3) the mechanisms underlying transition between normal homeostasis and disease [34, 46–48]. Indeed, reduced gravity constitutes a physical perturbation that alters cytoskeletal organization, cellular morphology, signal transduction, gene expression, metabolism, and differentiation in mammalian cells [32, 34, 46, 47]. Moreover, culturing cells under the microgravity environment of spaceflight has enhanced our understanding of cellular biomechanics, leading to major advances in bioengineering, including development of innovative bioreactor technology (the RWV) that allows establishment of 3-D organotypic (organoid) models of tissue-like assemblies from a variety of cells and their practical application as human surrogate models for disease research, therapeutic development, and regenerative medicine [33, 34, 46, 47].

4.3 Culture of 3-D Mammalian Tissue Aggregates in the RWV Bioreactor Ground-Based Spaceflight Analogue

The development of the dynamic RWV spaceflight analogue culture system has provided the opportunity to optimize experiments before spaceflight and validate findings after spaceflight, as well as enable translational advances in biomedical research for the general public. Perhaps the most widely used spaceflight analogue cell culture system is the RWV bioreactor (Synthecon, Houston, TX), which is arguably the best mimic of cellular spaceflight responses [33, 49]. The design and theory of the RWV bioreactor is thoroughly described in Chap. 2, and is based on the principle that organs and tissues function in a 3-D environment. This optimized form of suspension culture is used for growing 3-D cell cultures in vitro that recapitulate key aspects of the differentiated form and function of parental tissues in vivo, including 3-D spatial organization/polarity, cellular differentiation, multicellular complexity, and functionality [33, 34, 46, 48]. Key factors influencing the physiologically relevant cellular differentiation and development in the RWV include efficient mass transfer of nutrients and wastes in combination with culture conditions that provide a low fluid shear growth environment similar to that encountered in certain regions of the body [33, 34, 46, 47]. The dynamic RWV culture conditions allow cells to grow in three dimensions, aggregate based on natural cellular affinities (facilitating co-culture of multiple cell types), and to differentiate into 3-D tissue-like assemblies [25, 33, 34, 46, 47, 50].

While 3-D cells cultured in the RWV require more time to grow and are initially more costly as compared to monolayers, the inherent experimental flexibility and reproducibility of these human surrogate tissues are ideal for the design of hierarchical models with a modular functionality and multicellular complexity, such that different individual cell types can be mixed and matched to explore fundamental biological questions [25, 33, 34, 46, 47, 50]. RWV-derived 3-D tissue models can easily be exposed to infectious agents and external chemical compounds, thereby facilitating studies related to host–pathogen interactions or incorporation of models into drug, adjuvant, toxin, and vaccine screens [33, 51, 52] (also see patents in Chap. 1). A variety of RWV-derived 3-D organotypic cell culture models have been extensively characterized, and many different cell and tissue types from both cell lines and primary cells have been developed. This includes, but is not limited to, 3-D models of human small intestine, colon, lung, bladder, liver, placenta, neuronal, vaginal, tonsil, and lymphoma that have been used to investigate mechanisms of cellular differentiation, tissue morphogenesis, and transition to disease (e.g., infectious disease and cancer) [10–12, 33, 35, 46, 51, 53–70]. In addition, these studies have also contributed insight toward advancing our understanding of mechanisms leading to tissue loss and deconditioning as well as immune system dysfunction during prolonged stays in space [3, 45].

4.4 Conclusion

Cellular studies in true spaceflight or in the RWV spaceflight analogue bioreactor offer dynamic approaches to engineer high fidelity, physiologically relevant 3-D tissue models with a vast array of biomedical applications. Research findings generated from these organotypic in vitro cell culture models also reflect the benefits of architectural context previously described by Bissell and others that are necessary for the establishment of predictive models of human disease [31, 71–74]. In particular, these models have furthered our understanding of structure–function relationships and design principles of the cellular microenvironment and cellular biomechanics that are critical in establishment of in vitro models that better recapitulate in vivo responses as compared to conventional flat 2-D cultures, and have complemented the knowledge being gained from other 3-D cell culture approaches. The applications of this tissue engineering research are as diverse as the number of cell types that can be cultured using these platforms. The following chapters in this section of the book describe the use of spaceflight and the RWV bioreactor to better understand organogenesis and normal tissue development using cell lines, stem cells, and primary cells, as well as disease pathologies, including infectious disease, immunological disorders, and cancer. These studies have shown tremendous potential to accelerate our understanding of human physiology and susceptibility to disease and hold translational promise to benefit mankind on Earth. In addition, studying the response of mammalian cells to culture under microgravity and microgravity analogue conditions provides the opportunity to unveil underpinning mechanisms regulating spaceflight-induced alterations in human physiology, adaptation during long duration missions, and associated clinical problems for astronauts.

In acknowledgement of the significance and importance of the scientific potential for these platforms to advance human health, federal funding agencies including the National Aeronautics and Space Administration (NASA), National Institutes of Health (NIH), and the Defense Advanced Research Projects Agency (DARPA) have funded spaceflight and spaceflight analogue biomedical research. As stated by the NIH, "*the microgravity environment of spaceflight affords a new tool to investigate the influences of various forces on life*" that are often obscured on Earth by the presence of gravity and to understand how these forces are "*manifest in structural and functional processes in cells, tissues, and organ systems*" (http://grants.nih.gov/grants/guide/pa-files/PAR-09-120.html). Accordingly, cellular and molecular mechanisms that underlie disease are studied in the spaceflight environment, offering new opportunities to understand how cells operate in these conditions, and giving new fundamental mechanistic insight into the disease process that will lead to new strategies for treatment and prevention.

With the ISS, commercial spaceflight entities, and increasing international participation from spacefaring nations, we are in a Renaissance period for spaceflight (and spaceflight analogue) biomedical research that has tremendous potential for breakthrough advances to benefit life on Earth and exploration of space. It is indeed an exciting time to be a biomedical scientist!

References

1. Cogoli, A., Tschopp, A., & Fuchs-Bislin, P. (1984). Cell sensitivity to gravity. *Science, 225*(4658), 228–230.
2. Freed, L. E., Hollander, A. P., Martin, I., Barry, J. R., Langer, R., & Vunjak-Novakovic, G. (1998). Chondrogenesis in a cell-polymer-bioreactor system. *Experimental Cell Research, 240*(1), 58–65.
3. Freed, L. E., Langer, R., Martin, I., Pellis, N. R., & Vunjak-Novakovic, G. (1997). Tissue engineering of cartilage in space. *Proceedings of the National Academy of Sciences of the United States of America, 94*(25), 13885–13890.
4. Freed, L. E., & Vunjak-Novakovic, G. (1995). Cultivation of cell-polymer tissue constructs in simulated microgravity. *Biotechnology and Bioengineering, 46*(4), 306–313.
5. Carrier, R. L., Papadaki, M., Rupnick, M., Schoen, F. J., Bursac, N., Langer, R., et al. (1999). Cardiac tissue engineering: Cell seeding, cultivation parameters, and tissue construct characterization. *Biotechnology and Bioengineering, 64*(5), 580–589.
6. Molnar, G., Schroedl, N. A., Gonda, S. R., & Hartzell, C. R. (1997). Skeletal muscle satellite cells cultured in simulated microgravity. *In Vitro Cellular & Developmental Biology Animal, 33*(5), 386–391.
7. Pellis, N. R., Goodwin, T. J., Risin, D., McIntyre, B. W., Pizzini, R. P., Cooper, D., et al. (1997). Changes in gravity inhibit lymphocyte locomotion through type I collagen. *In Vitro Cellular & Developmental Biology Animal, 33*(5), 398–405.
8. Cooper, D., & Pellis, N. R. (1998). Suppressed PHA activation of T lymphocytes in simulated microgravity is restored by direct activation of protein kinase C. *Journal of Leukocyte Biology, 63*(5), 550–562.
9. Vandenburgh, H., Chromiak, J., Shansky, J., Del Tatto, M., & Lemaire, J. (1999). Space travel directly induces skeletal muscle atrophy. *FASEB Journal, 13*(9), 1031–1038.
10. Alcantara Warren, C., Destura, R. V., Sevilleja, J. E., Barroso, L. F., Carvalho, H., Barrett, L. J., et al. (2008). Detection of epithelial-cell injury, and quantification of infection, in the HCT-8 organoid model of cryptosporidiosis. *Journal of Infectious Diseases, 198*(1), 143–149.
11. Long, J. P., Pierson, S., & Hughes, J. H. (1998). Rhinovirus replication in HeLa cells cultured under conditions of simulated microgravity. *Aviation, Space, and Environmental Medicine, 69*(9), 851–856.
12. Nickerson, C. A., Goodwin, T. J., Terlonge, J., Ott, C. M., Buchanan, K. L., Uicker, W. C., et al. (2001). Three-dimensional tissue assemblies: Novel models for the study of *Salmonella enterica* serovar Typhimurium pathogenesis. *Infection and Immunity, 69*(11), 7106–7120.
13. Martin, I., Obradovic, B., Treppo, S., Grodzinsky, A. J., Langer, R., Freed, L. E., et al. (2000). Modulation of the mechanical properties of tissue engineered cartilage. *Biorheology, 37*(1–2), 141–147.
14. Rhee, H. W., Zhau, H. E., Pathak, S., Multani, A. S., Pennanen, S., Visakorpi, T., et al. (2001). Permanent phenotypic and genotypic changes of prostate cancer cells cultured in a three-dimensional rotating-wall vessel. *In Vitro Cellular & Developmental Biology Animal, 37*(3), 127–140.
15. Obradovic, B., Martin, I., Padera, R. F., Treppo, S., Freed, L. E., & Vunjak-Novakovic, G. (2001). Integration of engineered cartilage. *Journal of Orthopaedic Research, 19*(6), 1089–1097.
16. Lappa, M. (2003). Organic tissues in rotating bioreactors: Fluid-mechanical aspects, dynamic growth models, and morphological evolution. *Biotechnology and Bioengineering, 84*(5), 518–532.
17. Levenberg, S., Rouwkema, J., Macdonald, M., Garfein, E. S., Kohane, D. S., Darland, D. C., et al. (2005). Engineering vascularized skeletal muscle tissue. *Nature Biotechnology, 23*(7), 879–884.
18. Wang, R., Xu, J., Juliette, L., Castilleja, A., Love, J., Sung, S. Y., et al. (2005). Three-dimensional co-culture models to study prostate cancer growth, progression, and metastasis to bone. *Seminars in Cancer Biology, 15*(5), 353–364.

19. Boonyaratanakornkit, J. B., Cogoli, A., Li, C. F., Schopper, T., Pippia, P., Galleri, G., et al. (2005). Key gravity-sensitive signaling pathways drive T cell activation. *FASEB Journal, 19*(14), 2020–2022.
20. Marolt, D., Augst, A., Freed, L. E., Vepari, C., Fajardo, R., Patel, N., et al. (2006). Bone and cartilage tissue constructs grown using human bone marrow stromal cells, silk scaffolds and rotating bioreactors. *Biomaterials, 27*(36), 6138–6149.
21. Ohyabu, Y., Kida, N., Kojima, H., Taguchi, T., Tanaka, J., & Uemura, T. (2006). Cartilaginous tissue formation from bone marrow cells using rotating wall vessel (RWV) bioreactor. *Biotechnology and Bioengineering, 95*(5), 1003–1008.
22. Wu, X., Li, S. H., Lou, L. M., & Chen, Z. R. (2013). The effect of the microgravity rotating culture system on the chondrogenic differentiation of bone marrow mesenchymal stem cells. *Molecular Biotechnology, 54*(2), 331–336.
23. Sakai, S., Mishima, H., Ishii, T., Akaogi, H., Yoshioka, T., Ohyabu, Y., et al. (2009). Rotating three-dimensional dynamic culture of adult human bone marrow-derived cells for tissue engineering of hyaline cartilage. *Journal of Orthopaedic Research, 27*(4), 517–521.
24. Sung, S. Y., Hsieh, C. L., Law, A., Zhau, H. E., Pathak, S., Multani, A. S., et al. (2008). Coevolution of prostate cancer and bone stroma in three-dimensional coculture: Implications for cancer growth and metastasis. *Cancer Research, 68*(23), 9996–10003.
25. Crabbe, A., Sarker, S. F., Van Houdt, R., Ott, C. M., Leys, N., Cornelis, P., et al. (2010). Alveolar epithelium protects macrophages from quorum sensing-induced cytotoxicity in a three-dimensional co-culture model. *Cellular Microbiology, 13*, 469–481.
26. Brinley, A. A., Theriot, C. A., Nelman-Gonzalez, M., Crucian, B., Stowe, R. P., Barrett, A. D., et al. (2013). Characterization of Epstein-Barr virus reactivation in a modeled spaceflight system. *Journal of Cellular Biochemistry, 114*(3), 616–624.
27. David, J., Sayer, N. M., & Sarkar-Tyson, M. (2014). The use of a three-dimensional cell culture model to investigate host-pathogen interactions of Francisella tularensis in human lung epithelial cells. *Microbes and Infection, 16*(9), 735–745.
28. Barrila, J., Sarkar, S. F., Hansmeier, N., Briones, N., Park, J., Ott, C. M., et al. (2015). Microgravity uniquely alters the host-pathogen interaction between human intestinal epithelial cells and Salmonella enterica serovar Typhimurium. In *115th American Society for Microbiology General Meeting*. New Orleans, LA.
29. Nelson, C. M., & Bissell, M. J. (2005). Modeling dynamic reciprocity: Engineering three-dimensional culture models of breast architecture, function, and neoplastic transformation. *Seminars in Cancer Biology, 15*(5), 342–352.
30. Xu, R., Boudreau, A., & Bissell, M. J. (2009). Tissue architecture and function: Dynamic reciprocity via extra- and intra-cellular matrices. *Cancer Metastasis Reviews, 28*(1–2), 167–176.
31. Ingber, D. E. (2003). Mechanobiology and diseases of mechanotransduction. *Annals of Medicine, 35*(8), 564–577.
32. Ingber, D. (1999). How cells (might) sense microgravity. *FASEB Journal, 13*(Suppl), S3–S15.
33. Barrila, J., Radtke, A. L., Crabbe, A., Sarker, S. F., Herbst-Kralovetz, M. M., Ott, C. M., et al. (2010). Organotypic 3D cell culture models: Using the rotating wall vessel to study host-pathogen interactions. *Nature Reviews Microbiology, 8*(11), 791–801.
34. Duray, P. H., Hatfill, S. J., & Pellis, N. R. (1997). Tissue culture in microgravity. *Science and Medicine, 4*(3), 46–55.
35. Radtke, A. L., Wilson, J. W., Sarker, S., & Nickerson, C. A. (2010). Analysis of interactions of Salmonella type three secretion mutants with 3-D intestinal epithelial cells. *PLoS One, 5*(12), e15750.
36. De Weirdt, R., Crabbe, A., Roos, S., Vollenweider, S., Lacroix, C., van Pijkeren, J. P., et al. (2012). Glycerol supplementation enhances L. reuteri's protective effect against S. Typhimurium colonization in a 3-D model of colonic epithelium. *PLoS One, 7*(5), e37116.
37. O'Connor, K. C. (1999). Three-dimensional cultures of prostatic cells: Tissue models for the development of novel anti-cancer therapies. *Pharmaceutical Research, 16*(4), 486–493.

38. Margolis, L., Hatfill, S., Chuaqui, R., Vocke, C., Emmert-Buck, M., Linehan, W. M., et al. (1999). Long term organ culture of human prostate tissue in a NASA-designed rotating wall bioreactor. *Journal of Urology, 161*(1), 290–297.
39. O'Connor, K. C., Enmon, R. M., Dotson, R. S., Primavera, A. C., & Clejan, S. (1997). Characterization of autocrine growth factors, their receptors and extracellular matrix present in three-dimensional cultures of DU 145 human prostate carcinoma cells grown in simulated microgravity. *Tissue Engineering, 3*(2), 161–171.
40. Zhau, H. E., Goodwin, T. J., Chang, S. M., Baker, T. L., & Chung, L. W. (1997). Establishment of a three-dimensional human prostate organoid coculture under microgravity-simulated conditions: Evaluation of androgen-induced growth and PSA expression. *In Vitro Cellular & Developmental Biology Animal, 33*(5), 375–380.
41. Sambandam, Y., Townsend, M. T., Pierce, J. J., Lipman, C. M., Haque, A., Bateman, T. A., et al. (2014). Microgravity control of autophagy modulates osteoclastogenesis. *Bone, 61*, 125–131.
42. Capulli, M., Rufo, A., Teti, A., & Rucci, N. (2009). Global transcriptome analysis in mouse calvarial osteoblasts highlights sets of genes regulated by modeled microgravity and identifies a "mechanoresponsive osteoblast gene signature". *Journal of Cellular Biochemistry, 107*(2), 240–252.
43. Qiu, Q. Q., Ducheyne, P., & Ayyaswamy, P. S. (2001). 3D bone tissue engineered with bioactive microspheres in simulated microgravity. *In Vitro Cellular & Developmental Biology Animal, 37*(3), 157–165.
44. Martin, I., Shastri, V. P., Padera, R. F., Yang, J., Mackay, A. J., Langer, R., et al. (2001). Selective differentiation of mammalian bone marrow stromal cells cultured on three-dimensional polymer foams. *Journal of Biomedical Materials Research, 55*(2), 229–235.
45. Gueguinou, N., Huin-Schohn, C., Bascove, M., Bueb, J. L., Tschirhart, E., Legrand-Frossi, C., et al. (2009). Could spaceflight-associated immune system weakening preclude the expansion of human presence beyond Earth's orbit? *Journal of Leukocyte Biology, 86*(5), 1027–1038.
46. Unsworth, B. R., & Lelkes, P. I. (1998). Growing tissues in microgravity. *Nature Medicine, 4*(8), 901–907.
47. Freed, L. E., & Vunjak-Novakovic, G. (2002). Spaceflight bioreactor studies of cells and tissues. *Advances in Space Biology and Medicine, 8*, 177–195.
48. Vunjak-Novakovic, G., Searby, N., De Luis, J., & Freed, L. E. (2002). Microgravity studies of cells and tissues. *Annals of the New York Academy of Sciences, 974*, 504–517.
49. Wolf, D. A., & Schwarz, R. P. (1991). *Analysis of gravity-induced particle motion and fluid perfusion flow in the NASA-designed rotating zero-head space tissue culture vessel.*
50. Nickerson, C. A., & Ott, C. M. (2004). A new dimension in modeling infectious disease. *ASM News, 70*(4), 169–175.
51. Hjelm, B. E., Berta, A. N., Nickerson, C. A., Arntzen, C. J., & Herbst-Kralovetz, M. M. (2010). Development and characterization of a three-dimensional organotypic human vaginal epithelial cell model. *Biology of Reproduction, 82*(3), 617–627.
52. Durand, R. E., & Olive, P. L. (2001). Resistance of tumor cells to chemo- and radiotherapy modulated by the three-dimensional architecture of solid tumors and spheroids. *Methods in Cell Biology, 64*, 211–233.
53. Carterson, A. J., Honer zu Bentrup, K., Ott, C. M., Clarke, M. S., Pierson, D. L., Vanderburg, C. R., et al. (2005). A549 lung epithelial cells grown as three-dimensional aggregates: Alternative tissue culture model for Pseudomonas aeruginosa pathogenesis. *Infection and Immunity, 73*(2), 1129–1140.
54. Carvalho, H. M., Teel, L. D., Goping, G., & O'Brien, A. D. (2005). A three-dimensional tissue culture model for the study of attach and efface lesion formation by enteropathogenic and enterohaemorrhagic *Escherichia coli. Cellular Microbiology, 7*(12), 1771–1781.
55. Duray, P. H., Yin, S. R., Ito, Y., Bezrukov, L., Cox, C., Cho, M. S., et al. (2005). Invasion of human tissue ex vivo by Borrelia burgdorferi. *Journal of Infectious Diseases, 191*(10), 1747–1754.

56. Goodwin, T. J., Jessup, J. M., & Wolf, D. A. (1992). Morphologic differentiation of colon carcinoma cell lines HT-29 and HT-29KM in rotating-wall vessels. *Vitro Cellular & Developmental Biology, 28A*(1), 47–60.
57. Guo, P., Weinstein, A. M., & Weinbaum, S. (2000). A hydrodynamic mechanosensory hypothesis for brush border microvilli. *American Journal of Physiology Renal Physiology, 279*(4), F698–F712.
58. Hughes, J. H., & Long, J. P. (2001). Simulated microgravity impairs respiratory burst activity in human promyelocytic cells. *In Vitro Cellular & Developmental Biology Animal, 37*(4), 209–215.
59. Ingram, M., Techy, G. B., Saroufeem, R., Yazan, O., Narayan, K. S., Goodwin, T. J., et al. (1997). Three-dimensional growth patterns of various human tumor cell lines in simulated microgravity of a NASA bioreactor. *In Vitro Cellular & Developmental Biology Animal, 33*(6), 459–466.
60. Jessup, J. M., Frantz, M., Sonmez-Alpan, E., Locker, J., Skena, K., Waller, H., et al. (2000). Microgravity culture reduces apoptosis and increases the differentiation of a human colorectal carcinoma cell line. *In Vitro Cellular & Developmental Biology Animal, 36*(6), 367–373.
61. Long, J. P., Pierson, S., & Hughes, J. H. (1999). Suppression of Epstein-Barr virus reactivation in lymphoblastoid cells cultured in simulated microgravity. *In Vitro Cellular & Developmental Biology Animal, 35*(1), 49–54.
62. Margolis, L. B., Fitzgerald, W., Glushakova, S., Hatfill, S., Amichay, N., Baibakov, B., et al. (1997). Lymphocyte trafficking and HIV infection of human lymphoid tissue in a rotating wall vessel bioreactor. *AIDS Research and Human Retroviruses, 13*(16), 1411–1420.
63. Pei, M., He, F., Kish, V. L., & Vunjak-Novakovic, G. (2008). Engineering of functional cartilage tissue using stem cells from synovial lining: A preliminary study. *Clinical Orthopaedics and Related Research, 466*(8), 1880–1889.
64. Sainz, B., Jr., TenCate, V., & Uprichard, S. L. (2009). Three-dimensional Huh7 cell culture system for the study of hepatitis C virus infection. *Virology Journal, 6*, 103.
65. Schwarz, R. P., Goodwin, T. J., & Wolf, D. A. (1992). Cell culture for three-dimensional modeling in rotating-wall vessels: An application of simulated microgravity. *Journal of Tissue Culture Methods, 14*(2), 51–57.
66. Smith, Y. C., Grande, K. K., Rasmussen, S. B., & O'Brien, A. D. (2006). Novel three-dimensional organoid model for evaluation of the interaction of uropathogenic Escherichia coli with terminally differentiated human urothelial cells. *Infection and Immunity, 74*(1), 750–757.
67. Straub, T. M., Honer zu Bentrup, K., Orosz-Coghlan, P., Dohnalkova, A., Mayer, B. K., Bartholomew, R. A., et al. (2007). In vitro cell culture infectivity assay for human noroviruses. *Emerging Infectious Diseases, 13*(3), 396–403.
68. Honer zu Bentrup, K., Ramamurthy, R., Ott, C. M., Emami, K., Nelman-Gonzalez, M., Wilson, J. W., et al. (2006). Three-dimensional organotypic models of human colonic epithelium to study the early stages of enteric salmonellosis. *Microbes and Infection, 8*(7), 1813–1825.
69. Lamarca, H. L., Ott, C. M., Honer Zu Bentrup, K., Leblanc, C. L., Pierson, D. L., Nelson, A. B., et al. (2005). Three-dimensional growth of extravillous cytotrophoblasts promotes differentiation and invasion. *Placenta, 26*(10), 709–720.
70. Myers, T. A., Nickerson, C. A., Kaushal, D., Ott, C. M., Honer zu Bentrup, K., Ramamurthy, R., et al. (2008). Closing the phenotypic gap between transformed neuronal cell lines in culture and untransformed neurons. *Journal of Neuroscience Methods, 174*(1), 31–41.
71. Schmeichel, K. L., & Bissell, M. J. (2003). Modeling tissue-specific signaling and organ function in three dimensions. *Journal of Cell Science, 116*(Pt 12), 2377–2388.
72. Griffith, L. G., & Swartz, M. A. (2006). Capturing complex 3D tissue physiology in vitro. *Nature Reviews Molecular Cell Biology, 7*(3), 211–224.
73. Weigelt, B., & Bissell, M. J. (2008). Unraveling the microenvironmental influences on the normal mammary gland and breast cancer. *Seminars in Cancer Biology, 18*(5), 311–321.
74. Ingber, D. E. (1998). The architecture of life. *Scientific American, 278*(1), 48–57.

Chapter 5
Outpacing Infectious Disease: Mimicking the Host-Pathogen Microenvironment in Three-Dimensions

Aurélie Crabbé, Jennifer Barrila, C. Mark Ott, and Cheryl A. Nickerson

Historical Landmarks

1998—First infection of RWV-derived three-dimensional (3-D) cell culture models with any pathogen (Rhinovirus) [1].

2001—First infection of RWV-derived 3-D cell culture models with a bacterial pathogen (*Salmonella enterica* serovar Typhimurium) [2].

2008—First infection of RWV-derived 3-D cell culture models with a parasite (*Cryptosporidium parvum*) [3].

(continued)

A. Crabbé
The Biodesign Institute, Center for Infectious Diseases and Vaccinology, Arizona State University, Tempe, AZ, USA

Laboratory of Pharmaceutical Microbiology (LPM), Ghent University, Ghent, Belgium

J. Barrila
The Biodesign Institute, Center for Infectious Diseases and Vaccinology, Arizona State University, Tempe, AZ, USA

C. Mark Ott
Biomedical Research and Environmental Sciences Division, NASA Johnson Space Center, 2101 NASA Parkway, Mail Code: SK4, Houston, TX, USA

C.A. Nickerson
School of Life Sciences, The Biodesign Institute, Center for Infectious Diseases and Vaccinology, Arizona State University, Tempe, AZ, USA
e-mail: cheryl.nickerson@asu.edu

C.A. Nickerson et al. (eds.), *Effect of Spaceflight and Spaceflight Analogue Culture on Human and Microbial Cells*, DOI 10.1007/978-1-4939-3277-1_5

(continued)

2010—First RWV-derived 3-D immunocompetent co-culture model used for studying infectious disease process [4].

2010—First study using RWV-derived 3-D intestinal cell culture model that mechanistically confirmed clinical data showing that the *Salmonella* type three secretion system (T3SS) is not required for enteric infection in humans [5].

2012—First study to demonstrate that commensal bacteria can protect against bacterial infection in vitro using RWV-derived 3-D cell culture model [6].

2010—First study to profile the infection process in human cells during spaceflight [7].

2014—First transcriptomic study to reveal molecular differences in epithelial cells grown in 2-D versus RWV-derived 3-D cultures, before and after infection [8].

5.1 Introduction

Infectious diseases continue to rank as a leading cause of human morbidity and mortality worldwide, resulting in over seven million deaths/year [9]. The majority of these infections are caused by pathogens that attack at mucosal surfaces (e.g., *Streptococcus pneumoniae*, *Mycobacterium tuberculosis*, HIV, *Salmonella* spp., *Escherichia coli*, *Shigella*, etc.). The intestinal, lung, and vaginal mucosa represent the major portals of entry for most pathogens and are the most common sites of infection. These mucosal tissues are characterized by an architectural complexity that varies depending on their physical location and biological functionality. The 3-D microenvironment within each tissue—which includes multiple cell types (e.g., epithelial cells, fibroblasts, immune cells), mucins, extracellular matrix (ECM) components (proteins, polysaccharides, biochemical signals, peptides, hormones), and commensal microbiota—plays an important role in regulating the bidirectional crosstalk between host and pathogen that modulates disease initiation, progression, and outcome [10–16]. In particular, the tensile forces between the cell cytoskeleton, cell surface adhesion molecules, and the surrounding ECM plays a critical role in regulating a diverse array of processes ranging from cellular proliferation and differentiation to immune functions and cell survival [17–25]. In order to accurately model this complex in vivo network of interactions, an in vitro system must be able to recapitulate multiple aspects of this dynamic 3-D microenvironment. The incorporation of these advanced functional models into both fundamental and preclinical studies will lead to improved strategies for preventing and treating infectious disease.

For decades, most researchers have used the approach of culturing host cells as conventional two-dimensional (2-D) monolayers on plastic or glass to facilitate our understanding of host responses to infection. The technique has provided a great deal of mechanistic insight into host responses to infection and is relatively inexpensive and easy to learn. However, the lack of complexity in 2-D models leads to the loss of important phenotypic and functional characteristics that are critical for predicting in vivo responses [10, 20, 24–35]. This problem is largely reflected in the simple fact that despite decades of scientific progress, 90 % of drug candidates fail to make it to market due to problems associated with side effects, toxicity and/or a failure to translate early successes observed in cell culture and animal models to humans [36]. In recent years, an increasing number of studies have employed a range of 3-D tissue models to study infectious disease. A variety of methodologies and culture systems can be applied to engineer these models, including transwell systems, hollow fiber bioreactors, implantation into 3-D matrix scaffolds, microfluidic cell perfusion culture systems, and rotating culture bioreactors [10]. Each of these systems has advantages and disadvantages with regards to the required levels of expertise, cost, culture time, and compatibility with downstream analyses. For the purposes of this chapter, we will focus primarily on 3-D models of human tissue developed using the rotating wall vessel (RWV) bioreactor for infectious disease research (Fig. 5.1).

5.1.1 The RWV Bioreactor

The RWV is a cylindrical, rotating bioreactor that provides an optimized form of suspension culture for the development of 3-D organotypic tissue aggregates [10]. Figure 5.1 shows a diagram of the slow-turning lateral vessel (STLV)—one of the two RWV designs that are commonly used for cell culture. The other type of RWV, the high-aspect ratio vessel (HARV, not shown), differs from the STLV in its aeration source and shape. However, the operational principals are the same for both types of vessels. During culture, the bioreactor is completely filled with culture medium and rotates about the horizontal axis, leading to a solid body rotation of the media. As a result, the sedimentation of cells within the RWV is offset by the rotating fluid, creating a constant, gentle fall of the cells through the medium within a restricted orbit, thereby allowing the cells to grow and differentiate under physiologically relevant fluid shear conditions. This quiescent environment allows the cells to aggregate and assemble in three-dimensions based on their natural cellular affinities. The physiologically relevant levels of fluid shear that are present during culture are also similar to what is encountered by pathogens in certain regions of the body, including in between the brush border microvilli of epithelial cells and in utero [37]. This advanced bioreactor technology has been used in combination with a variety of cell lines and/or primary cells to engineer 3-D models of human tissue including small intestine, colon, lung, vaginal epithelium, placenta, bladder, lymphoma, and liver (Fig. 5.2) [10].

In order to generate 3-D models using the RWV, cells must first be grown as conventional 2-D monolayers until they reach the appropriate density. Cells are then removed from the tissue culture flask, re-suspended in the appropriate medium and

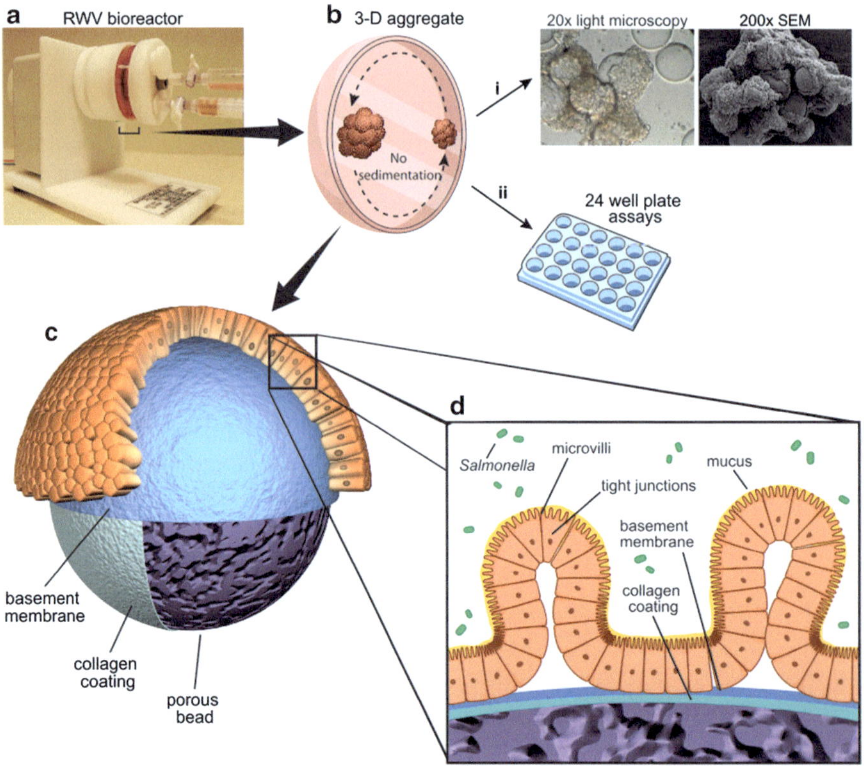

Fig. 5.1 Depiction and operation of the RWV cell culture system. (**a**) Rotating wall vessel (RWV) bioreactor containing cell culture media and porous microcarrier beads. (**b**) Sedimentation of cells within the RWV is offset by the rotating fluid, creating a constant free fall of the cells through the culture medium, allowing the cells to form 3-D aggregates under conditions of physiological fluid shear. 3-D aggregates are taken out of the bioreactor for analysis including light microscopy and scanning electron microscopy (SEM) (*i*), or seeded evenly in a 24 well plate format for various profiling and infection assays (*ii*). (**c**) Enlarged representation of intestinal epithelial cells attached to microcarrier beads of a 3-D aggregate. After intestinal epithelial cells adhere to the beads, they begin to differentiate and display in vivo-like characteristics such as narrow columnar cellular morphology, basement membrane, tight junctions, mucus, and microvilli [2, 5, 40]. (**d**) A further enlarged depiction of a single bead within the 3-D aggregate displaying characterized cellular surface structures and components, and infected with *Salmonella* [5]

then incubated with porous ECM-coated microcarrier beads or other scaffolding for attachment, and then introduced into the RWV bioreactor (Fig. 5.1). Alternatively, depending on the cell type and experimental needs, cells can be introduced into the bioreactor and allowed to grow and/or aggregate without attachment to a scaffold. The size, material, and porosity of microcarrier beads used for culture can be selected based on individual experimental needs. The advantage to using a porous scaffold (as opposed to a non-porous surface) is that the cells are able to respond to chemical and molecular gradients in three-dimensions, similar to what occurs in the parental tissue in vivo.

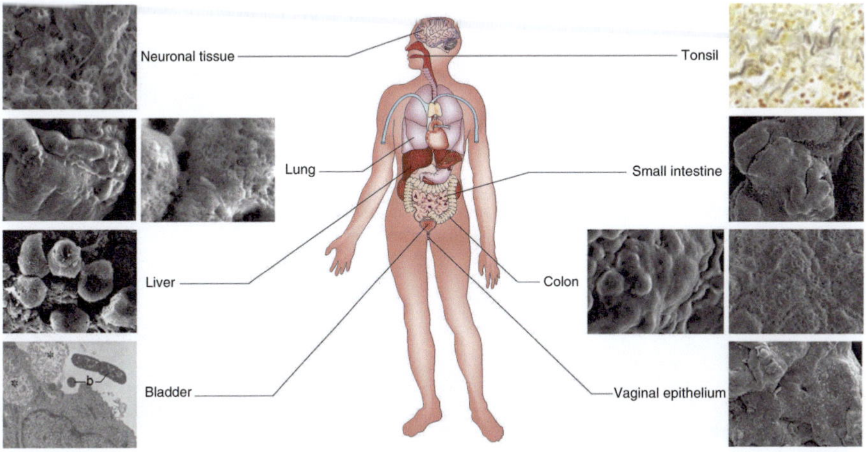

Fig. 5.2 Rotating wall vessel (RWV)-derived cell culture models of human tissue used for infection studies. A variety of RWV-derived 3-D models are currently being used for infectious disease research. This includes, but is not limited to, models of: neuronal tissue—scanning electron micrograph (SEM) of 3-D SH-SY5Y cell aggregates showing well-developed neurite extensions—this model is currently being used to study HIV-associated dementia (2000×); liver—SEM of primary human liver cells (1500×)—RWV liver models have been applied to study hepatitis C virus infections; lung—SEM of 3-D A549 lung cell aggregates (2000×, *left image*), SEM of 3-D A549 lung cell aggregates infected with *Pseudomonas aeruginosa* (7500×, *right image*); bladder—TEM of 3-D 5637 bladder cell aggregates showing close association (b) of uropathogenic *E. coli* with superficial urothelial cells (*asterisks* indicate regions displaying loss of structural integrity); vaginal epithelium—SEM of 3-D V19I vaginal epithelial cells (300×), this model is currently being used to study several sexually transmitted infectious agents; colon—SEM of 3-D HT-29 aggregates (2000×, *left image*), SEM of 3-D HT-29 aggregates infected with *Salmonella* Typhimurium (2000×, *right image*); small intestine—SEM of 3-D Int-407 aggregates which have been used to study the early stages of human enteric salmonellosis (500×; 3-D Int-407 model development and imaging done in the lab of the corresponding author); tonsillar tissue–light micrograph showing primary tonsillar cells 10 days after infection with *Borrelia burgdorferi* [10]

The growth and differentiation of RWV-derived 3-D models typically takes between 2 and 4 weeks depending on the cell type(s) and the complexity of the model. During the model development phase, cultures are typically sampled at select points over time and profiled for structural and/or functional characteristics that are representative of the in vivo host tissue. This may include the profiling for cellular morphology, cellular organization (formation of unilayers versus stacking), the expression and localization of polarity markers and tight junctional proteins, and multiple cell types, as well as other tissue-specific markers or functional assays that reflect the in vivo tissue (Fig. 5.3). Once cultures are established, the most common way to use them is to remove the 3-D aggregates from the RWV bioreactor and seed them evenly in multi-well plates (Fig. 5.1). Most RWV-derived cultures retain their differentiated state once removed from the bioreactor for multiple days, thereby allowing for a broad range of experiments to be conducted over time. However, it should be emphasized that this timing needs to be confirmed with the development of each new model. Once the model is ready and validated, it can be used for a wide

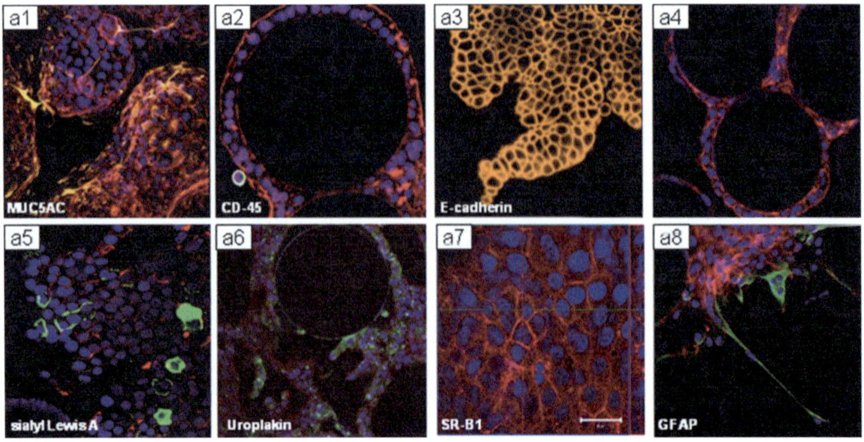

Fig. 5.3 Rotating wall vessel (RWV)-derived 3-D cell culture models display in vivo-like differentiation properties critical for pathogenesis. Immunohistochemical profiling of RWV-derived 3-D models from various human cell lines shows expression/organization of biomarkers and cellular organization relevant to those found in normal tissues in vivo. Shown are 3-D models established from the following cell lines that have been or are currently being used in infection studies: (**A1**) A549 lung epithelium co-cultured with U937 monocytes—MUC5AC mucin marker (*yellow*); (**A2**) A549 lung epithelium co-cultured with U937 monocytes—CD45 mononuclear phagocytic cell marker (*white/yellow*); (**A3**) HT-29 colonic epithelium—E-cadherin tight junctional marker (*orange*); (**A4**) Int-407 small intestinal epithelial cells exhibit unilayer morphology; (**A5**) Int-407 small intestinal epithelium—sialyl Lewis A, an M cell marker (*green*); (**A6**) HTB9-5637 bladder epithelium—Uroplakin–urothelial specific antigen (*green*); (**A7**) Huh7 (liver)—HCV Receptor SR-B1 (*red*); (**A8**) SH-SY5Y neuronal cells co-cultured with HTB-14 astrocytes and THP-1 monocytes—GFAP (glial fibrillary acidic protein) intermediate filament protein marker (*green*). *Blue*—DAPI (nuclear staining); *Red*—Phalloidin (actin cytoskeleton)—except in (**A7**). (Development and imaging of 3-D models shown in (**A2**–**A6**) and (**A8**) were done in the lab of C.A. Nickerson (**A1**: [120]); (**A7**: [94]) [10]

variety of studies, including (1) standard infection assays—microbial adherence, invasion, and intracellular survival; (2) transcriptomic, proteomic, and metabolomic analyses; (3) cytokine profiling, and (4) testing the efficacy, mechanism-of-action, and/or toxicity of drug candidates.

5.2 RWV-Derived 3-D Organotypic Cell Cultures Used as Human Surrogates for Infectious Disease Research

5.2.1 3-D Models of the Intestine

In vivo-like features of intestinal tissue mimicked in 3-D cell culture models. Intestinal infections are among the most common infections of individuals in developing countries and a leading cause of malnutrition and death in children under 5 years old (http://www.who.int/mediacentre/factsheets/fs330/en/). The intestine is a critical interface between environment and host, serving as the major portal of entry for nutrients, microbes, and drugs/therapeutics into the body [13]. Moreover, 70 % of the immune system is located in the intestine, making it the most important immune organ [38, 39]. A major obstacle in the development of novel strategies against enteric infections is the lack of representative and practical in vitro models of human gastrointestinal (GI) mucosa that faithfully recapitulate the tissue microenvironment of the parental tissue, including 3-D architecture, multicellular complexity, and functionality. Consequently, there is urgent need for human intestinal mucosal models that are structural and functional equivalents of the intestinal mucosa in order to dissect the dynamic interactions between host and pathogen that lead to enteric infectious disease.

Three-dimensional cell culture models of the small and large (colon) intestinal epithelium have been developed with the RWV bioreactor technology using porous collagen 1-coated microcarrier beads as scaffolds for cell growth [2, 5, 6, 10, 40]. These 3-D intestinal models were shown to mimic important physiological properties of the parental tissue that were not observed in conventional 2-D monolayer cultures grown on plastic, and which are of key importance for infectious disease studies. Specifically, RWV-derived 3-D models of small intestinal epithelium derived from the Int-407 cell line (originally derived from fetal small intestinal epithelium, since contaminated with HeLa cells) or the HT-29 cell line (colon adenocarcinoma) resulted in the localized expression of tight junctional markers (β-catenin, E-cadherin, Zonula Occludens 1/ZO-1) at cell–cell boundaries, apical secretion of mucins (MUC2, MUC5AC), apical localization of the microvillar protein villin as well as abundant brush border microvilli, basal expression of ECM protein collagen IV, organization of cells as a single layer (for Int-407 models), and expression of markers of multiple epithelial cell types (including enterocytes, goblet cells, M/M-like cells, and Paneth cells, thus indicating a pluripotent-like potential of cells cultured in the RWV), as well as decreased expression of the carcinoma markers cytokeratins 18 and 19 [2, 5, 40]. Therefore, aspects of the in vivo barrier function, single cell layer columnar cell organization, polarity, and multicellular complexity found in the normal parental tissue in vivo are mimicked in RWV-derived 3-D intestinal models. In contrast, growth of these same cell lines as conventional 2-D monolayers resulted in poorly differentiated cell culture models, which lacked many of the in vivo-like characteristics recapitulated by their respective 3-D models [2, 5, 40].

While a majority of the RWV-derived 3-D cell culture models used for infection studies were generated by growing cell lines on the surface of ECM-coated porous microcarrier beads, a limited number of studies used an alternative scaffold for cell growth. For example, a 3-D colorectal model was developed by growing HCT-8 cells on the surface of hydrated cell culture sheets of small intestinal submucosa [41]. In a similar fashion as the 3-D intestinal models described above, the 3-D colorectal model exhibited enhanced physiological localization of tight junctional proteins and apical localization of villin. Formation of brush border microvilli at the apical surface was confirmed at the phenotypic level using transmission electron microscopy (TEM). In addition, 3-D HCT-8 cells demonstrated enhanced production of carbohydrate digestion enzymes alkaline phosphatase and maltase as compared to monolayer controls of the same cell line. On the other hand, formation of multiple stacked epithelial cell layers was observed in the 3-D HCT-8 cultures, which does not reflect the single layer cell organization observed in the normal intestine in vivo, but rather, is reflective of conditions observed in abnormal/tumorigenic tissues.

New Insights into Host Responses to Infection Using 3-D Intestinal Models

The two major functions of the intestinal epithelium that are essential to mimic in an in vitro cell culture model are: (1) the physical barrier function to protect against commensal and pathogenic organisms and (2) the generation of innate immune responses following pathogen recognition. As described above, the 3-D intestinal models exert architectural and phenotypic integrity, including highly localized expression of tight junctions, ECM proteins, and villus-like structures which results in relevant in vitro modeling of the in vivo barrier function [2, 3, 5, 10, 40]. As a result, 3-D cell culture models of small intestine (Int-407) and colon epithelial cells (HCT-8) demonstrated enhanced resistance to infection with *S.* Typhimurium [2] and with the protozoan parasite *Cryptosporidium parvum* [3], respectively, as compared to conventional monolayers. Two hours post-infection, 3-D small intestinal cultures challenged with *S.* Typhimurium exhibited a minimal loss of integrity, no induction of apoptosis, and a faster recovery of cell structure (Fig. 5.4). In contrast, monolayers infected for the same amount of time showed a major loss of integrity and a strong induction of apoptosis (between 70 % and 90 % of cells) [2] (Fig. 5.4). In vivo, *S.* Typhimurium-induced damage to small intestinal epithelium is believed to be transient and repairs quickly, with a rapid restoration of the mucosal barrier [10]. This finding is relevant to the interactions of *Salmonella enterica* with the small intestinal epithelium in vivo during the natural course of infection, since the vast majority of cases of *Salmonella*-induced gastroenteritis are not reported (less than 5 % reported), which would not be the case if ~90 % of cells

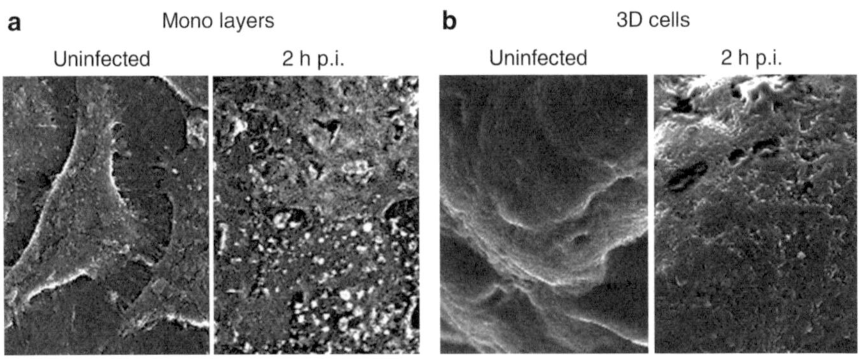

Fig. 5.4 SEM of Int-407 monolayers and 3-D Int-407 aggregates infected with *Salmonella* Typhimurium. (**a**) SEM of uninfected Int-407 monolayers (control) and infected Int-407 monolayers at 2 h postinfection (p.i.); (**b**) SEM of uninfected 3-D Int-407 cells (control) and infected 3-D Int-407 cells at 2 h p.i. [40] (All images 2000×)

in the intestinal unilayer were undergoing apoptosis. Therefore, the observed response of the 3-D small intestinal epithelial cells to *S.* Typhimurium infection shows strong similarities to the in vivo scenario.

When 3-D colon epithelial cells (derived from the HT-29 cell line) were infected with *S.* Typhimurium, a higher number of bacteria (~6-fold) were found to adhere to the surface of the 3-D cells as compared to the same cell type cultured as 2-D monolayers [40]. However, of the bacteria that had adhered, a much lower number of bacteria ultimately invaded into the 3-D cells relative to monolayer cultures 2 h post-infection (~9-fold lower). Intriguingly, despite the lower invasion levels into the 3-D cells, at 24 h post-infection there was greater structural damage that had occurred to the tissue as compared to monolayers of the same cell type (Fig. 5.5), which is the opposite of what was observed for the 3-D

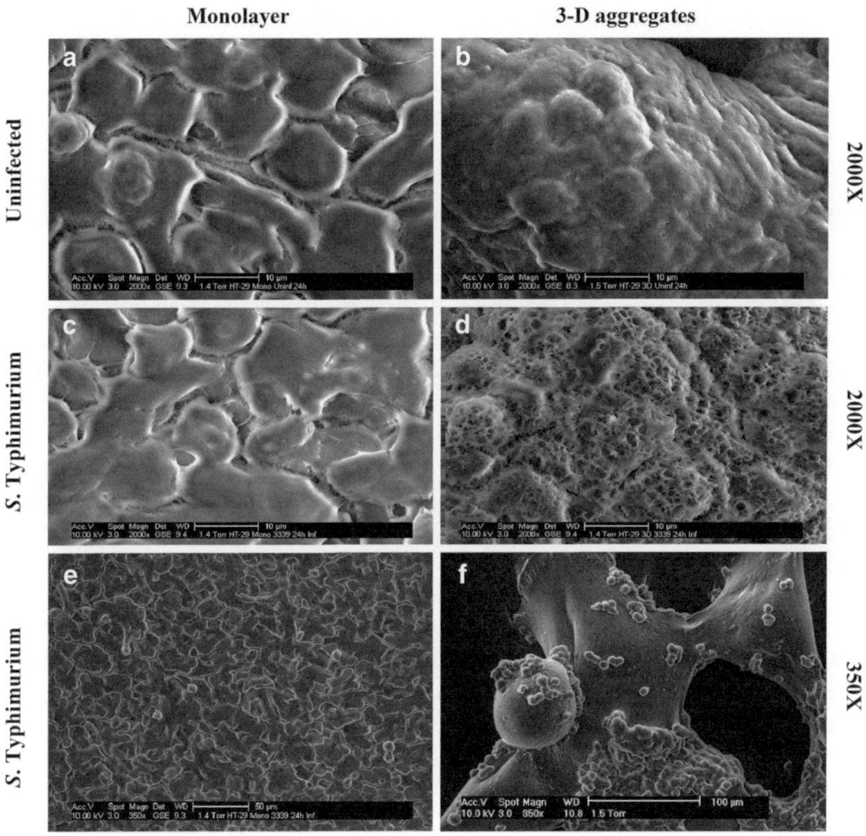

Fig. 5.5 Infection with *Salmonella* Typhimurium causes structural changes in 3-D HT-29 colon epithelial cells but not in monolayers. Uninfected HT-29 monolayers (**a**) and 3-D control cultures (**b**) show no evidence of damage. *Salmonella* infected HT-29 monolayers (**c**, **e**) and 3-D aggregates (**d**, **f**) at 24 h post-infection (pi) reveals the extent of damage to 3-D cells, but not monolayers, upon infection (**b**) [40]

small intestinal epithelial cells [40] (Fig. 5.4). As with the 3-D small intestinal model, these findings were also in agreement with in vivo infection profiles, as the colon (and not the small intestine) is the primary site for non-typhoidal *Salmonella* spp. infection and host damage [42].

The second characteristic of intestinal epithelium that is important to mimic in vitro is the protective and regulated inflammatory response following pathogen encounter, which coordinates host defense during the infection process. While both 3-D small intestinal epithelial cells and monolayers secrete pro-inflammatory cytokines in response to *S.* Typhimurium infection, production of pro-inflammatory cytokines such as tumor necrosis factor (TNF), interleukin-6 (IL-6), IL-1α, and IL-1β, is much higher in monolayers compared to the 3-D cells [2]. In contrast, 3-D small intestinal epithelial cells secrete higher levels of anti-inflammatory cytokines (such as IL-1-receptor antagonist) than monolayers. The production of lower levels of pro-inflammatory cytokines and higher levels of anti-inflammatory cytokines by 3-D cells in response to *S.* Typhimurium challenge could be related to the observed lower level of apoptosis and maintenance of structural integrity at early post-infection time points compared to monolayers, a phenotype that is relevant to the in vivo infection process of *S.* Typhimurium [43].

New Insights into Bacterial Pathogenesis Using 3-D Intestinal Models

It is hypothesized that when highly differentiated 3-D cell culture models of the parental tissue are used for infectious disease studies, a research outcome that is more relevant to the in vivo scenario will be obtained. This hypothesis has been confirmed for the enteric bacterial pathogen *Salmonella enterica* serovar Typhimurium [5, 10, 40]. According to the prevailing paradigm, *S.* Typhimurium invasion into intestinal epithelial cells is an essential step in the onset of infection and is mediated through the pathogenicity island 1 (SPI-1)-encoded type III secretion system (T3SS) [44–57]. The paradigm that the SPI-1 T3SS is essential for *S.* Typhimurium invasion has been largely based on the use of conventional 2-D monolayers of various cell lines [54–57]. However, clinical reports from human enteric salmonellosis outbreaks and data obtained with animal models demonstrated that strains lacking a functional SPI-1 T3SS could cause severe gastrointestinal disease, indicating that the SPI-I-encoded T3SS is not required for establishment of enteric infection in animals or humans [58–61]. Indeed, experiments performed on mice infected with SPI-I mutants of *S.* Typhimurium showed that this bacterium can invade intestinal epithelial cells other than enterocytes, including antigen-sampling M-cells [58]. The highly differentiated 3-D HT-29 cell culture model was able to replicate aspects of these in vivo findings, as an *S.* Typhimurium *invA* mutant (lacking a functional SPI-1 system) could invade the 3-D cells as effectively as the wild type strain [5, 40]. In mechanistic follow up studies using a series of *Salmonella* mutant strains carrying single, double, and triple knockout mutations, it was shown that invasion into the 3-D HT-29 cells occurred in the absence of all characterized *S.* Typhimurium T3SSs (SPI-1 T3SS, SPI-2 T3SS, and the flagellar pore) [5] (Fig. 5.6).

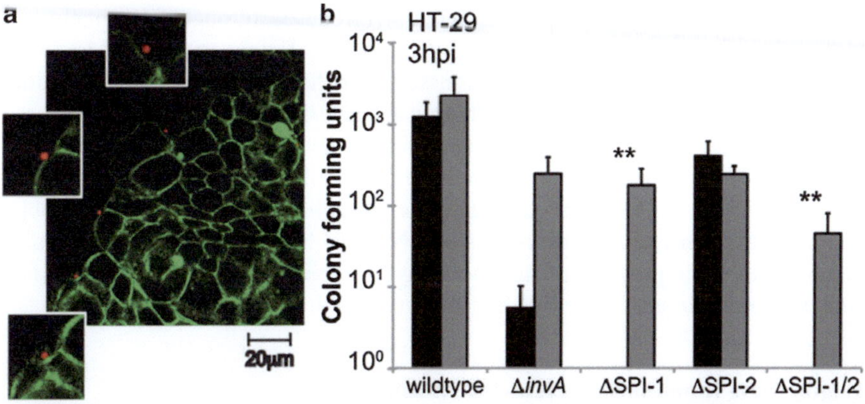

Fig. 5.6 SPI-1, SPI-2, and the flagellar secretory system are not required for *Salmonella* invasion into 3-D intestinal cells. (**a**) Representative single image from z-stack confocal immunofluorescence microscopy images (1006objective) of 3-D HT-29 aggregates exposed to red fluorescent beads (2 µm) for 1 hour, washed, fixed 3 hours post exposure, and stained with phalloidin antibody to visualize host cell actin (green) - showing that 3-D HT-29 cells do not actively take up particles. (**b**) HT-29 monolayers (*black bars*) and 3-D aggregates (*gray bars*) infected with wildtype, Δ*invA* ΔSPI-1, ΔSPI-2, and ΔSPI-1/2: *Salmonella* mutants for 1 h at an m.o.i. of 10. To measure bacterial invasion, host cells were lysed at 3 h post-infection (hpi) and intracellular bacteria were enumerated and plotted as colony forming units CFU obtained for each bacterium. Data represents the average of at least three independent experiments from separate batches of cells ($N=3$). Limit of detection is 10 CFU. *Asterisks* and *double asterisks* indicate statistical significance of $p<0.05$ and 0.001, respectively, comparing 3-D aggregates to monolayers for each bacterial strain [5]

Moreover, neither the *S.* Typhimurium wild type or T3SS mutant strains (SPI-1, SPI-2, SPI-1/SPI-2, and *flhDC*) preferentially targeted different epithelial cell types identified in the models (enterocytes, M/M-like cells, and Paneth cells) which could contribute to the establishment of an effective infection in the 3-D model and potentially in vivo. However, each T3SS was required for intracellular growth and survival, as the different T3SS mutants were unable to replicate inside the cells. Taken together, the in vitro 3-D colon model replicated aspects of the in vivo *Salmonella* infection process and provided novel insights as well, which highlights the potential of this organotypic model to serve as a valuable platform for infectious disease research and development of novel treatment approaches.

With our expanding knowledge on the role of the gut microbiome in tissue homeostasis and onset of infectious diseases [62–64], incorporating key microbial community members into in vitro cell culture models is of importance to replicate the in vivo infectious disease process. The RWV-derived 3-D HT-29 colon model was used to study whether host susceptibility to *S.* Typhimurium is altered in the presence of a gut commensal bacterium, *Lactobacillus reuteri* [6]. In humans, *L. reuteri* has been shown to protect against various gastrointestinal infections, but the mechanism of protection is unknown [65]. Using the 3-D HT-29 model, it was demonstrated that

L. reuteri's antimicrobial metabolite reuterin (produced through fermentation of glycerol) could lower the adhesion and invasion of *S.* Typhimurium [6]. Moreover, when *S.* Typhimurium colonization was assessed in the presence of reuterin-producing *L. reuteri*, a greater reduction in adhesion and invasion was observed as compared to addition of a non-reuterin-producing strain. Of note, the pure reuterin molecule and the glycerol ferment were toxic for 3-D HT-29 cells after long-term (24 h) exposure, which warrants further investigation on the applicability of these antimicrobial components against gastrointestinal disease. Nevertheless, these foundational studies demonstrate the importance of incorporating microbial community members when studying interactions of enteric pathogens with intestinal epithelial cells.

Engineering Immunocompetent 3-D Intestinal Models

As mentioned, the incorporation of multiple cell types into an in vitro model system is essential for predicting in vivo responses to infection. For example, during a *Salmonella* infection, macrophages are a critical immune cell type that is targeted by the pathogen during infection, and intracellular survival of *Salmonella* in these cells is essential for bacterial virulence [66–68]. However, most intestinal epithelial cell cultures used to model human enteric salmonellosis do not typically include macrophages. This is an important consideration, as cultures of either epithelial cells or macrophages alone can respond differently to infection with pathogens or treatment with their toxins as compared to co-culture models containing both of these cell types. Recently an immunocompetent 3-D co-culture model of human colonic epithelial cells (HT-29) and functional macrophages (U937) capable of phagocytosis was developed and applied towards the study of colonization profiles of *Salmonella enterica* pathovars with different host adaptations and disease phenotypes (Yang et al, manuscript in preparation). Differences were observed in the adherence, invasion, and intracellular survival of all *Salmonella* serovars in the co-culture model (epithelial cells and macrophages) relative to the monotypic model (epithelial cells only), indicating the contribution of macrophages in the infection process. Moreover, the data showed that the 3-D immunocompetent co-culture model could distinguish between challenge with select pathovars in ways that differed from the monotypic model, thus reinforcing the importance of using in vitro infection models that recapitulate the multicellular complexity normally encountered by *Salmonella* during enteric infection.

5.2.2 3-D Models of the Lung

In Vivo-Like Features of the Lung Mimicked in 3-D Cell Culture Models

Another major mucosal site that is prone to bacterial infections is the respiratory system. Three-dimensional cell culture models of lung tissue have been developed using the RWV bioreactor technology and applied to infection studies with respiratory pathogens or their toxins [4, 69]. In a similar fashion as the 3-D intestinal epithelial cells, 3-D alveolar (A549 adenocarcinoma cell line) epithelial cells grown on the surface of collagen I-coated microcarrier beads expressed markers indicative of tight junctional formation (ZO-1, occludin, E-cadherin, β-catenin), polarity (apical expression of villin, intercellular adhesion molecule 1 (ICAM-1); and basolateral expression of laminin and collagen IV), and mucus production (MUC5AC and MUC1) [4, 69]. Further in agreement with the 3-D intestinal epithelial model, a physiologically relevant single cell layer was observed as well as a decreased expression of carcinoma markers [2, 4, 69]. When these same cell lines were grown as 2-D monolayers, most markers of differentiation were expressed at low levels (except for cancer markers) or in a manner that is not reflective of the in vivo situation.

The more in vivo-like phenotype of 3-D A549 epithelial cells compared to monolayers was confirmed at the transcriptomic level [8]. Among 2500 differentially expressed genes (>2-fold change) between 3-D and 2-D A549 cultures, pathways involved in cell signaling, membrane trafficking, and ECM production and remodeling (such as mucins and matrix metalloproteinases) were induced in 3-D cultures, while pathways related to advancement of cell cycle and tumorigenic oncogenes were downregulated [8]. The gene expression data set confirms in vivo-like traits of the 3-D A549 cell culture model [8, 69].

The RWV-derived 3-D A549 lung model was further advanced by incorporating the most important innate immune cell of the lung, the macrophage [4]. This model was generated by co-culturing U937 monocytes with already differentiated 3-D A549 cultures. As a result, a 3-D immunocompetent model was obtained that exerted key phenotypic and architectural characteristics of the parental tissue. Specifically, U937 monocytes spontaneously differentiated into phagocytic macrophage-like cells that were localized at the apical surface of the 3-D A549 cells in a physiologically relevant ratio (Fig. 5.7). Furthermore, the addition of monocytes to the 3-D A549 epithelial cells did not negatively impact the tight junctional formation, polarity, and mucus content of the model. In contrast, the presence of macrophage-like cells on the surface of 3-D A549 epithelial cells further enhanced the apical expression of ICAM-1 compared to monotypic 3-D A549 cultures (Fig. 5.7), a phenotype that is indicative of a high differentiation status [70, 71].

New Insights into Infectious Agent Pathogenesis Using 3-D Lung Models

The 3-D A549 lung model was used to study the infectious disease process of the major respiratory pathogen *Pseudomonas aeruginosa* and of the biological agent *Francisella tularensis* [8, 69]. Both *P. aeruginosa* and *F. tularensis* invaded the 3-D A549 epithelial cells to a lower level compared to monolayer controls. In addition, 3-D A549 cells were more resistant to *P. aeruginosa* infection as they maintained structural integrity for longer periods of time than 2-D monolayers [69]. These data are relevant to the initial interactions of *P. aeruginosa* with the healthy

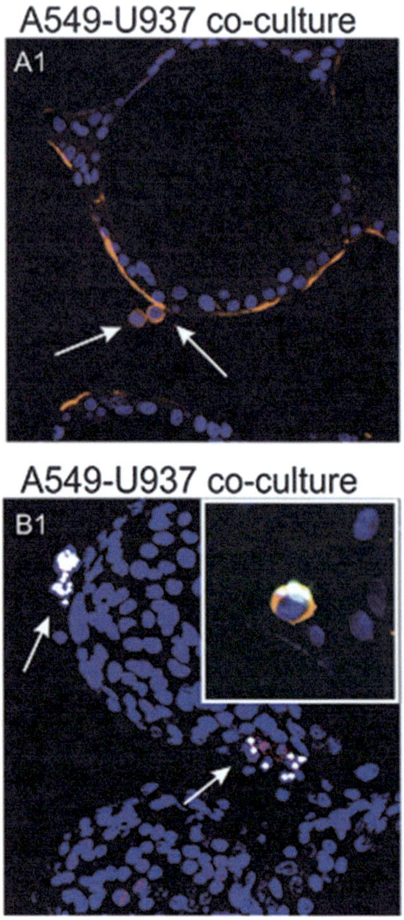

Fig. 5.7 ICAM-1 expression and phagocytosis activity of 3-D A549-U937 co-cultures. (**A1**). ICAM-1 expression on co-cultured alveolar epithelium and monocytes/macrophages in the 3-D A549-U937 co-culture. (**B1**). Phagocytosis of 2 μm fluorescent beads by the 3-D A549-U937 co-culture. Macrophage-like cells in the 3-D co-culture (**B1**) are labeled with the CD45 antibody (*orange*) showing specific uptake of beads (*white–green*) by this cell population. Z-stack assembly is presented. Cell nuclei are *stained blue* (DAPI) and images are based on 400× magnifications. *White arrows* point out macrophage-like cells [120]. With permission from Wiley

host in vivo, since polarized lung epithelial cells are less susceptible to *P. aeruginosa* infection compared to non-polarized cells [11, 72]. Indeed, the occurrence of non-polarized epithelial cells in vivo is often a result of epithelial layer injury, and enhances accessibility of the basolateral cell surface and transient apical expression of receptors that mediate *P. aeruginosa* binding and initiation of disease [73–78]. Moreover, expression of the cystic fibrosis transmembrane conductance regulator (CFTR) protein, which is believed to be essential for clearance of *P. aeruginosa* in vivo [79, 80], is dependent on cell polarity [74, 81, 82]. Besides the presence of polarized epithelial cells in the 3-D A549 model, the in vivo-like expression of tight junctional proteins and production of mucus mimics the barrier function of the respiratory mucosa and could further contribute to the lower colonization of 3-D A549 cells by *P. aeruginosa* compared to 2-D monolayers [11, 69]. Hence, these data emphasize that the characteristics of differentiation present in RWV-derived 3-D lung epithelial models are relevant for studying the infectious disease process of clinically significant microorganisms.

New Insights into Host Responses to Infection Using 3-D Lung Models

The 3-D A549 lung epithelial model was utilized to study the production of inflammatory mediators in response to *P. aeruginosa* infection [69]. A major difference in cytokine production between 3-D cultures and monolayers infected with *P. aeruginosa* was the stronger production of anti-inflammatory interleukin-10 (IL-10) in 3-D A549 cells compared to monolayers. One possibility is that the higher levels of IL-10 in 3-D cultures act as a negative-feedback system to compensate for induction of pro-inflammatory cytokines like IL-12. In contrast, the monolayer controls did not strongly induce IL-10 production, which could suggest a weaker negative-feedback mechanism. Mimicking the in vivo inflammatory cascades in vitro is, among others, of importance in the context of chronic lung infections in patients with cystic fibrosis, where *P. aeruginosa* is believed to continuously stimulate the production of cytokines, contributing to destructive lung disease [83].

As described in the previous section, 3-D A549 epithelial cells were more resistant to *F. tularensis* infection compared to monolayers [8]. In order to investigate whether differential host responses to infection between 3-D cells and monolayers could be causative for the observed differences in colonization, global transcriptomic profiling was performed following infection. Based on the transcriptional profile, three host pathways were further investigated as potential causes for higher resistance of 3-D cells to infection, i.e. matrix metalloproteinases, complement system, and CDC42. While all three pathways were induced in *F. tularensis*-infected 3-D cells, inhibition of these pathways using targeted inhibitors did not outbalance observed differences in bacterial invasion between both cell culture models. Therefore, further investigation of host pathways in response to infection might shed light on the observed differences in colonization profiles.

Like all other tissues in the human body, the respiratory mucosa is comprised of numerous different cell types. Recapitulating aspects of the in vivo multicellular complexity in vitro can allow mimicking complex intercellular interactions that orchestrate the response of each single cell type to pathogens and their virulence factors. In this regard, the 3-D immunocompetent model that is comprised of A549 epithelial cells and U937 macrophages was used to study host responses to a quorum sensing molecule that is produced by *P. aeruginosa* during the natural infection process [4]. The quorum sensing molecule had been reported to exert cytotoxic effects in monotypic cultures of monocytes and macrophages [84, 85]. In contrast, when monocytes/macrophages were co-cultured with 3-D A549 epithelial cells, no such cytotoxicity was observed [4]. The authors demonstrated that 3-D A549 epithelial cells could degrade the quorum sensing molecules before they exerted cytotoxic effects on monocytes/macrophages in the 3-D co-culture (Fig. 5.8). These findings demonstrate the importance of incorporating multiple cell types into in vitro 3-D cultures, to enhance the physiological relevance of the host response to infection.

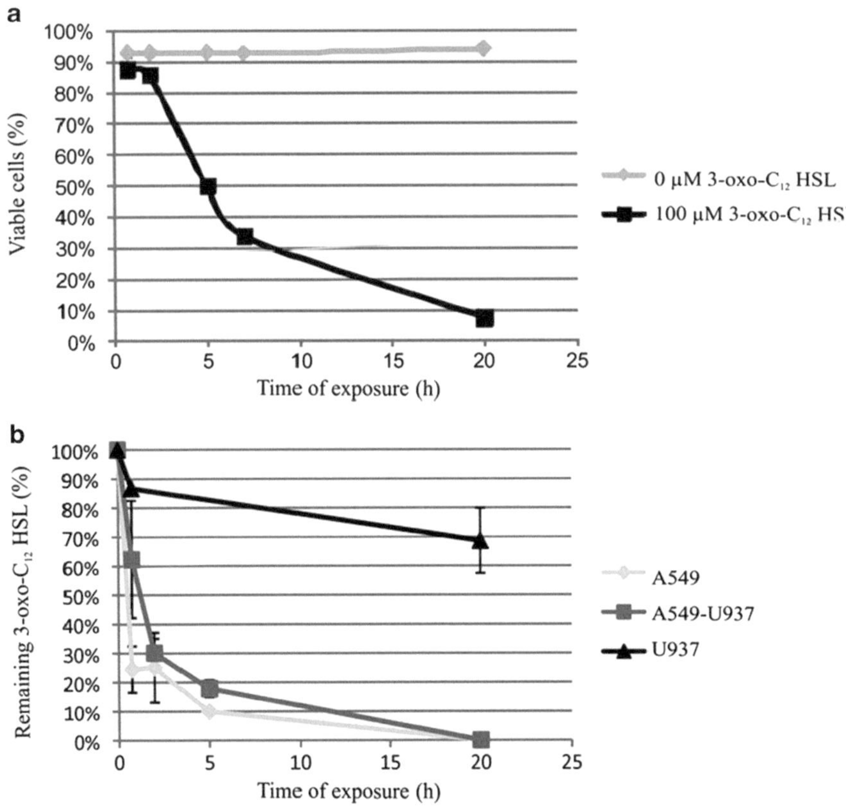

Fig. 5.8 (**a**) Viability of U937 cells as a function of time following exposure to 0 μM (DMSO control) and 100 μM 3-oxo-C_{12} HSL. Standard deviations for U937 exposed to 0 μM and 100 μM were lower than 1 % and 2 %, respectively, and are therefore not visible on the chart ($n=2$). (**b**). Concentration of 3-oxo-C_{12} HSL as a function of time after exposure to 3-D A549, 3-D A549-U937, and U937 cultures. Results are presented as the ratio of the remaining concentration to the original 3-oxo-C_{12} HSL concentration (100 μM). *Error bars* represent the mean ± standard deviation ($n=2$) [120]. With permission from Wiley

5.2.3 3-D Models of the Female Reproductive System

In Vivo-Like Features of the Female Reproductive Tract Mimicked in 3-D Cell Culture Models

Three-dimensional epithelial cell models of vaginal and endocervical tissues have been developed using the RWV bioreactor technology [86–89]. Vaginal epithelial cells (V191 immortalized cell line) grown on collagen I-coated microcarrier beads progressively developed through an early and late developmental stage [86]. Specifically, the early stage was characterized by the presence of a single cell layer on the microcarrier beads with phenotypic traits of the early vaginal canal and fornix of postnatal BALB/c mice, including sparse apical distribution of microvilli. In the late developmental phase, 3-D vaginal epithelial cell aggregates showed characteristics of polarized and differentiated vaginal tissue, including flattened multiple cell layers with densely packed lawns of microvilli, microfolds (i.e., rugae), and secretory vesicles/mucus deposits on the apical cell surface and between connecting cells. Immunohistochemical profiling confirmed production of the major vaginal mucins MUC1 and MUC4, and in vivo-like expression of tight junctional proteins ZO-1 and E-cadherin. Furthermore, involucrin, a terminal differentiation marker of keratinocytes, was expressed at the cell surface of 3-D vaginal epithelial cells. In a similar fashion as for the 3-D intestinal and lung epithelial cells, monolayers of vaginal epithelial cells did not exhibit most of the in vivo-like characteristics of their 3-D counterparts. Taken together, these data suggest that the phenotype of 3-D vaginal epithelial cells is representative of normal human vaginal tissue.

The 3-D endocervical epithelial cell model (A2EN cell line) also exhibited hallmarks of the human parental tissue, including a single layer of cobblestone-shaped cells that expressed microvilli, tight junctions, and secretory vesicles [89]. In addition, genes encoding the major endocervical mucins (MUC1, MUC5, MUC5AC, MUC6, and MUC16) and antimicrobial peptides (SLP1, CCL20, LL37, hBD-1, hBD-2, hBD-3, hBD-4) all displayed higher expression in 3-D compared to monolayer cells. The listed phenotypic traits are reflective of the human endocervical tissue, once more highlighting the potential of the RWV bioreactor technology to generate tissue culture models with striking similarities to the parental tissue.

New Insights into Host Responses to Infection Using 3-D Models of the Female Reproductive Tract

The RWV-derived 3-D vaginal model was used for testing the cytotoxicity of Nonoxynol-9 (N-9), the active ingredient of the over-the-counter spermicide Conceptrol [86]. This spermicide showed initial potential as an anti-HIV-1 microbicide both in vitro (monolayers and tissue explants) and in animal models. However, Conceptrol was ineffective in clinical trials, and even increased incidence of HIV-1 transmission in some cases [90, 91]. Indeed, further research demonstrated in vivo toxicity and in vitro epithelial barrier disruption [90, 92]. The human 3-D vaginal epithelial model was able to parallel the in vivo toxicity of N-9 [86], and could thus potentially serve as an interesting platform for candidate microbicide testing before advancing to clinical trials (Fig. 5.9).

In an effort to understand the microbial flora that plays a role in the initiation of bacterial sexually transmitted vaginosis, the 3-D vaginal epithelial model was exposed to bacterial members of the healthy and vaginosis-associated microbiome [87]. In particular, *Atopobium vaginae*, a vaginosis-associated commensal bacterium initiated a robust pro-inflammatory immune response through induction

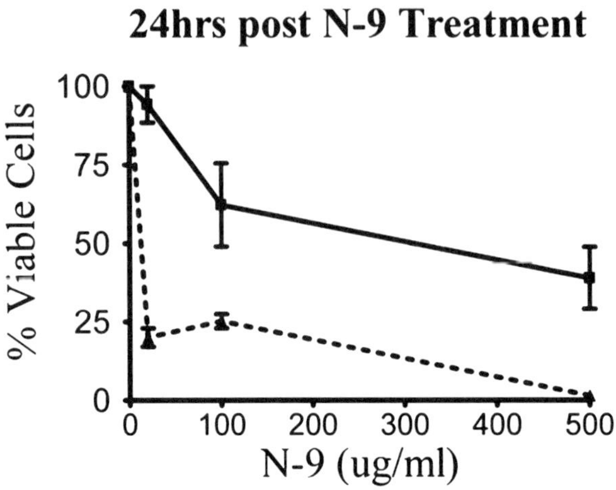

Fig. 5.9 Validation of 3-D vaginal epithelial model for microbicide evaluation. Nonoxynol-9 dose dependent viability curve 24 h post-treatment of 3-D vaginal epithelial cells compared to confluent monolayer cultures of the same cells. *Solid lines* represent the 3-D vaginal epithelial model, whereas *dashed lines* represent confluent monolayers of vaginal epithelial cells. *Error bars* are shown as standard error of the mean (±SEM) [86]

of cytokines and antimicrobial peptides. Since this pro-inflammatory response can affect barrier function, colonization of vaginal epithelium with *A. vaginae* could potentially contribute to the enhanced susceptibility to sexually transmitted infections.

Given the in vivo-like expression of mucins and antimicrobial peptide genes in the 3-D endocervical model, this model was utilized for studying the expression of these innate immunity markers in response to various bacterial and viral products [89]. In addition, the host transcriptional response to *Mycobacterium genitalium*, a commonly reported sexually transmitted bacterial pathogen, was studied using the 3-D endocervical model [88]. This approach demonstrated the induction of MUC1, MUC4, and MUC16 in 3-D endocervical epithelial cells regardless of the microbial product tested [89]. In agreement with these data, expression of MUC1 was induced in response to *M. genitalium* even though other mucins were not differentially expressed [87]. With regard to antimicrobial peptide production in response to microbial products or *M. genitalium*, unique antimicrobial peptide immunosignatures of the 3-D endocervical epithelial cells were identified depending on the microbial product/bacterium to which the cells were exposed. Genes encoding antimicrobial peptides that were induced in response to *M. genitalium* infection included three human β-defensin genes (hBD-1, hBD-2, and hBD-4) and the secreted leukocyte peptidase inhibitor (SLP1). In addition, other markers of innate immunity were induced, including cytokines and chemokines, and this specific acute infection signature could shed light into the infectious disease mechanisms of *M. genitalium*. Of particular interest is the induction of IL-7, which plays a role in the entry and replication of HIV in immune cells and could thus potentially provide clues why patients with *M. genitalium* are more susceptible to HIV and other sexually transmitted diseases.

5.2.4 3-D Models of Other Tissue for Infection Studies

In addition to the 3-D cell culture models of the human intestine, lung and female reproductive tract, which are the most commonly used RWV-derived mucosal cell culture models for host-pathogen interaction studies, models of other tissues have been developed and applied for infectious disease research. For example, several 3-D tissue culture models enabled the cultivation and study of pathogens that lacked suitable in vitro or in vivo model systems [10]. For example, a 3-D tonsillar model was applied for culturing of *Borrelia burgdorferi*, the etiological agent of Lyme disease for which no ex vivo tissue system was available [93]. *B. burgdorferi* invaded the 3-D tonsillar tissue model readily and was able to mount a productive intracellular infection. In a similar fashion, a 3-D liver model offered new research possibilities for hepatitis C and E viral infections, as these pathogens lacked adequate cell culture and small-animal infection models [94, 95]. Specifically, differentiated and polarized 3-D liver models enabled viral entry and replication. A recently developed 3-D neuronal model allowed the establishment of persistent infection of Varicella Zoster Virus (VZV), a neurotropic human alphaherpesvirus that lacked long-duration in vitro infection models and animal models (as it is a human-specific pathogen) [96]. The RWV cell culture system was also used to generate a Human Immunodeficiency Virus (HIV) infection model, based on surgically excised human lymphoid tissue blocks [97]. These examples illustrate that 3-D models of various tissues replicate phenotypic characteristics of the parental tissue that are essential for the pathogen to replicate and initiate infectious diseases.

The RWV has also been utilized to study herpes viral replication and reactivation [1, 98–100]. Studying viral replication and reactivation in microgravity-analogue culture systems is of interest since it has been well-documented that reactivation of Epstein–Barr virus (EBV) occurs during spaceflight, as evidenced through increased viral load and lytic viral transcripts in saliva [101–109]. EBV is one of the most common human viruses, is causative for mononucleosis and associated with particular types of cancer (e.g., Burkitt's lymphoma) and autoimmune diseases (e.g., dermatomyositis, arthritis) [110]. Unveiling the mechanisms of viral reactivation during spaceflight is important to ensure astronaut health during space travel, and may provide novel insight into mechanisms of viral pathogenesis for the general public. The first publication on viral replication in RWV-exposed 3-D cell cultures reported enhanced Rhinovirus replication as compared to static and roller bottle cultures [1]. With regard to viral reactivation, Long et al. first reported impaired viral reactivation of EBV-positive lymphoblastoid cell lines (P3HR-1 and Daudi) when exposed to RWV conditions as compared to conventional monolayers [98, 99]. Transfer of RWV-cultured 3-D lymphoblastoid cells to static conditions could not restore reactivation, and chemical induction of viral reactivation was less effective in RWV compared to static cultures [98]. Recently, Brinley et al. further confirmed that microgravity-analogue conditions presumably do not induce EBV reactivation [100]. Specifically, radiation exposure, and not RWV culture, was a major contributing factor that enhanced expression of the EBV lytic antigen ZEBRA in an EBV-positive cell line (Raji). Interestingly, EBV-positive cells exposed to RWV conditions decreased DNA repair and experienced greater DNA damage and production of reactive oxygen species (ROS) following radiation exposure compared to uninfected cells. Furthermore, an EBV-negative cell line (BJAB), but not the EBV-positive cell line (Raji), experienced decreased viability and increased apoptosis when exposed to RWV growth conditions as compared to growth in conventional monolayers [100]. Taken together, these findings suggest that EBV infection could preclude natural defense mechanisms against DNA damage, which could lead to a greater sensitivity to DNA-damaging agents, such as spaceflight irradiation.

In addition to use of RWV-derived 3-D models for cultivation and study of microorganisms that are difficult to cultivate, other 3-D tissue culture models of mucosal origin were also used to study the infectious disease process of bacteria that can easily be cultured. One example is a 3-D model of human bladder urothelium that was used to gain mechanistic insights into uropathogenic *Escherichia coli* (EPEC) infections. EPEC colonized and induced in vivo-like exfoliation (an innate host defense mechanism that clears bacteria) of the 3-D bladder cells with less damage compared to monolayer cultures infected with the same strain [111]. A UPEC virulence factor (alpha-hemolysin) was identified as a causative factor for the observed exfoliation, emphasizing the usefulness of in vivo-like 3-D models to understand the role of bacterial virulence factors in the various steps of the infectious disease process.

In addition to the described 3-D models that were generated using the RWV bioreactor technology and were applied for dissecting infectious disease mechanisms, a plethora of other models have been developed. These include 3-D models of various stem cells, pancreatic cells, cardiac muscle, bone, cartilage, cornea, periodontal ligament, and prostate [112–119], which could also be explored to advance the fields of infectious diseases, drug testing, and beyond.

5.3 Conclusions

The studies outlined in this chapter represent major advances towards the establishment of advanced in vitro model systems that are able to recapitulate the complex 3-D microenvironment found in vivo. The establishment and characterization of a variety of 3-D tissues using innovative RWV bioreactor technology and their practical application as human surrogates for host–microbe interactions have provided specific examples of how the study of microbial pathogenesis and infectious disease can be advanced by using appropriate, biologically meaningful 3-D models. Studies have shown that RWV-derived 3-D tissue models respond to microbial pathogens (and their toxins) as well as commensals in key ways that reflect the in vivo condition, and which cannot be recapitulated by traditional in vitro cell culture models. The ability of these models to serve as faithful predictors of the human response to microbicides has also been demonstrated. Collectively, these studies have facilitated: (1) the meaningful dissection of the molecular mechanisms of infectious disease, (2) the study of pathogens that lack suitable cell culture and animal models, and (3) the discovery of novel mechanisms underlying the transition between tissue homeostasis and infectious disease. Ongoing studies are focused on the integration of additional cell types, novel bioscaffolds, and microbial consortia into RWV-derived 3-D models. In addition, the incorporation of additional physiologically relevant biomechanical forces—e.g., stretching, compression, heterogeneous fluid shear—into these models will provide yet another level of complexity reflective of the parental tissue in vivo.

The enhanced knowledge derived from using such advanced physiological models in host-pathogen studies will help drive development of more rapid and accurate predictive models for disease risks and outcomes and next generation therapeutic modalities to combat infectious disease.

Questions for Future Research

- Build 3-D models of animal tissues to study zoonotic infections
- Advance physiological relevance of 3-D models, including incorporation of additional components of the tissue microenvironment, including different immune cell types, primary cells, bioscaffolds, microbial consortia, other biophysical forces, etc.
- What cellular pathways regulate the transition from 2-D to 3-D in RWV-derived models? Are these responses conserved across multiple cell types?
- Can RWV-derived models be engineered with enough complexity to one day replace animal testing?
- Can RWV-derived models be engineered with enough complexity to be used for tissue/organ replacement?

References

1. Long, J. P., Pierson, S. S., & Hughes, J. H. (1998). Rhinovirus replication in HeLa cells cultured under conditions of simulated microgravity. *Aviation, Space, and Environmental Medicine, 69*, 851–856.
2. Nickerson, C. A., et al. (2001). Three-dimensional tissue assemblies: Novel models for the study of *Salmonella enterica* serovar Typhimurium pathogenesis. *Infection and Immunity, 69*(11), 7106–7120.

3. Alcantara Warren, C., et al. (2008). Detection of epithelial-cell injury, and quantification of infection, in the HCT-8 organoid model of cryptosporidiosis. *The Journal of Infectious Diseases, 198*(1), 143–149.
4. Crabbé, A., et al. (2011). Alveolar epithelium protects macrophages from quorum sensing-induced cytotoxicity in a three-dimensional co-culture model. *Cellular Microbiology, 13*(3), 469–481.
5. Radtke, A. L., et al. (2010). Analysis of interactions of *Salmonella* type three secretion mutants with 3-D intestinal epithelial cells. *PLoS One, 5*(12), e15750.
6. De Weirdt, R., et al. (2012). Glycerol supplementation enhances *L. reuteri's* protective effect against *S.* Typhimurium colonization in a 3-D model of colonic epithelium. *PLoS One, 7*(5), e37116.
7. Barrila, J., et al. (2015) Microgravity uniquely alters the host-pathogen interaction between human intestinal epithelial cells and *Salmonella enterica* serovar Typhimurium in 115th American Society for Microbiology General Meeting, New Orleans, LA.
8. David, J., Sayer, N. M., & Sarkar-Tyson, M. (2014). The use of a three-dimensional cell culture model to investigate host-pathogen interactions of *Francisella tularensis* in human lung epithelial cells. *Microbes and Infection, 16*(9), 735–745.
9. GBD 2013 Mortality and Causes of Death Collaborators. (2015). Global, regional, and national age-sex specific all-cause and cause-specific mortality for 240 causes of death, 1990–2013: A systematic analysis for the Global Burden of Disease Study 2013. *Lancet, 385*(9963), 117–171.
10. Barrila, J., et al. (2010). Organotypic 3D cell culture models: Using the rotating wall vessel to study host-pathogen interactions. *Nature Reviews Microbiology, 8*(11), 791–801.
11. Crabbé, A., Ledesma, M. A., & Nickerson, C. A. (2014). Mimicking the host and its microenvironment in vitro for studying mucosal infections by *Pseudomonas aeruginosa. Pathogens and Disease, 71*(1), 1–19.
12. Singh, B., et al. (2012). Human pathogens utilize host extracellular matrix proteins laminin and collagen for adhesion and invasion of the host. *FEMS Microbiology Reviews, 36*(6), 1122–1180.
13. Lopetuso, L. R., et al. (2014). The gastrointestinal microbiome – Functional interference between stomach and intestine. *Best Practice & Research Clinical Gastroenterology, 28*(6), 995–1002.
14. Duerkop, B. A., Vaishnava, S., & Hooper, L. V. (2009). Immune responses to the microbiota at the intestinal mucosal surface. *Immunity, 31*(3), 368–376.
15. Bals, R., & Hiemstra, P. S. (2004). Innate immunity in the lung: How epithelial cells fight against respiratory pathogens. *The European Respiratory Journal, 23*(2), 327–333.
16. Kraehenbuhl, J. P., & Corbett, M. (2004). Immunology. Keeping the gut microflora at bay. *Science, 303*(5664), 1624–1625.
17. Ingber, D. E., & Tensegrity, I. I. (2003). How structural networks influence cellular information processing networks. *Journal of Cell Science, 116*(Pt 8), 1397–1408.
18. Ingber, D. E., & Tensegrity, I. (2003). Cell structure and hierarchical systems biology. *Journal of Cell Science, 116*(Pt 7), 1157–1173.
19. Ingber, D. E. (2003). Mechanobiology and diseases of mechanotransduction. *Annals of Medicine, 35*(8), 564–577.
20. Bissell, M. J., Rizki, A., & Mian, I. S. (2003). Tissue architecture: The ultimate regulator of breast epithelial function. *Current Opinion in Cell Biology, 15*(6), 753–762.
21. Xu, R., Boudreau, A., & Bissell, M. J. (2009). Tissue architecture and function: Dynamic reciprocity via extra- and intra-cellular matrices. *Cancer and Metastasis Reviews, 28*(1–2), 167–176.
22. Nelson, C. M., & Bissell, M. J. (2006). Of extracellular matrix, scaffolds, and signaling: Tissue architecture regulates development, homeostasis, and cancer. *Annual Review of Cell and Developmental Biology, 22*, 287–309.
23. Hagios, C., Lochter, A., & Bissell, M. J. (1998). Tissue architecture: The ultimate regulator of epithelial function? *Philosophical Transactions of the Royal Society of London. Series B, Biological Sciences, 353*(1370), 857–870.

24. Griffith, L. G., & Swartz, M. A. (2006). Capturing complex 3D tissue physiology in vitro. *Nature Reviews Molecular Cell Biology, 7*(3), 211–224.
25. Schmeichel, K. L., & Bissell, M. J. (2003). Modeling tissue-specific signaling and organ function in three dimensions. *Journal of Cell Science, 116*(Pt 12), 2377–2388.
26. Cukierman, E., Pankov, R., & Yamada, K. M. (2002). Cell interactions with three-dimensional matrices. *Current Opinion in Cell Biology, 14*(5), 633–639.
27. Yamada, K. M., & Cukierman, E. (2007). Modeling tissue morphogenesis and cancer in 3D. *Cell, 130*(4), 601–610.
28. Nickerson, C. A., & Ott, C. M. (2004). A new dimension in modeling infectious disease. *ASM News, 70*(4), 169–175.
29. Nickerson, C. A., Richter, E. G., & Ott, C. M. (2007). Studying host-pathogen interactions in 3-D: Organotypic models for infectious disease and drug development. *Journal of Neuroimmune Pharmacology, 2*(1), 26–31.
30. Mueller-Klieser, W. (1997). Three-dimensional cell cultures: From molecular mechanisms to clinical applications. *The American Journal of Physiology, 273*(4 Pt 1), C1109–C1123.
31. Abbott, A. (2003). Biology's new dimension. *Nature, 424*, 870–872.
32. Debnath, J., & Brugge, J. S. (2005). Modelling glandular epithelial cancers in three-dimensional cultures. *Nature Reviews Cancer, 5*(9), 675–688.
33. Roskelley, C. D., Desprez, P. Y., & Bissell, M. J. (1994). Extracellular matrix-dependent tissue-specific gene expression in mammary epithelial cells requires both physical and biochemical signal transduction. *Proceedings of the National Academy of Sciences of the United States of America, 91*(26), 12378–12382.
34. Smalley, K. S., Lioni, M., & Herlyn, M. (2006). Life isn't flat: Taking cancer biology to the next dimension. *In Vitro Cellular & Developmental Biology. Animal, 42*(8–9), 242–247.
35. Freshney, R. I. (Ed.). (2010). *Culture of animal cells: A manual of basic technique and specialized applications* (6th ed.). Hoboken, NJ: Wiley.
36. Selick, H. E., Beresford, A. P., & Tarbit, M. H. (2002). The emerging importance of predictive ADME simulation in drug discovery. *Drug Discovery Today, 7*(2), 109–116.
37. Guo, P., Weinstein, A. M., & Weinbaum, S. (2000). A hydrodynamic mechanosensory hypothesis for brush border microvilli. *American Journal of Physiology. Renal Physiology, 279*(4), F698–F712.
38. Furness, J. B., Kunze, W. A., & Clerc, N. (1999). Nutrient tasting and signaling mechanisms in the gut. II. The intestine as a sensory organ: Neural, endocrine, and immune responses. *The American Journal of Physiology, 277*(5 Pt 1), G922–G928.
39. Castro, G. A., & Arntzen, C. J. (1993). Immunophysiology of the gut: A research frontier for integrative studies of the common mucosal immune system. *The American Journal of Physiology, 265*(4 Pt 1), G599–G610.
40. Honer zu Bentrup, K., et al. (2006). Three-dimensional organotypic models of human colonic epithelium to study the early stages of enteric salmonellosis. *Microbes and Infection, 8*(7), 1813–1825.
41. Carvalho, H. M., et al. (2005). A three-dimensional tissue culture model for the study of attach and efface lesion formation by enteropathogenic and enterohaemorrhagic *Escherichia coli*. *Cellular Microbiology, 7*(12), 1771–1781.
42. Darwin, K. H., & Miller, V. L. (1999). Molecular basis of the interaction of *Salmonella* with the intestinal mucosa. *Clinical Microbiology Reviews, 12*(3), 405–428.
43. Rastogi, D., Ratner, A. J., & Prince, A. (2001). Host-bacterial interactions in the initiation of inflammation. *Paediatric Respiratory Reviews, 2*(3), 245–252.
44. Galan, J. E. (2001). *Salmonella* interactions with host cells: Type III secretion at work. *Annual Review of Cell and Developmental Biology, 17*, 53–86.
45. Kuhle, V., & Hensel, M. (2004). Cellular microbiology of intracellular *Salmonella enterica*: Functions of the type III secretion system encoded by *Salmonella* pathogenicity island 2. *Cellular and Molecular Life Sciences, 61*(22), 2812–2826.
46. Pullinger, G. D., et al. (2007). Systemic translocation of *Salmonella enterica* serovar Dublin in cattle occurs predominantly via efferent lymphatics in a cell-free niche and requires type

III secretion system 1 (T3SS-1) but not T3SS-2. *Infection and Immunity, 75*(11), 5191–5199.
47. Raffatellu, M., et al. (2005). SipA, SopA, SopB, SopD, and SopE2 contribute to *Salmonella enterica* serotype typhimurium invasion of epithelial cells. *Infection and Immunity, 73*(1), 146–154.
48. Baumler, A. J., et al. (1997). Synergistic effect of mutations in invA and lpfC on the ability of *Salmonella typhimurium* to cause murine typhoid. *Infection and Immunity, 65*(6), 2254–2259.
49. Bueno, S. M., et al. (2010). *Salmonella* pathogenicity island 1 differentially modulates bacterial entry to dendritic and non-phagocytic cells. *Immunology, 130*(2), 273–287.
50. Morgan, E., et al. (2004). Identification of host-specific colonization factors of *Salmonella enterica* serovar Typhimurium. *Molecular Microbiology, 54*(4), 994–1010.
51. Watson, P. R., et al. (1995). Characterization of intestinal invasion by *Salmonella typhimurium* and *Salmonella dublin* and effect of a mutation in the invH gene. *Infection and Immunity, 63*(7), 2743–2754.
52. Tenor, J. L., et al. (2004). *Caenorhabditis elegans*-based screen identifies *Salmonella* virulence factors required for conserved host-pathogen interactions. *Current Biology, 14*(11), 1018–1024.
53. Santos, R. L., et al. (2001). Animal models of *Salmonella* infections: Enteritis versus typhoid fever. *Microbes and Infection, 3*(14–15), 1335–1344.
54. Zhang, S., et al. (2003). Molecular pathogenesis of *Salmonella enterica* serotype typhimurium-induced diarrhea. *Infection and Immunity, 71*(1), 1–12.
55. Sun, J., et al. (2005). Crosstalk between NF-kappaB and beta-catenin pathways in bacterial-colonized intestinal epithelial cells. *American Journal of Physiology. Gastrointestinal and Liver Physiology, 289*(1), G129–G137.
56. Ye, Z., et al. (2007). *Salmonella* effector AvrA regulation of colonic epithelial cell inflammation by deubiquitination. *The American Journal of Pathology, 171*(3), 882–892.
57. Zhou, D., & Galan, J. (2001). *Salmonella* entry into host cells: The work in concert of type III secreted effector proteins. *Microbes and Infection, 3*(14–15), 1293–1298.
58. Clark, M. A., et al. (1996). Invasion of murine intestinal M cells by *Salmonella typhimurium* inv mutants severely deficient for invasion of cultured cells. *Infection and Immunity, 64*(10), 4363–4368.
59. Rescigno, M., et al. (2001). Dendritic cells shuttle microbes across gut epithelial monolayers. *Immunobiology, 204*(5), 572–581.
60. Murray, R. A., & Lee, C. A. (2000). Invasion genes are not required for *Salmonella enterica* serovar typhimurium to breach the intestinal epithelium: Evidence that salmonella pathogenicity island 1 has alternative functions during infection. *Infection and Immunity, 68*(9), 5050–5055.
61. Hu, Q., et al. (2008). *Salmonella enterica* serovar Senftenberg human clinical isolates lacking SPI-1. *Journal of Clinical Microbiology, 46*(4), 1330–1336.
62. Kaiko, G. E., & Stappenbeck, T. S. (2014). Host-microbe interactions shaping the gastrointestinal environment. *Trends in Immunology, 35*(11), 538–548.
63. Lozupone, C. A., et al. (2012). Diversity, stability and resilience of the human gut microbiota. *Nature, 489*(7415), 220–230.
64. Integrative HMP (iHMP) Research Network Consortium. (2014). The Integrative Human Microbiome Project: Dynamic analysis of microbiome-host omics profiles during periods of human health and disease. *Cell Host & Microbe, 16*(3), 276–289.
65. Weizman, Z., Asli, G., & Alsheikh, A. (2005). Effect of a probiotic infant formula on infections in child care centers: Comparison of two probiotic agents. *Pediatrics, 115*(1), 5–9.
66. Hurley, D., et al. (2014). *Salmonella*-host interactions – Modulation of the host innate immune system. *Frontiers in Immunology, 5*, 481.
67. Behnsen, J., et al. (2015). Exploiting host immunity: The *Salmonella* paradigm. *Trends in Immunology, 36*(2), 112–120.
68. Ruby, T., et al. (2012). *Salmonella*'s long-term relationship with its host. *FEMS Microbiology Reviews, 36*(3), 600–615.

69. Carterson, A. J., et al. (2005). A549 lung epithelial cells grown as three-dimensional aggregates: Alternative tissue culture model for *Pseudomonas aeruginosa* pathogenesis. *Infection and Immunity, 73*(2), 1129–1140.
70. Christensen, B. B., et al. (1998). Establishment of new genetic traits in a microbial biofilm community. *Applied and Environmental Microbiology, 64*, 2247–2255.
71. Kang, B. H., et al. (1993). Intercellular adhesion molecule-1 expression on the alveolar epithelium and its modification by hyperoxia. *American Journal of Respiratory Cell and Molecular Biology, 9*(4), 350–355.
72. Engel, J., & Eran, Y. (2011). Subversion of mucosal barrier polarity by *Pseudomonas aeruginosa. Frontiers in Microbiology, 2*, 114.
73. Fleiszig, S. M., et al. (1997). Epithelial cell polarity affects susceptibility to *Pseudomonas aeruginosa* invasion and cytotoxicity. *Infection and Immunity, 65*(7), 2861–2867.
74. Plotkowski, M. C., et al. (1999). *Pseudomonas aeruginosa* internalization by human epithelial respiratory cells depends on cell differentiation, polarity, and junctional complex integrity. *American Journal of Respiratory Cell and Molecular Biology, 20*(5), 880–890.
75. Kazmierczak, B. I., Mostov, K., & Engel, J. N. (2004). Epithelial cell polarity alters Rho-GTPase responses to *Pseudomonas aeruginosa. Molecular Biology of the Cell, 15*(2), 411–419.
76. Bucior, I., Mostov, K., & Engel, J. N. (2010). *Pseudomonas aeruginosa*-mediated damage requires distinct receptors at the apical and basolateral surfaces of the polarized epithelium. *Infection and Immunity, 78*(3), 939–953.
77. Heiniger, R. W., et al. (2010). Infection of human mucosal tissue by *Pseudomonas aeruginosa* requires sequential and mutually dependent virulence factors and a novel pilus-associated adhesin. *Cellular Microbiology, 12*(8), 1158–1173.
78. Bucior, I., Pielage, J. F., & Engel, J. N. (2012). *Pseudomonas aeruginosa* pili and flagella mediate distinct binding and signaling events at the apical and basolateral surface of airway epithelium. *PLoS Pathogens, 8*(4), e1002616.
79. Pier, G. B. (2012). The challenges and promises of new therapies for cystic fibrosis. *The Journal of Experimental Medicine, 209*(7), 1235–1239.
80. Pier, G. B., et al. (1996). How mutant CFTR may contribute to *Pseudomonas aeruginosa* infection in cystic fibrosis. *American Journal of Respiratory and Critical Care Medicine, 154*(4 Pt 2), S175–S182.
81. Morris, A. P., et al. (1994). Polarization-dependent apical membrane CFTR targeting underlies cAMP-stimulated Cl- secretion in epithelial cells. *The American Journal of Physiology, 266*(1 Pt 1), C254–C268.
82. Loffing-Cueni, D., et al. (2001). Trafficking of GFP-tagged DeltaF508-CFTR to the plasma membrane in a polarized epithelial cell line. *American Journal of Physiology. Cell Physiology, 281*(6), C1889–C1897.
83. Hartl, D., et al. (2012). Innate immunity in cystic fibrosis lung disease. *Journal of Cystic Fibrosis, 11*(5), 363–382.
84. Tateda, K., et al. (2003). The *Pseudomonas aeruginosa* autoinducer N-3-oxododecanoyl homoserine lactone accelerates apoptosis in macrophages and neutrophils. *Infection and Immunity, 71*(10), 5785–5793.
85. Kaufmann, G. F., et al. (2008). The quorum quenching antibody RS2-1G9 protects macrophages from the cytotoxic effects of the *Pseudomonas aeruginosa* quorum sensing signalling molecule N-3-oxo-dodecanoyl-homoserine lactone. *Molecular Immunology, 45*(9), 2710–2714.
86. Hjelm, B. E., et al. (2010). Development and characterization of a three-dimensional organotypic human vaginal epithelial cell model. *Biology of Reproduction, 82*(3), 617–627.
87. Doerflinger, S. Y., Throop, A. L., & Herbst-Kralovetz, M. M. (2014). Bacteria in the vaginal microbiome alter the innate immune response and barrier properties of the human vaginal epithelia in a species-specific manner. *The Journal of Infectious Diseases, 209*(12), 1989–1999.
88. McGowin, C. L., et al. (2013). Mycoplasma genitalium infection activates cellular host defense and inflammation pathways in a 3-dimensional human endocervical epithelial cell model. *The Journal of Infectious Diseases, 207*(12), 1857–1868.

89. Radtke, A. L., Quayle, A. J., & Herbst-Kralovetz, M. M. (2012). Microbial products alter the expression of membrane-associated mucin and antimicrobial peptides in a three-dimensional human endocervical epithelial cell model. *Biology of Reproduction, 87*(6), 132.
90. van de Wijgert, J. H., & Shattock, R. J. (2007). Vaginal microbicides: Moving ahead after an unexpected setback. *AIDS, 21*(18), 2369–2376.
91. Hiller, S. L., Moench, T., & Shattock, R. J. (2005). In vitro and in vivo: The story of nonoxynol 9. *Journal of Acquired Immune Deficiency Syndromes, 39*, 1–8.
92. Ayehunie, S., et al. (2006). Organotypic human vaginal-ectocervical tissue model for irritation studies of spermicides, microbicides, and feminine-care products. *Toxicology In Vitro, 20*(5), 689–698.
93. Duray, P. H., et al. (2005). Invasion of human tissue ex vivo by *Borrelia burgdorferi. The Journal of Infectious Diseases, 191*(10), 1747–1754.
94. Sainz, B., Jr., TenCate, V., & Uprichard, S. L. (2009). Three-dimensional Huh7 cell culture system for the study of Hepatitis C virus infection. *Virology Journal, 6*, 103.
95. Berto, A., et al. (2013). Replication of hepatitis E virus in three-dimensional cell culture. *Journal of Virological Methods, 187*(2), 327–332.
96. Goodwin, T. J., et al. (2013). Three-dimensional normal human neural progenitor tissue-like assemblies: A model of persistent varicella-zoster virus infection. *PLoS Pathogens, 9*(8), e1003512.
97. Margolis, L. B., et al. (1997). Lymphocyte trafficking and HIV infection of human lymphoid tissue in a rotating wall vessel bioreactor. *AIDS Research and Human Retroviruses, 13*(16), 1411–1420.
98. Long, J. P., & Hughes, J. H. (2001). Epstein-Barr virus latently infected cells are selectively deleted in simulated-microgravity cultures. *In Vitro Cellular & Developmental Biology. Animal, 37*(4), 223–230.
99. Long, J. P., Pierson, S., & Hughes, J. H. (1999). Suppression of Epstein-Barr virus reactivation in lymphoblastoid cells cultured in simulated microgravity. *In Vitro Cellular & Developmental Biology. Animal, 35*(1), 49–54.
100. Brinley, A. A., et al. (2013). Characterization of Epstein-Barr virus reactivation in a modeled spaceflight system. *Journal of Cellular Biochemistry, 114*(3), 616–624.
101. Cohrs, R. J., et al. (2008). Asymptomatic reactivation and shed of infectious varicella zoster virus in astronauts. *Journal of Medical Virology, 80*(6), 1116–1122.
102. Mehta, S. K., et al. (2004). Stress-induced subclinical reactivation of varicella zoster virus in astronauts. *Journal of Medical Virology, 72*(1), 174–179.
103. Mehta, S. K., et al. (2013). Reactivation of latent viruses is associated with increased plasma cytokines in astronauts. *Cytokine, 61*(1), 205–209.
104. Mehta, S. K., & Pierson, D. L. (2007). Reactivation of latent herpes viruses in cosmonauts during a Soyuz taxi mission. *Microgravity Science and Technology, XIX*(5/6), 215–218.
105. Pierson, D. L., Mehta, S. K., & Stowe, R. P. (2007). Reactivation of latent herpes viruses in astronauts. In R. Ader (Ed.), *Psychoneuroimmunology* (pp. 851–868). New York: Academic.
106. Pierson, D. L., et al. (2005). Epstein-Barr virus shedding by astronauts during space flight. *Brain, Behavior, and Immunity, 19*(3), 235–242.
107. Stowe, R. P., et al. (2001). Immune responses and latent herpesvirus reactivation in spaceflight. *Aviation, Space, and Environmental Medicine, 72*(10), 884–891.
108. Stowe, R. P., Pierson, D. L., & Barrett, A. D. (2001). Elevated stress hormone levels relate to Epstein-Barr virus reactivation in astronauts. *Psychosomatic Medicine, 63*(6), 891–895.
109. Stowe, R. P., et al. (2000). Stress-induced reactivation of Epstein-Barr virus in astronauts. *Neuroimmunomodulation, 8*(2), 51–58.
110. Cohen, J. I. (2000). Epstein-Barr virus infection. *New England Journal of Medicine, 343*(7), 481–492.
111. Smith, Y. C., et al. (2006). Novel three-dimensional organoid model for evaluation of the interaction of uropathogenic *Escherichia coli* with terminally differentiated human urothelial cells. *Infection and Immunity, 74*(1), 750–757.

112. Teo, A., et al. (2014). A novel perfused rotary bioreactor for cardiomyogenesis of embryonic stem cells. *Biotechnology Letters, 36*(5), 947–960.
113. Margolis, L., et al. (1999). Long term organ culture of human prostate tissue in a NASA-designed rotating wall bioreactor. *The Journal of Urology, 161*(1), 290–297.
114. Hoz, L., et al. (2012). Cementum protein 1 (CEMP1) induces differentiation by human periodontal ligament cells under three-dimensional culture conditions. *Cell Biology International, 36*(2), 129–136.
115. Muhitch, J. W., et al. (2000). Characterization of aggregation and protein expression of bovine corneal endothelial cells as microcarrier cultures in a rotating-wall vessel. *Cytotechnology, 32*(3), 253–263.
116. Samuelson, L., & Gerber, D. A. (2013). Improved function and growth of pancreatic cells in a three-dimensional bioreactor environment. *Tissue Engineering. Part C, Methods, 19*(1), 39–47.
117. Sheyn, D., et al. (2010). The effect of simulated microgravity on human mesenchymal stem cells cultured in an osteogenic differentiation system: A bioinformatics study. *Tissue Engineering. Part A, 16*(11), 3403–3412.
118. Frith, J. E., Thomson, B., & Genever, P. G. (2010). Dynamic three-dimensional culture methods enhance mesenchymal stem cell properties and increase therapeutic potential. *Tissue Engineering. Part C, Methods, 16*(4), 735–749.
119. Freed, L. E., Vunjak-Novakovic, G., & Langer, R. (1993). Cultivation of cell-polymer cartilage implants in bioreactors. *Journal of Cellular Biochemistry, 51*(3), 257–264.
120. Crabbe, A., et al. (2011). Alveolar epithelium protects macrophages from quorum sensing-induced cytotoxicity in a three-dimensional co-culture model. *Cell Microbiology, 13*, 469–481.

Chapter 6
Use of In Vitro Cell Culture Models to Understand the Cellular and Molecular Basis of Immune Dysfunction During Spaceflight

Svantje Tauber, Buqing Yi, Alexander Choukèr, and Oliver Ullrich

Historical Landmarks

1973—Konstantinova et al. report reduced reactivity of lymphoid blood cells from crew members of "Soyuz" space missions 6–8 as compared to the status before flights [1].

1983–1998—NASA Spacelab allowed for systematic investigation of the effects of space conditions on cell cultures [2].

1983—Cogoli et al. conducted in vitro experiments with peripheral blood lymphocytes in real microgravity during the first Spacelab Mission. They found the proliferative response to mitogenic stimulation decreased by 93 % [3].

1991—The rotating wall vessel (RWV) bioreactor as a model system for simulated microgravity culture of cells was developed at NASA-Johnson Space Center. It is today still one of the most frequently used ground-based model systems in microgravity research using cell cultures [4].

1992—Wolfgang Briegleb develops and describes the fast-rotating clinostat, a ground-based system to model microgravity which is today a common tool in microgravity research with cell cultures [5].

1998—Lewis et al. documented spaceflight-induced alteration of the cytoskeleton on a molecular level when they observed that microtubules of Jurkat T lymphocytes were shortened and extended from poorly defined organizing centers during Space Shuttle flight [6].

(continued)

S. Tauber (✉) • O. Ullrich
Division of Cellbiology, Institute of Anatomy, University of Zurich,
Winterthurerstr. 190, CH-8057 Zurich, Switzerland
e-mail: svantje.tauber@anatom.uzh.ch

B. Yi • A. Choukèr
Department of Anesthesiology, Stress and Immunology Lab,
Hospital of the University of Munich, Marchioninstraße 15, 81366 Munich, Germany

C.A. Nickerson et al. (eds.), *Effect of Spaceflight and Spaceflight Analogue Culture on Human and Microbial Cells*, DOI 10.1007/978-1-4939-3277-1_6

(continued)

1998—The random positioning machine (RPM), which had been used to model microgravity in gravitropism-studies with plants before, was firstly used in cell culture experiments with human immune cells [7, 8].

2001/2005—Analyses on the mRNA level revealed that microgravity affects the mRNA expression of cytoskeletal genes [9] and of genes that are essentially involved in T cell activation and regulated by transcription factors NF-kappaB, CREB, ELK, AP-1, and STAT [10].

2013—Adrian et al. showed that the oxidative burst reaction of alveolar macrophages is triggered by gravity. Real time quantification of the burst reaction revealed that simulated microgravity (2D clinostat) and real microgravity (parabolic flight) significantly reduced the release of reactive oxygen species, while hypergravity conditions enhanced their release [11].

6.1 Introduction

Cell culture models are commonly used to study aspects of spaceflight-induced alterations to the immune system and have been extremely useful in evaluating biological and physical perturbations underlying these changes [12–15]. These in vitro systems allow investigations into cellular and molecular mechanisms of immune cell function and interrogation of signal transduction pathways that are altered in the microgravity environment. However, the limitations of in vitro cell cultures should be taken into consideration as well. These limitations include the use of tumorigenic or transformed cell lines and the inherently reductionistic nature of in vitro cell culture models as compared to the in vivo scenario (e.g., inability to faithfully model the complexity of the biological, physical, and chemical microenvironment found in whole organisms) [16]. While the influence of spaceflight on the immune system of the whole organism cannot be fully recapitulated by in vitro cell culture models, the latter offers a highly controlled and experimentally flexible platform to study and understand mechanisms that underlie in vivo immune system dysfunction.

6.2 Immune Cell Responses to True Spaceflight and Ground-Based Spaceflight Analogue Culture

The first cell culture study in the microgravity environment of spaceflight addressed the effect of spaceflight conditions on T lymphocyte proliferation, which plays a critical role in functional adaptive immune responses. In this study, leukocytes were separated from astronauts' blood samples and cultured in vitro during a Space Shuttle mission. Subsequently, these cells were treated with mitogens to induce the proliferation of lymphocytes. Spaceflight cultured T cells exhibited a dramatic reduction in their ability to proliferate [3], a finding in alignment with the compromised delayed type hypersensitivity (DTH) cell mediated immune response reported in astronauts during spaceflight [17]. Interestingly, it has been shown that T cells can proliferate as well as their ground controls if they are placed in a 1*g* centrifuge. This finding indicates that aspects of the microgravity environment of spaceflight can impact the proliferation of T cells, however, it remains to be determined if this was a direct or indirect effect of microgravity (conventional wisdom suggests the latter) [6].

These experiments led to a great interest in the influence of gravity on cellular and molecular levels of immune function. A blunted immune system obviously represents a concern and potential limitation for human long duration missions outside of low earth orbit, e.g. to Mars [18], and detailed knowledge about the underling cellular and molecular mechanisms that mediate spaceflight and microgravity-induced changes are essential to mitigate risk to crew health. Since the 1980s, experiments aiming to elucidate the effects and mechanisms of gravitational changes on immune function have been performed on board the Space Shuttle, the International Space Station (ISS), the Mir space station, sounding rockets, parabolic flights, and unmanned satellites [17, 19–27]. In addition to these platforms, which provide real microgravity, ground-based model systems that simulate some aspects of microgravity (e.g., lack of sedimentation and low fluid shear) have been established, such as the NASA designed rotating wall vessel (RWV) bioreactor, clinorotation, 3D-random positioning machine (RPM), and diamagnetic levitation [5, 28–30]. These models offer an opportunity to complement aspects of cell culture experiments in space, with the RWV bioreactor and clinorotation commonly considered as the methods that best replicate biological spaceflight responses in mammalian cell cultures [30]. These ground-based microgravity analogue systems have helped to reveal underlying mechanisms that regulate normal cellular homeostasis or transition to disease phenotypes that are not observed during conventional culture conditions, and allow for higher experimental frequencies at much lower costs. A broad set of data regarding the effects of microgravity on a variety of immune cell types and functions has been reported within the last 3 decades using both real and simulated microgravity conditions [14, 15, 24, 31, 32].

The majority of these studies have focused on the investigation of signal transduction pathways eliciting the altered reactivity of T lymphocytes in microgravity. Bechler and colleagues found that T lymphocyte proliferation could be recovered by enabling T cell–macrophage interactions during spaceflight. The authors hypothesized that their findings suggested that an inhibited interaction between macrophages and lymphocytes in the microgravity environment affected lymphocyte proliferation [33]. Subsequent investigations looked at the mechanisms of inhibited activation of T cells in response to mitogen-activation. It was shown that the primary events during stimulation of T cells with ConA, patching and capping of the ConA-receptors in the membrane, were not altered by microgravity conditions [19]. In contrast, the translocation of protein kinase C (PKC) to cell particulate fraction, as a downstream signaling process, was reduced [34, 35]. By exposing T lymphocytes to simulated microgravity conditions in an RPM followed by gene expression analysis using qPCR, it was found that signal transduction through NF-kappaB, cAMP response element binding protein (CREB), and transcription factor AP-1 was affected by simulated microgravity conditions [10]. A separate study using microarray gene expression profiling revealed that primary human T cells in microgravity (on the ISS) showed a decreased transactivation of Rel/NF-κB, CREB, and serum response factor (SRF) gene targets when compared to samples from a 1*g* on board-centrifuge following mitogen stimulation. Further analyses of these samples suggested that the TNF pathway is affected strongly and early in microgravity [22]. Recently, key proteins of the membrane proximal T cell activation cascade were quantified upon stimulation with ConA during a sounding rocket experiment (MASER-12 by ESA). Results from this study suggested that microgravity caused quantitative changes only in a few signaling components whereas most of the investigated signaling components remained quantitatively unaffected [36].

In addition to the process of activation, chromatin regulation [20], epigenetic regulation [37], cell cycle regulation [38], and the expression profile of microRNAs [39] have also been reported to be differentially impacted during microgravity culture of T lymphocytes. Furthermore, expression of cytokines including interleukin-1, interleukin-2, interferon-gamma [12] and cytokine-receptors, including the IL-2 receptor [29], have also been reported to be altered in response to microgravity conditions.

Not only lymphocytes, but also a major cell type of the innate immune defense and the connector between innate and adaptive immunity—the monocyte-macrophage-system—is affected by microgravity. A central module in activating the adaptive immune response and recruiting other immune cells of the innate immune system is the secretion of cytokines by macrophages. It was shown in real microgravity that the IL-1 secretion of cell culture monocytes is inhibited compared to normal gravity [33, 40]. In contrast, a study with the bone marrow macrophage cell line B6MP102 revealed enhanced LPS-induced tumor necrosis factor-alpha and IL-1 secretion in microgravity during spaceflight [41]. In a more recent study it was shown that microgravity reduces the LPS-induced expression of IL-6, TNF-alpha, and IL-10, and enhances the expression of IL-1β in monocytes of Space Shuttle flight crewmembers. Levels of IL-8 were also changed, however, not all of the microgravity-experiments showed the same direction of change [42].

The capacity to engulf bacteria and to induce an oxidative burst reaction is part of the crucial immune effector functions of monocytes. These functions were investigated in the blood of astronauts before and after 5–11 days of spaceflight, all of them appeared to be impaired after spaceflight [43]. A recent study demonstrated that the release of reactive oxygen species (ROS) during the oxidative burst reaction depends on gravity conditions: ROS release was reduced in microgravity, enhanced in hypergravity and responds rapidly and reversibly to altered gravity within seconds [11]. In another study, reduced levels of human leucocyte antigen HLA-DR were measured in human monocytes after spaceflight, which could be indicative of a reduced capacity to present antigens [42].

Another main feature of immune cells—not only of monocytes and macrophages, but also of neutrophils—is the capability to migrate and to adhere to tissues. These processes are a prerequisite to reach the sites of infection and eliminate pathogens. Migration and adhesion are strongly dependent on the expression profile of surface molecules, and they also require intact cytoskeletal reorganization. Meloni and colleagues were able to show that simulated gravity in an RPM alters the distribution of cytoskeletal filaments and decreases the motility of J-111 monocytes as measured by a migration-assay using colloid gold-coated cover slides [44]. In an experiment on the ISS, changes in cytoskeletal architecture and impaired motility of monocytes were observed in microgravity when compared to $1g$ in-flight controls [44, 45]. During the SIMBOX (Science in Microgravity Box) mission on Shenzhou-8, severely disturbed actin cytoskeleton, disorganized tubulin, and distinctly reduced expression of CD18, CD36, and MHC-II were detected after 5 days in microgravity [46]. A chemotactic assay of neutrophils after short term flight of astronauts has revealed that the cell response decreased by tenfold after spaceflight [47]. In lymphocytes, simulated microgravity in the ground-based RWV culture system inhibited locomotion by type I collagen [48, 49]. In accordance with these findings, the culture of T lymphocytes in hypergravity ($10\times g$) led to enhanced motility as measured by time-lapse bright-field microscopy [50]. A recent study also showed that short duration microgravity leads to decreased expression levels of the adhesion relevant surface-protein CD62L in monocytes [42].

In addition to the above mentioned effects of microgravity on the cytoskeleton in monocytes [45, 46], the cytoskeleton of other cell types has also been shown to be affected by altered gravity. These cell types include, cultured glial cells [51], nerve cells [52, 53], chondrocytes [54], and endothelial cells [55]. In endothelial cells, remodeling of the cytoskeleton occurred immediately [55, 56], and the mRNA-expression of several cytoskeletal proteins was altered after only 22 s of microgravity [56].

Likewise, apoptotic pathways have also been shown to be gravity-sensitive in several immune cell types. For example, human NK cells cultured in simulated microgravity in the RWV exhibited increased occurrence of apoptosis [57]. In astrocytes, simulated microgravity-induced apoptosis was verified by means of morphological hallmarks, as well as Caspase 7 and DNA-fragmentation [58]. A recent study investigated lymphocytes cultured on board the ISS and revealed that after 48 h, several signs of apoptosis were abundant: including DNA fragmentation, enhanced mRNA levels of the apoptosis-related markers p53, calpain, and an increase of 5-LOX activity [59]. In addition, Ohnishi et al. found an enhanced level of p53 protein in rat muscle and skin tissue during spaceflight [60].

6.3 Conclusion

Taken together, in the last 30 years in vitro cell culture models have been used extensively to study the functional and molecular aspects of microgravity-induced effects on immune cells. Although the results of these cell culture experiments cannot represent the influence of microgravity on the immune system of the whole organism, cell culture experiments offer a highly reproducible and standardized platform that allows identifying and investigating mechanisms that underlie in vivo immune cell dysfunction. Studies using both real and simulated microgravity have revealed that on the cellular level, immune cells are affected in their respective functions and distinct intracellular signal transduction processes are altered upon exposure to microgravity. In the future, cell culture models (including those of increasing complexity) will continue to contribute as a useful tool to a further understanding of the biological response of human cells to gravitational changes and help to elucidate the processes and mechanisms underlying immune dysfunction during spaceflight.

Questions for Future Research

1. What changes does spaceflight induce in immune cell dynamics, including alterations in quantities of immune cell types, gene and protein expression, and function of innate, mucosal, adaptive, and progenitor immune cells?
2. How does spaceflight alter the tissue microenvironment that could change microbiota composition? Would this change in the crew microbiome impact immune status and disease risk?
3. How does the response of immune cells cultured in spaceflight compare with immunological data from astronauts?
4. Can the data from spaceflight immunological experiments be translated to clinical applications on Earth?

References

1. Konstantinova, I. V., Antropova, E. N., Legen'kov, V. I., & Zazhirei, V. D. (1973). Reactivity of lymphoid blood cells in the crew of "Soiuz-6", "Soiuz-7" and "Soiuz-8" spacecraft before and after flight. *Kosmicheskaia Biologiia i Meditsina, 7*(6), 35–40.
2. Cogoli, A., & Gmunder, F. K. (1991). Gravity effects on single cells: Techniques, findings, and theory. *Advances in Space Biology and Medicine, 1*, 183–248.
3. Cogoli, A., Tschopp, A., & Fuchs-Bislin, P. (1984). Cell sensitivity to gravity. *Science, 225*(4658), 228–230.
4. Schwarz, R.P. & Wolf, D.A. (1991). *Rotating bio-reactor cell culture apparatus*.
5. Briegleb, W. (1992). Some qualitative and quantitative aspects of the fast-rotating clinostat as a research tool. *ASGSB Bulletin, 5*(2), 23–30.
6. Lewis, M. L., Reynolds, J. L., Cubano, L. A., Hatton, J. P., Lawless, B. D., & Piepmeier, E. H. (1998). Spaceflight alters microtubules and increases apoptosis in human lymphocytes (Jurkat). *FASEB Journal, 12*(11), 1007–1018.
7. Walther, I., Pippia, P., Meloni, M. A., Turrini, F., Mannu, F., & Cogoli, A. (1998). Simulated microgravity inhibits the genetic expression of interleukin-2 and its receptor in mitogen-activated T lymphocytes. *FEBS Letters, 436*(1), 115–118.
8. Hoson, T., Kamisaka, S., Masuda, Y., Yamashita, M., & Buchen, B. (1997). Evaluation of the three-dimensional clinostat as a simulator of weightlessness. *Planta, 203*(1), S187–S197.
9. Lewis, M. L., Cubano, L. A., Zhao, B., Dinh, H. K., Pabalan, J. G., Piepmeier, E. H., et al. (2001). cDNA microarray reveals altered cytoskeletal gene expression in space-flown leukemic T lymphocytes (Jurkat). *FASEB Journal, 15*(10), 1783–1785.

10. Boonyaratanakornkit, J. B., Cogoli, A., Li, C. F., Schopper, T., Pippia, P., Galleri, G., et al. (2005). Key gravity-sensitive signaling pathways drive T cell activation. *FASEB Journal, 19*(14), 2020–2022.
11. Adrian, A., Schoppmann, K., Sromicki, J., Brungs, S., von der Wiesche, M., Hock, B., et al. (2013). The oxidative burst reaction in mammalian cells depends on gravity. *Cell Communication and Signaling, 11*, 98.
12. Cogoli, A., & Cogoli-Greuter, M. (1997). Activation and proliferation of lymphocytes and other mammalian cells in microgravity. *Advances in Space Biology and Medicine, 6*, 33–79.
13. Moore, D., Bie, P., & Oser, H. (1996). *Biological and medical research in space: An overview of life sciences research in microgravity*. Berlin: Springer.
14. Cubano, L. A., & Maldonado, H. M. (2006). Immune cells under altered gravity conditions. *Boletín de la Asociación Médica de Puerto Rico, 98*(3), 223–228.
15. Hauschild, S., Tauber, S., Lauber, B., Thiel, C. S., Layer, L. E., & Ullrich, O. (2014). T cell regulation in microgravity–the current knowledge from in vitro experiments conducted in space, parabolic flights and ground-based facilities. *Acta Astronautica, 104*(1), 365–377.
16. Rothman, S. (2001). *Lessons from the living cell: The limits of reductionism*. New York: McGraw-Hill.
17. Taylor, G. R., & Janney, R. P. (1992). In vivo testing confirms a blunting of the human cell-mediated immune mechanism during space flight. *Journal of Leukocyte Biology, 51*(2), 129–132.
18. Gueguinou, N., Huin-Schohn, C., Bascove, M., Bueb, J. L., Tschirhart, E., Legrand-Frossi, C., et al. (2009). Could spaceflight-associated immune system weakening preclude the expansion of human presence beyond Earth's orbit? *Journal of Leukocyte Biology, 86*(5), 1027–1038.
19. Cogoli, M., Bechler, B., Cogoli, A., Arena, N., Barni, S., Pippia, P., et al. (1992). Lymphocytes on sounding rockets. *Advances in Space Research, 12*(1), 141–144.
20. Sciola, L., Cogoli-Greuter, M., Cogoli, A., Spano, A., & Pippia, P. (1999). Influence of microgravity on mitogen binding and cytoskeleton in Jurkat cells. *Advances in Space Research, 24*(6), 801–805.
21. Cogoli, A., & Tschopp, A. (1985). Lymphocyte reactivity during spaceflight. *Immunology Today, 6*(1), 1–4.
22. Chang, T. T., Walther, I., Li, C. F., Boonyaratanakornkit, J., Galleri, G., Meloni, M. A., et al. (2012). The Rel/NF-kappaB pathway and transcription of immediate early genes in T cell activation are inhibited by microgravity. *Journal of Leukocyte Biology, 92*, 1133–1145.
23. Buravkova, L. B., Rykova, M. P., Grigorieva, V., & Antropova, E. N. (2004). Cell interactions in microgravity: Cytotoxic effects of natural killer cells in vitro. *Journal of Gravitational Physiology, 11*(2), 177–P180.
24. Cogoli, A. (1996). Gravitational physiology of human immune cells: A review of in vivo, ex vivo and in vitro studies. *Journal of Gravitational Physiology, 3*(1), 1–9.
25. Paulsen, K., Thiel, C., Timm, J., Schmidt, P. M., Huber, K., Tauber, S., et al. (2010). Microgravity-induced alterations in signal transduction in cells of the immune system. *Acta Astronautica, 67*(9–10), 1116–1125.
26. Kaufmann, I., Feuerecker, M., Salam, A., Schelling, G., Thiel, M., & Chouker, A. (2011). Adenosine A2(A) receptor modulates the oxidative stress response of primed polymorphonuclear leukocytes after parabolic flight. *Human Immunology, 72*(7), 547–552.
27. Novoselova, E. G., Lunin, S. M., Khrenov, M. O., Parfenyuk, S. B., Novoselova, T. V., et al. (2014). Changes in immune cell signalling, apoptosis and stress response functions in mice returned from the BION-M1 mission in space. *Immunobiology, 220*(4), 500–509.
28. Hemmersbach, R., von der Wiesche, M., & Seibt, D. (2006). Ground-based experimental platforms in gravitational biology and human physiology. *Signal Transduction, 6*(6), 381–387.
29. Schwarzenberg, M., Pippia, P., Meloni, M. A., Cossu, G., Cogoli-Greuter, M., & Cogoli, A. (1999). Signal transduction in T lymphocytes--a comparison of the data from space, the free fall machine and the random positioning machine. *Advances in Space Research, 24*(6), 793–800.

30. Herranz, R., Anken, R., Boonstra, J., Braun, M., Christianen, P. C., de Geest, M., et al. (2013). Ground-based facilities for simulation of microgravity: Organism-specific recommendations for their use, and recommended terminology. *Astrobiology, 13*(1), 1–17.
31. Ullrich, O., & Thiel, C. S. (2012). Gravitational force - triggered stress in cells of the immune system. In A. Choukèr (Ed.), *Stress challenges and immunity in space*. Heidelberg: Springer.
32. Sonnenfeld, G. (2002). The immune system in space and microgravity. *Medicine and Science in Sports and Exercise, 34*(12), 2021–2027.
33. Bechler, B., Cogoli, A., Cogoli-Greuter, M., Muller, O., Hunzinger, E., & Criswell, S. B. (1992). Activation of microcarrier-attached lymphocytes in microgravity. *Biotechnology and Bioengineering, 40*(8), 991–996.
34. Schmitt, D. A., Hatton, J. P., Emond, C., Chaput, D., Paris, H., Levade, T., et al. (1996). The distribution of protein kinase C in human leukocytes is altered in microgravity. *FASEB Journal, 10*(14), 1627–1634.
35. Hatton, J. P., Gaubert, F., Cazenave, J. P., & Schmitt, D. (2002). Microgravity modifies protein kinase C isoform translocation in the human monocytic cell line U937 and human peripheral blood T-cells. *Journal of Cellular Biochemistry, 87*(1), 39–50.
36. Tauber, S., Hauschild, S., Crescio, C., Secchi, C., Paulsen, K., Pantaleo, A., et al. (2013). Signal transduction in primary human T lymphocytes in altered gravity—Results of the MASER-12 suborbital space flight mission. *Cell Communication and Signaling, 11*(1), 32.
37. Singh, K. P., Kumari, R., & Dumond, J. W. (2010). Simulated microgravity-induced epigenetic changes in human lymphocytes. *Journal of Cellular Biochemistry, 111*(1), 123–129.
38. Thiel, C. S., Paulsen, K., Bradacs, G., Lust, K., Tauber, S., Dumrese, C., et al. (2012). Rapid alterations of cell cycle control proteins in human T lymphocytes in microgravity. *Cell Communication and Signaling, 10*(1), 1.
39. Mangala, L. S., Zhang, Y., He, Z., Emami, K., Ramesh, G. T., Story, M., et al. (2011). Effects of simulated microgravity on expression profile of microRNA in human lymphoblastoid cells. *Journal of Biological Chemistry, 286*(37), 32483–32490.
40. Limouse, M., Manie, S., Konstantinova, I., Ferrua, B., & Schaffar, L. (1991). Inhibition of phorbol ester-induced cell activation in microgravity. *Experimental Cell Research, 197*(1), 82–86.
41. Chapes, S. K., Morrison, D. R., Guikema, J. A., Lewis, M. L., & Spooner, B. S. (1994). Production and action of cytokines in space. *Advances in Space Research, 14*(8), 5–9.
42. Crucian, B., Stowe, R., Quiriarte, H., Pierson, D., & Sams, C. (2011). Monocyte phenotype and cytokine production profiles are dysregulated by short-duration spaceflight. *Aviation, Space, and Environmental Medicine, 82*(9), 857–862.
43. Kaur, I., Simons, E. R., Castro, V. A., Ott, C. M., & Pierson, D. L. (2005). Changes in monocyte functions of astronauts. *Brain, Behavior, and Immunity, 19*(6), 547–554.
44. Meloni, M. A., Galleri, G., Pippia, P., & Cogoli-Greuter, M. (2006). Cytoskeleton changes and impaired motility of monocytes at modelled low gravity. *Protoplasma, 229*(2–4), 243–249.
45. Meloni, M. A., Galleri, G., Pani, G., Saba, A., Pippia, P., & Cogoli-Greuter, M. (2011). Space flight affects motility and cytoskeletal structures in human monocyte cell line J-111. *Cytoskeleton (Hoboken), 68*(2), 125–137.
46. Paulsen, K., Tauber, S., Goelz, N., Simmet, D. M., Engeli, S., Birlem, M., et al. (2013). Severe disruption of the cytoskeleton and immunologically relevant surface molecules in a human macrophageal cell line in microgravity—Results of an in vitro experiment on board of the Shenzhou-8 space mission. *Acta Astronautica, 94*(1), 277–292 (January–February 2014).
47. Stowe, R. P., Sams, C. F., Mehta, S. K., Kaur, I., Jones, M. L., Feeback, D. L., et al. (1999). Leukocyte subsets and neutrophil function after short-term spaceflight. *Journal of Leukocyte Biology, 65*(2), 179–186.
48. Pellis, N. R., Goodwin, T. J., Risin, D., McIntyre, B. W., Pizzini, R. P., Cooper, D., et al. (1997). Changes in gravity inhibit lymphocyte locomotion through type I collagen. *In Vitro Cellular & Developmental Biology Animal, 33*(5), 398–405.
49. Sundaresan, A., Risin, D., & Pellis, N. R. (2002). Loss of signal transduction and inhibition of lymphocyte locomotion in a ground-based model of microgravity. *In Vitro Cellular & Developmental Biology Animal, 38*(2), 118–122.

50. Galimberti, M., Tolic-Norrelykke, I. M., Favillini, R., Mercatelli, R., Annunziato, F., Cosmi, L., et al. (2006). Hypergravity speeds up the development of T-lymphocyte motility. *European Biophysics Journal, 35*(5), 393–400.
51. Uva, B. M., Masini, M. A., Sturla, M., Prato, P., Passalacqua, M., Giuliani, M., et al. (2002). Clinorotation-induced weightlessness influences the cytoskeleton of glial cells in culture. *Brain Research, 934*(2), 132–139.
52. Uva, B. M., Strollo, F., Ricci, F., Pastorino, M., Mason, J. I. & Masini, M. A. (2005). Morpho-functional alterations in testicular and nervous cells submitted to modelled microgravity. *Journal of Endocrinological Investigation, 28*(11 Suppl Proceedings), 84–91.
53. Rosner, H., Wassermann, T., Moller, W., & Hanke, W. (2006). Effects of altered gravity on the actin and microtubule cytoskeleton of human SH-SY5Y neuroblastoma cells. *Protoplasma, 229*(2–4), 225–234.
54. Aleshcheva, G., Sahana, J., Ma, X., Hauslage, J., Hemmersbach, R., Egli, M., et al. (2013). Changes in morphology, gene expression and protein content in chondrocytes cultured on a random positioning machine. *PLoS One, 8*(11), e79057.
55. Carlsson, S. I., Bertilaccio, M. T., Ballabio, E., & Maier, J. A. (2003). Endothelial stress by gravitational unloading: Effects on cell growth and cytoskeletal organization. *Biochimica et Biophysica Acta, 1642*(3), 173–179.
56. Grosse, J., Wehland, M., Pietsch, J., Ma, X., Ulbrich, C., Schulz, H., et al. (2012). Short-term weightlessness produced by parabolic flight maneuvers altered gene expression patterns in human endothelial cells. *FASEB Journal, 26*(2), 639–655.
57. Li, Q., Mei, Q., Huyan, T., Xie, L., Che, S., Yang, H., et al. (2013). Effects of simulated microgravity on primary human NK cells. *Astrobiology, 13*(8), 703–714.
58. Uva, B. M., Masini, M. A., Sturla, M., Tagliafierro, G., & Strollo, F. (2002). Microgravity-induced programmed cell death in astrocytes. *Journal of Gravitational Physiology, 9*(1), 275–P276.
59. Battista, N., Meloni, M. A., Bari, M., Mastrangelo, N., Galleri, G., Rapino, C., Dainese, E., et al. (2012). 5-Lipoxygenase-dependent apoptosis of human lymphocytes in the International Space Station: Data from the ROALD experiment. *FASEB Journal, 26*(5), 1791–1798.
60. Ohnishi, T., Takahashi, A., Wang, X., Ohnishi, K., Ohira, Y., & Nagaoka, S. (1999). Accumulation of a tumor suppressor p53 protein in rat muscle during a space flight. *Mutation Research, 430*(2), 271–274.

Chapter 7
Using a Spaceflight Three-Dimensional Microenvironment to Probe Cancer–Stromal Interactions

Ruoxiang Wang, Gina Chia-Yi Chu, Haiyen E. Zhau, and Leland W.K. Chung

Historical Landmarks

1889—Paget, proposed "seed and soil" hypothesis, indicating the permissive and determinant role of mesenchymal stroma in cancer metastasis [1].

1983—Chung and Cunha, demonstrated the permissive and determinant role of urogenital sinus mesenchyme in regulating prostate epithelial cell growth and differentiation [2].

1997—Pathak et al., showed that human prostate cancer cells can transform normal mouse stromal cells by cytogenetic analysis [3].

2001—Rhee et al., demonstrated that permanent and non-random phenotypic and genotypic changes can be induced in prostate cancer epithelial cells upon co-culture in 3-dimensional rotary wall vessel with either prostate or bone stromal cells [4].

2005—Wang et al., summarized the research experience with 3-dimensional co-culture of cancer and stromal cells in space [5].

2008—Sung et al., demonstrated the co-evolution of cancer and the interacting stromal cells during prostate cancer progression and metastasis [6].

R. Wang (✉) • G.C.-Y. Chu • H.E. Zhau • L.W.K. Chung (✉)
Department of Medicine, Uro-Oncology Research Program, Samuel Oschin Comprehensive Cancer Institute, Cedars-Sinai Medical Center, 8750 Beverly Blvd. Atrium 103, Los Angeles, CA 90048, USA
e-mail: Ruoxiang.wang@cshs.org; Leland.chung@cshs.org

C.A. Nickerson et al. (eds.), *Effect of Spaceflight and Spaceflight Analogue Culture on Human and Microbial Cells*, DOI 10.1007/978-1-4939-3277-1_7

7.1 Introduction

Early spaceflight research focused on how to protect astronauts from loss of gravity and from strong cosmic radiation, to which they were exposed during spaceflight. A number of projects were conducted in space and in space stations [7–10] to understand lymphocyte interactions in space and how embryonic and mesenchymal stem cells differentiate in microgravity conditions compared to ground-based gravity. The results of these investigations helped clarify the effect of spaceflight on cytoskeletal structure and cell motility, and on cell growth and differentiation [11]. These studies are of particular interest for regenerative medicine in terms of organ regeneration and tissue repair. In recent years, studies conducted in spaceflight and in space stations shared much common designs with earlier experiments conducted under 3-dimensional (3-D) conditions on ground [12, 13]. Such studies help determine the effects of physical environments, including flow and shear forces, on cell growth and differentiation. In the field of cancer research, microgravity-simulated growth conditions also helped assess the co-evolution of tumor and stroma [14] and probe irreversible changes in cancer and benign cells responding to inductive cues from interactive cell types maintained under rotary wall vessel (RWV) bioreactor conditions, which have been used to recapitulate the microgravity spaceflight conditions [4, 5].

Cancer is a life-threatening disease mainly because of the effects of cancer metastasis. Whereas primary and localized tumors can be treated fairly successfully with surgical resection, chemotherapy, and radiation therapy, metastatic cancer cells are often unstable, and can acquire plasticity to become highly resistant to therapy [15]. Although the exact pathophysiologic cause of lethality has yet to be determined, metastatic tumor growth in secondary and tertiary sites is thought to impair the function of the affected normal host tissue and organs, causing wasting, cachexia, and dwindling vitality [16–18]. To prolong survival and improve the quality of life of cancer patients, a fundamental understanding of the biology of cancer cells and their interactive microenvironment is needed, in order to design new approaches tackling the spread of cancer cells and overcoming their development of resistance. While a number of mechanisms for metastatic cancer progression have been proposed, our recent experimental data support a new concept, in which recruitment and interaction of bystander dormant cells by metastasis-initiating cancer cells (MICs) are central to the development of cancer metastasis. We discovered that MICs could recruit and reprogram or transform bystander cells to participate in carcinogenesis both under 3-D co-culture conditions and during in vivo xenograft growth [19]. We further obtained experimental evidence that MICs can also recruit and reprogram circulating tumor cells (CTCs) to express markers of aggressive prostate cancer. One common denominator of these studies is the use of 3-D co-culture, either with RWV [4, 6, 20] or in a 3-D suspension [19]. These results are in agreement with those observed in in vivo xenograft tumor formation, where we observed that MICs reprogrammed non-tumorigenic and non-metastatic cancer cells to form metastatic growth in the host mouse skeleton [19].

This review will focus on the following relevant areas of cancer and cancer metastasis research, particularly the utilization of 3-D RWV and 3-D suspension culture for evaluating cancer cell behaviors in vitro: (1) the roles of mesenchymal–epithelial interactions in controlling glandular epithelial cell differentiation and functional maturation; (2) the contribution of mesenchymal-derived stromal cells to prostate cancer progression and bone metastasis; (3) the concept that MICs significantly recruit and reprogram bystander dormant cells to participate in cancer metastasis; (4) the intracellular cell signal network that accounts for MIC recruitment and induction of bystander dormant cells; and (5) the use of spaceflight to probe cancer–stromal interaction under 3-D conditions in order to define the mechanism of malignant progression and prostate cancer metastasis.

7.2 Critical Role of Mesenchyme-Derived Stromal Cells in Regulating Glandular Epithelial Cell Growth, Differentiation, and Function

A large literature exists defining the importance of mesenchyme-derived stromal cells that regulate glandular epithelial growth, differentiation, and function [2, 21–25]. During organogenesis and cytodifferentiation of the prostate gland, for example, fetal urogenital mesenchyme (UGM) has been used as an inductor to recapitulate the development of the urogenital tract, including the prostate gland, the seminal vesicles, the bulbourethral gland, and the urinary bladder [26, 27]. UGM was shown to drive epithelial cell differentiation in a reciprocal manner and this induction cue could be either permissive or determinant depending on the cell types and the ages of the fetal tissues used to construct the tissue recombination. A natural extension of this technology was to use UGM as an inductor to drive the growth, differentiation, and subsequent development of castration-resistant prostate cancer (CRPC) in either prostate cancer or prostate stem cell preparations, which can be subjected to detailed morphologic, biochemical, and molecular interrogation [28–30].

Embryonic organogenesis and functional maturation of the glandular prostate is a process controlled by androgenic hormones, and the same could be true for oncogenesis and development of the prostate cancer in later life. Chimeric tissue recombination studies were proven to be critical to mechanistic elucidation of regulatory mechanism [31–33]. Because of the success of using tissue recombinants to study the mechanisms of steroid hormone action, many variations of these techniques have subsequently been developed and applied to normal epithelial or tumor cells in recombination with normal mesenchymal stroma or cancer-associated fibroblasts. Genetically engineered (gene overexpression or knockdown) epithelial or stromal components could be used in the recombination or co-inoculation of cancer cells with stromal fibroblasts of different lineages. The general conclusions from these studies are: (1) UGMs are robust and can be used successfully to construct tissue recombinants with normal fetal or adult tissues, or with cancer cells. (2) Similar to UGM cells, adult prostate stromal fibroblasts isolated from benign or prostate cancer tissues are also robust to drive prostate cancer cell growth and progression, when co-inoculated as chimeric organoids in the host of athymic mice. (3) Androgen receptor residing in the stromal fibroblasts, but not the epithelium, is responsible for driving epithelial prostate growth in a ligand-dependent manner. (4) Similar to prostate stromal fibroblasts, bone marrow-derived mesenchymal stem/stromal cells have the potential to induce the growth and transdifferentiation of prostate cancer cells both in co-culture in vitro and in xenograft chimeric tumor formation in vivo. Derivative cancer cells isolated from the chimeric tumor through ex vivo culture displayed androgen-independent characteristics and could form metastatic tumors to distant sites such as bone and soft tissues [34–37]. Interestingly, the reprogramming of prostate cancer cells by stromal fibroblasts occurred exclusively through chimeric tumor formation, or under 3-D-RWV or 3-D suspension co-culture conditions. This observation strongly suggests that topographical factors induce a unique gene expression profile that alters cell contact or adhesion [24, 38], thus determining or fundamentally changing the fate of cancer cells. This conclusion also means that in order to recapitulate the histogenesis, cytodifferentiation, and transdifferentiation of tissues or cells in normal organ development or during cancer development and progression, it is essential to co-culture the multiplex cell types under 3-D-RWV or 3-D suspension conditions.

7.3 Mesenchymal Stromal Fibroblasts Are Responsible for Driving Prostate Cancer Progression and Bone Metastasis

The underlying mechanisms by which mesenchymal stromal cells modulate the growth, survival, and progression of prostate cancer cells are hijacked from normal organ development, where stromal fibroblasts play a determining role in prostate epithelial growth and differentiation [31, 39, 40]. Cancer metastasis is a highly inefficient and rare event [41, 42], and in most cases, it takes a substantial amount of time for cancer cells to establish metastasis. Cancer metastasis can occur when the host selectively accommodates the growth of cancer cell colonies. In the case of *clonal selection*, cancer metastasis occurs by having the "fittest" clone expand at the metastatic sites [43, 44]. Cancer metastasis can also occur as the consequence of incremental genetic changes, small steps that make cancer cells gradually more likely to metastasize, through a process of *clonal evolution and adaptation* [45, 46]. Multiple epigenetic changes can be acquired to culminate into selective growth and survival advantages that ultimately lead to the establishment of metastatic foci at distant organs. We believe that these acquired changes are more important elements of cancer metastasis than the original genetic abnormalities. This reasoning is based in part on interracial comparative studies of human cancers, including prostate cancer, where cancer incidence seems to be determined by dietary and environmental factors rather than genetics of the host germ lines [47–50]. On the other hand, given that accumulated genetic changes could ultimately drive cancer metastasis, it is difficult to explain why metastatic deposits at multiple sites do not necessarily share an identical genetic makeup but instead display high intercellular heterogeneity [48, 51–53]. Our previous work and current studies suggest that the genetic variety may be explained by tumor–host interactions accounting for organ-specific cancer metastasis.

As depicted by the "seed and soil" theory of cancer metastasis [1], dissipating tumor cells colonize specific tissue and organs only when there is a favorable mesenchymal stromal microenvironment. In humans, the skeletal system is a frequent target organ of cancer metastasis, and bone metastatic tumors are mostly incurable [54, 55]. The interaction between tumor and bone marrow mesenchymal stem/stromal cells is thus a rational research subject. To establish a co-inoculation protocol, we first demonstrated that the LNCaP human prostate cancer cells or bone and prostate stromal fibroblasts were non-tumorigenic and non-metastatic, as inoculating each individual cell types alone did not cause xenograft tumor formation nor observed metastasis in athymic male mice. However, when co-inoculated with a bone stromal cell line with or without endogenous androgen, the LNCaP cells would become tumorigenic and metastatic [36, 56]. A number of *tumorigenic* or *metastatic* sublines, called C4, C4-2, and C4-2B cells, have been established from ex vivo culture of LNCaP chimeric tumors. Subsequent characterization demonstrated that these derivative cells represent a unique panel of lineage-related prostate cancer cell lines for the study of prostate cancer progression to the CRPC state. These studies clearly indicated that LNCaP cells can acquire additional malignancies and convert from a non-tumorigenic state into CRPC with bone and soft tissue metastatic potential, while selective mesenchymal stromal cells of the tumor microenvironment are likely to play a permissive or determinant role in conferring the malignant progression and the cancer epithelial cells.

Comparative genomic analysis was conducted to assess contribution of genetic changes to tumorigenic and metastatic prostate cancer progression. Whereas majorities of genomic abnormalities in derivative sublines could be traced back to the original cytogenetic profiles of the parental LNCaP cells [36, 56], sublines derived from chimeric tumors of castrated hosts were found to harbor additional genetic, epigenetic, and behavioral changes, in correlation to the strong phenotypic CRPC features. In parallel studies, we observed that LNCaP cells co-cultured in 3-D-RWV with adult prostate or bone stromal cells were induced to undergo reprogramming into derivative cells that were androgen-independent and had bone and soft tissue metastatic potential [4, 6, 36] (Fig. 7.1). Cytogenetic analysis demonstrated that, compared to their parental LNCaP cells, lineage-related LNCaP sublines underwent chromosomal reorganization and lost the Y chromosome in almost all cases. Comparative studies by many laboratories further identified additional phenotypic divergences between these cells, and molecular comparison identified various genes whose differential expression was correlated to phenotypic progression of the derivative sublines [57–59], while the acquired genetic and phenotypic changes were found to be stable and permanent. Although the mechanism has yet to be defined, cancer cells appeared capable of acquiring genomic and genetic abnormalities through the process of interacting with selective mesenchymal stromal cells. These studies collectively suggest the importance of 3-D-RWV conditions that can recapitulate the in vivo reprogramming of LNCaP cells grown as chimeric tumors with inductive bone stromal cells in mice. Whether any of the alterations are causal factors for tumorigenic and metastatic progression remains to be determined, but these studies consolidated the idea that mesenchymal stromal cells, when grown as chimeras in 3-D conditions with LNCaP cells maintained in a castrated state, directed the progression of prostate cancer to develop CRPC phenotypes with the propensity for bone and soft tissue metastases.

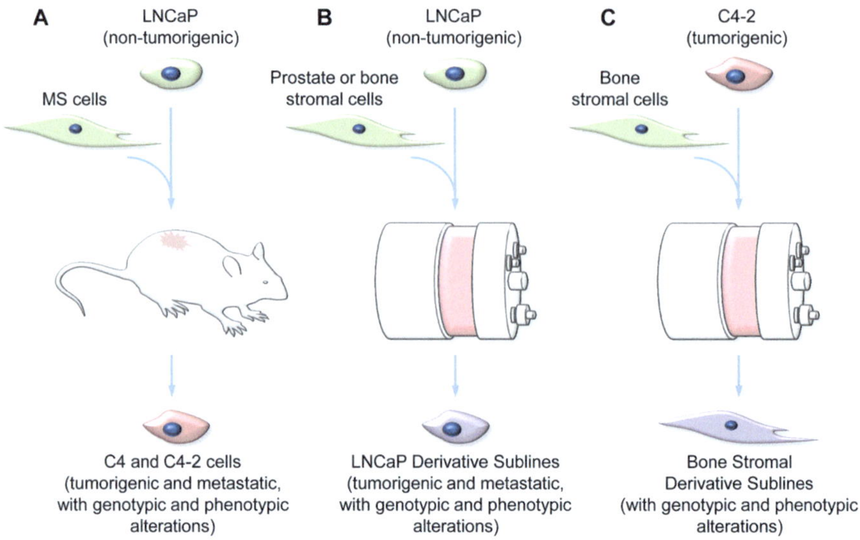

Fig. 7.1 3-D cancer–stromal interaction driving prostate cancer malignant progression. These schemas summarize the importance of 3-D cancer-stromal cell contact to prostate cancer progression and metastasis. (**A**) When inoculated separately, neither LNCaP prostate cancer cells nor the MS bone stromal cells would form subcutaneous tumor growth in male athymic mice (not shown). Co-inoculation of these two cell types in mix, however, could cause xenograft tumor formation. Derivative cancer cell clones isolated by ex vivo culture of the chimeric tumor (e.g., the C4 and C4-2 cells) have been determined by cytogenetics to be of the LNCaP cell lineage, but are with significant tumorigenic and metastatic potential. (**B**) Similar results could be replicated in vitro, where non-metastatic LNCaP cells can be reprogrammed by stromal cells derived either from the prostate or the bone when were subjected to 3-D-RWV co-culture. (**C**) In a reciprocal manner, the tumorigenic C4-2 derivative cells were used in the co-culture where at the end of this co-culture, multiple genotypic and phenotypic abnormalities could be detected in the participating bone stromal cells. It appeared that mesenchymal stromal cells could be reprogrammed by metastatic human PCa cells. Cancer–stromal interaction not only leads to cancer progression, but can also elicit a malignant transformation of the bystander cells

7.4 Metastasis-Initiating Cancer Cells (MICs) Can Recruit and Reprogram Host Bystander Dormant Cells to Participate in Cancer Metastasis

Conventional studies of cancer-stromal interaction focus on the tropic role of stroma-derived soluble factors, which may exert paracrine effects encouraging the growth and survival of cancer cells. In response to cancer cell invasion, the adjacent cancer-associated stromal cells resident of the tumor microenvironment could be activated to become "reactive stroma" [60–62], with elevated growth factor production. Blockade of soluble factor communication, however, has only limited efficacy against tumor progression and metastasis, indicating that additional mechanisms, such as autocrine and intracrine regulation may mediate cancer–stromal interaction. In addition, cancer metastasis may involve the participation of bystander dormant cells, including resident host mesenchymal stem cells, endothelial cells, or inflammatory cells at metastatic sites, and CTCs and disseminated tumor cells (DTCs) which are known to facilitate the establishment of an equilibrium with tumor cells at primary and the metastatic sites [38, 63, 64].

We developed a series of genetically tagged cells in order to assess the involvement of "bystander or dormant" cells in cancer-stroma co-culture. Highly bone metastatic prostate cancer cells were used as MICs. In one such study, MICs were co-inoculated with bystander dormant cancer cells which were tagged genetically with a red-fluorescent protein (RFP) in mouse skeleton at a ratio of 1:1000. Alone, either MICs or dormant cells failed to form metastasizing tumors in mice; but when they were co-inoculated in mouse skeleton, robust formation of intraosseous tumors was observed [19]. Based on the genetic tags, we were able to isolate both MICs and dormant cells from the chimeric tumors and investigate separately their gene expression profiles and tumorigenic and metastatic behaviors (Fig. 7.2). The results of these studies showed that: (1) Bystander dormant prostate cancer cells can be recruited and transformed by MICs to become tumorigenic and metastatic. (2) Dormant cells after reprogramming would express a distinctive gene expression profile, which was shared with the MICs. Reprogrammed dormant cells express receptor activator of NF-κB ligand (RANKL) and its RANK receptor, c-Met (a protooncogene that recognizes a hepatocyte growth factor, HGF), phosphorylated c-Met (p-c-Met), and neuropilin-1 (NPR-1, a co-receptor of VEGF). In addition, the dormant cells gained the ability to undergo epithelial-mesenchymal transition (EMT), expressed markedly elevated stem cell, neuroendocrine cell and bone cell phenotypes known to facilitate the ability of cancer cells to develop skeletal metastases [19, 65–68]. (3) Gene expression and behavioral changes in the dormant cells were switched on by coordinated upregulation of several key transcription factors (TFs). We observed that an activated RANK-mediated signaling network promoted a feed forward loop involving the induction of RANKL and c-Met, but repression of androgen signaling pathways. Site-directed mutagenesis and transcription factor interference assays identified common TFs, c-Myc/Max and AP4, as critical regulatory nodes. Specifically, RANKL-RANK signaling activated a number of master regulator TFs that control EMT (Twist1, Slug, Zeb1, Zeb2), stem cell properties (Sox2, Myc, Oct3/4, and Nanog), neuroendocrine differentiation (Sox9, HIF-1α, and FoxA2), and osteomimicry (c-Myc/Max, Sox2, Sox9, HIF1α, and Runx2). Conversely, abrogating RANK or its downstream c-Myc/Max or c-Met signaling network could inhibit skeletal metastasis in mice. RANKL-expressing LNCaP cells recruited and induced neighboring dormant cells to express RANKL and c-Met, with enhanced p-c-Met level. These initially non-tumorigenic dormant cells, after being retrieved from the tumors, were now able to colonize and grow in bone. In clinical prostate cancer tissues, we found that an activated RANK-mediated signal network, including RANKL, RANK, p-c-Met, and NRP-1, could be used to predict the survival of prostate cancer patients [69–71]. (4) We found that intricate interactions between MICs

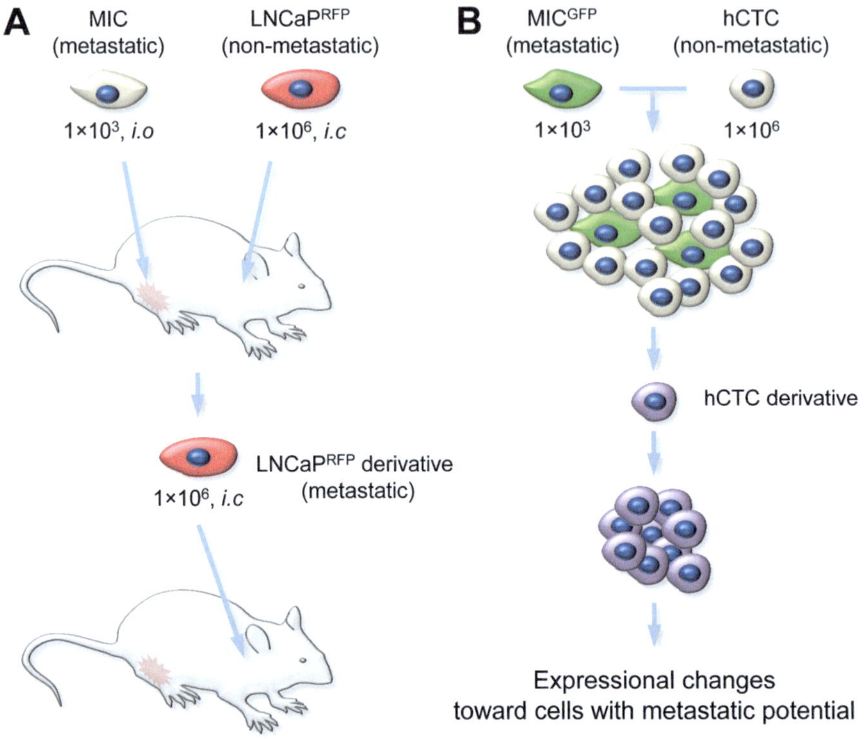

Fig. 7.2 MICs recruitment and reprogramming of bystander cells in 3-D suspension culture. These schemas summarize a novel mechanism of prostate cancer progression and metastasis. (**A**) When administered separately, neither MICs by low-dose intraosseous (*i.o*) inoculation nor red fluorescent LNCaP cells ($LNCaP^{RFP}$) by high-dose intracardiac (*i.c*) inoculation could cause bone tumor formation in male athymic mice (not shown). Successive inoculation of these two cell types to the same mouse, however, could lead to frequent tibial bone tumors. Subsequent ex vivo culture of the bone tumor detected the presence of red fluorescent cells. The red fluorescent tumor cell clones isolated from the tumor had acquired malignant potential, since these clones could now cause xenograft tumor formation by itself in the absent of MICs. (**B**) Similar results could be obtained in vitro with 3-D suspension co-culture. Suspension co-culture with GFP-tagged MICs (MIC^{GFP}), for instance, could reprogram CTCs, which were pre-amplified from blood samples of clinical prostate cancer patients (hCTC)

and the bystander dormant cells occur both in mice and in chimeric tumors grown in 3-D suspension culture. Once again, we noted that 3-D and in vivo chimerical growth were the critical determinant of efficient recruitment and reprogramming that ultimately led to cancer metastasis [5, 19]. This experimental system can be further expanded to assess the recruitment and reprogramming of other cell types from hosts (such as CTCs, host inflammatory cells, mesenchymal stem cells, host endothelial cells, and macrophages) in experimental mice or under 3-D suspension co-culture (Fig. 7.2).

7.5 Probing Cancer-Stromal Co-evolution with 3-D Co-culture in the RWV

To understand how tumor and stroma interact in vivo and the cellular and molecular bases of interaction, we used 3-D-RWV co-culture model to probe the molecular changes of prostate cancer and prostate and bone stromal cells. The RWV model is unique in that many of the phenotypic and genotypic changes found in the co-cultured cancer and stromal cells failed to occur when these cells were grown in conventional 2-D culture. Human bone stromal cells, after 3-D co-culture with human prostate cancer cells in vitro, underwent permanent cytogenetic and gene expression changes with reactive oxygen species serving as mediators [4, 6, 62]. The evolved stromal cells were highly inductive of human PCa growth in mice, and expressed increased levels of extracellular matrix (versican and tenascin) and chemokine genes (BDNF, CCL5, CXCL5, and CXCL16). These genes were validated in clinical tissue and/or serum specimens and could be used as predictors for invasive and bone metastatic prostate cancer to a remarkable degree. These results, combined with our previous observations, support the use of 3-D-RWV models to study permanent genetic and behavioral changes in prostate cancer cells after being co-cultured with prostate or bone stromal cells as 3-D prostate organoids or grown as tumor xenografts in mice. These observations collectively suggest that the co-evolution of cancer and stromal cells depends on 3-D growth conditions to accelerate cancer growth and metastasis. Biomarkers generated under 3-D co-culture conditions mimicked clinical prostate cancer in patients, again supporting the value of RWV models as a discovery engine for biomarker identification and validation. Figure 7.3 illustrates a hypothetical approach in which biomarkers can be identified from cells subjected to either in vivo growth or 3-D culture and validated in clinical specimens.

7.6 Cancer–Stromal Interaction in Space

We were fortunate to be given an opportunity to investigate cancer–stromal interaction in space, as a participant in NASA-sponsored spaceflight science projects [5, 14]. Our intent was to probe if genetic and phenotypic alterations of cancer and stroma in a space environment would allow us to define the plasticity of cancer and stroma during their co-evolution. Unfortunately, this opportunity was lost due to the tragic Shuttle Columbia disaster in 2003. We did, however, obtain some valuable data transmitted from this spaceflight during the flight showing the growth of tumor and stroma in space (based on the pH of the conditioned media and oxygen consumption of the culture). The data also indicated that active tumor–stroma interactions occurred in space, as golf-ball-sized chimeric organoids were formed in spaceflight, much larger than the organoids formed in parallel experiments on ground [5, 14]. Based on these results, it is possible that in the future we could probe genetic

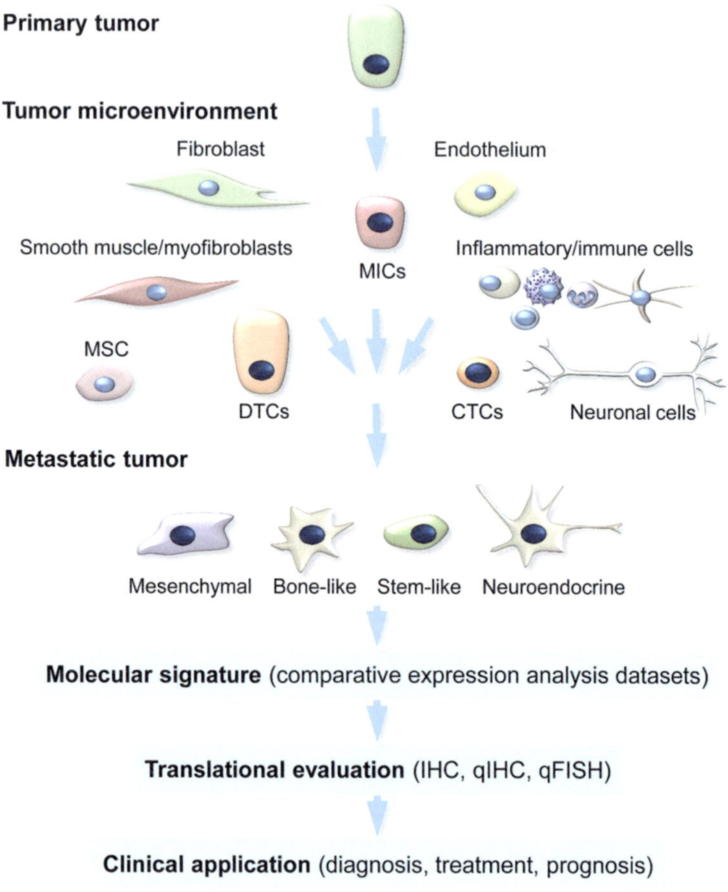

Fig. 7.3 3-D co-culture model for elucidating the mechanism of prostate cancer progression and metastasis. This schema outlines a research strategy based on our current understanding of the recruitment and reprogramming mechanism. Different from primary tumor, metastatic prostate cancer is depicted as the product of extensive and complex interaction between cancer and resident cells of the tumor microenvironment. In addition to reciprocal communication through soluble factors, resident cells of the tumor microenvironment may physically join the malignant force to become active participants in prostate cancer progression and metastasis. 3-D-RWV and 3-D suspension co-culture will be ideal experimental system for such studies. Subsequently, genotypic and phenotypic changes in the participating cells could be explored as molecular signature of the cancer–stromal interaction. The signature could have clinical applications if validated on clinical specimens with immunohistochemical (IHC), quantitative IHC (qIHC), and quantitative fluorescence in situ hybridization (qFISH) methods

and phenotypic changes of tumor and stroma and identify the factors released by these interactive cellular compartments. The nature of tumor–stroma contacts in space could favor the development of plasticity in prostate cancer and stromal cells. Ultimately we hope that further investigation of tumor–stromal interactions in the space environment could generate valuable genetic and epigenetic information on the participating cell types. The information could be translated clinically to improve prostate cancer diagnosis, prognosis, and treatment (see Fig. 7.3). These possibilities have already been realized in our 3-D-RWV co-culture studies where tumor–stroma secreted factors have matched clinical specimens from patients [6].

7.7 Potential Molecular Mechanisms Underlying Cancer–Stromal Interaction under 3-D Conditions

The molecular mechanisms underlying the co-evolution of cancer and stromal cells have not yet been fully elucidated. Besides the well-recognized permissive and determinant role of mesenchymal stromal cells, our accumulated experience with human prostate cancer xenograft tumor formation points to the potential involvement of soluble growth factors, extracellular matrices, and cell fusion as candidate mechanisms promoting cancer–stromal interaction. In addition, the temporal and spatial relationship between cancer and the host microenvironment is expected to be coordinately regulated to explain the progression and the development of bone metastasis in some patients. By understanding the signal network promoting cancer progression and metastasis, new therapeutic approaches can be developed.

7.7.1 Soluble Factor Paracrine Communication

We and others have identified reactive oxygen species, cytokines, chemokines, and other growth factors as key players mediating the communication between cancer and stromal cells. In a series of studies, for instance, our laboratory found that β2-microglobulin, a co-receptor of the MHC complex, potently promoted EMT and the ability of prostate cancer cells to express bone-like properties [67, 72–76]. Serving as an immune modulator with growth factor property, β2-microglobulin is a promising therapeutic target in advanced prostate cancer [67, 77–79]. Targeting β2-microglobulin proved to be highly effective for inhibiting the growth of both human liquid [79–81] and solid tumors. In our laboratory, we found that anti-β2-microglobulin antibody is an effective agent antagonizing the growth of both androgen-sensitive and androgen-resistant human prostate cancer cell lines [67, 72–76]. This antibody was highly effective against the growth of chimeric prostate tumors comprised of prostate cancer and bone stroma and grown as organoids under 3-D-RWV conditions. Moreover, due to the special gating effects of this antibody in iron transport through transferrin and transferrin receptor complex with an MHC-like protein, hemochromatosis gene (HFE), the cytotoxic effects of anti-β2-microglobulin antibody in prostate tumors can be greatly enhanced by radiation therapy [75]. Further translation of these results showed that anti-β2-microglobulin antibody effectively caused the regression of previously formed prostate cancer bone metastasis in mouse skeleton [67, 75]. In another respect, results from these studies demonstrated that the 3-D co-culture of prostate cancer and stroma is an effective model that faithfully reflects the therapeutic response of tumors in mice.

7.7.2 Adhesion-Mediated Signaling from Extracellular Matrices

The mesenchymal stromal compartment is structured with rich extracellular matrices, which can function to confer cell motility and survival via extracellular matrices–integrin interactions. In addition, the matrices can sequester and store soluble growth factors which, upon release by matrix-degrading enzymes, can alter cell functions through growth factor-receptor recognition [82, 83]. Cancer cells have remarkable abilities to make both the extracellular matrix components and the degrading enzymes, both could be promoting the ability of cancer cells to sense and respond to environmental cues by concentrating growth factors and by increasing their adhesion, survival, and motility [83–85]. On the other hand, extracellular matrices could also be provided by host cells such as host stromal fibroblasts and infiltrating inflammatory cells. Our laboratory discovered that ROS produced by the prostate stroma can promote the differential expression of a specific protein called a disintegrin and metalloprotease domain 9 (ADAM9) [86–90]. ADAM9 expression is associated with increased malignant potential and metastasis by promoting the growth and survival of prostate cancer cells. ADAM9 can be effectively targeted by a lentiviral construct encoding an anti-ADAM9 plasmid DNA. This treatment strategy was found to be highly effective in mice as well as in 3-D culture. It is likely that this treatment therapy interrupted cancer–integrin interaction by blockading the supply of cancer and cells in the tumor microenvironment with the appropriate extracellular matrix fragments and growth factors, upregulating p21 and impairing the progression of prostate cancer cells through the cell cycle [86–90].

7.7.3 The Recruitment and Reprogramming of Bystander Dormant or Host Cells to Participate in Cancer Metastasis

We hypothesize significant 3-D cell contact takes place between MICs, tumor cells with significant metastatic potential, and bystander cells, tumor or host cells with dormant presence in the tumor microenvironment at metastatic sites or in circulation as CTCs or DTCs. Using genetically tagged cell lines, we observed that MICs could reprogram the bystander dormant cells, or CTCs to participate in metastatic bone colonization through the activation of specific TFs. We found that the activation of many master regulators could fundamentally change the phenotypes and behaviors of the bystanders or CTCs, allowing them to express genes downstream from these regulatory TFs [19]. These induced phenotypes via cellular interactions are permanent and could be a major source of metastatic cancer cells (see above).

7.7.4 Cancer-Stromal Cell Fusion

Using red and green fluorescence proteins as differentiating markers to track cell fates in co-culture, we have found that prostate cancer cells placed in direct contact could fuse spontaneously with prostate or bone marrow stromal cells to form cancer-stromal cell hybrids [91]. Most perished eventually, possibly due to mitotic catastrophe, but some of the fusion hybrids survived to form derivative clones with acquired malignant potential. Our unpublished results showed that prostate cancer cells could fuse with a wide array of cell types of mesenchymal lineage, suggesting that cancer-stromal cell fusion is a general phenomenon during cancer–stromal interaction. Cancer cells are thus fusogenic, capable of hybridizing indiscriminatingly with bystander cells and various resident stromal cells in the tumor microenvironment. Should certain target cells be activated from the resting state and enter the cell cycle, the subsequent fusion hybrid would have a chance to escape mitotic catastrophe to form a derivative population bearing the admixed genomic makeup of the cancer and stromal parents to increase the phenotypic heterogeneity characteristic of tumor cells [91].

Although biologic cell fusion is a critical mechanism for fertilization, embryonic development, and functional maturation [92–94], investigation of the impact of cancer cell fusion on tumor formation and metastasis has just begun, thanks to the advent of a new fluorescence protein tagging technology allowing convenient real-time observation of the interaction between cancer and stromal cells under 3-D conditions. Cancer-stromal cell fusion has now been documented in many cancer types [95], and well-characterized 3-D culture models are mandatory for future investigation.

7.8 Space Research on Cancer–Stromal Interaction

Our accumulated research has established a permissive and determinant role of the mesenchymal stromal compartment in prostate cancer bone metastasis. The interaction may result additionally in the recruitment and reprogramming of bystander dormant cells or CTCs to participate in metastatic colonization of the skeleton. It is possible that a host of other cells including inflammatory cells, endothelial cells, and mesenchymal cells of different development lineages can also participate in these interactions. On the other hand, cancer–stromal interaction may result in cancer cell annihilation by the interacting stromal cells, through mitotic catastrophe following cancer-stromal cell fusion. New studies of cancer–stromal interaction incorporating these new insights may expand our understanding of the origin of tumor cell heterogeneity and help devise more effective therapies targeting the cells contributing to metastatic colonization. Based on our observations of the permanent reprogramming of prostate cancer phenotypes in 3-D growth, we believe that space research on cancer-stromal interaction will help us understand tumor growth and metastasis in the host microenvironment.

In the space age, humans will increasingly travel in space, and be exposed to cosmic-galactic radiation under microgravity or other partial gravity conditions. The long-term effect of manned missions on human health remains to be fully assessed, as comparative models show different conclusions [9]. The cancer incidence in veteran astronauts has been reported by some researchers to be lowered as compared to the general population, while the cancer mortality rate was not increased [95].

However, others have noted that the probability of causation of cancer is far more complex and spaceflight may lead to a greater risk of cancer [96]. Indeed, it remains unclear whether microgravity conditions promote or inhibit the metastatic progression of existing cancers. In this respect, many space studies under microgravity conditions have detected behavioral changes when established tumor cell lines were used as subjects. At the subcellular level, the subject cells were seen to have altered cytoskeletal organizations [97–99] and suppressed signal transduction [100–102]. Cultured under microgravity in space, cell lines were found to have marked gene expressional changes compared to controls on ground [103]. It would be intriguing to determine whether the observed changes in monoculture may serve as initiating factors triggering trophic cancer–stromal interaction.

7.9 Conclusion

We explored the possibility of studying the mechanism of prostate cancer bone metastasis under 3-D-RWV conditions, which mimics certain aspects of a microgravity setting. Cancer metastasis appears to be the net result of interaction between invading malignant cells and resident cells of the mesenchymal lineage, such as inflammatory and immune cells, endothelial cells, and stromal fibroblasts. Specifically, we found that cancer cells with MIC property could recruit bystander cells or dormant cancer cells and convert them into active participants in cancer metastasis. The growth and behaviors of cancer and stromal cells are affected by soluble and insoluble trophic factors upon cancer–stromal interaction, while additional unknown mechanisms may also mediate the cancer-stromal co-evolution. A comprehensive elucidation of the mediating mechanisms is crucial for identifying critical therapeutic targets blocking malicious cancer–stromal interaction. One rational research strategy is to examine cancer–stromal interaction in 3-D co-culture. In this respect, we found that while the 3-D-RWV could be considered an optimal experimental model, microgravity in spaceflight or in a space station is even more optimal for studying tumor–stroma, tumor–tumor, and tumor–extracellular matrices interactions. We propose that similar studies of cancer–stromal interaction could now be conducted in space to expedite the identification of the critical steps leading to irreversible (as opposed to reversible) modifications of the target cells that changing from non-tumorigenic or non-metastatic state to highly aggressive and metastatic progression. Use of 3-D microgravity interaction in the past has helped us to define the co-evolution of cancer and stroma, understand permanent vs. transient changes in cancer cell gene expression and behavior, the recruitment and reprogramming of bystander dormant cells, CTCs or host cells, and tumor–stroma cell fusion. Future experimental 3-D microgravity, spaceflight or space station models coupled with fundamental biology conducted in the laboratory using 3-D suspension cultures and live animals could reveal additional biomarkers and novel therapeutic targets for more effective diagnosis and treatment of men with metastatic cancers.

Questions for Future Research

1. Does microgravity facilitate the metastatic progression of existing tumors?
2. What are the critical molecular events that mediate trophic cancer–stromal interaction?
3. How could these critical events be explored as targets for improving the prognosis of aggressive prostate cancer and therapies for advanced and metastatic prostate cancer?

References

1. Paget, S. (1889). The distribution of secondary growths in cancer of the breast. *Lancet, 1*, 571–573.
2. Chung, L. W., & Cunha, G. R. (1983). Stromal-epithelial interactions: II. Regulation of prostatic growth by embryonic urogenital sinus mesenchyme. *Prostate, 4*(5), 503–511.
3. Pathak, S., Nemeth, M. A., Multani, A. S., Thalmann, G. N., von Eschenbach, A. C., & Chung, L. W. (1997). Can cancer cells transform normal host cells into malignant cells? (Translated from eng). *British Journal of Cancer, 76*(9), 1134–1138 (in eng).
4. Rhee, H. W., Zhau, H. E., Pathak, S., Multani, A. S., Pennanen, S., Visakorpi, T., et al. (2001). Permanent phenotypic and genotypic changes of prostate cancer cells cultured in a three-dimensional rotating-wall vessel. *In Vitro Cellular & Developmental Biology. Animal, 37*(3), 127–140.
5. Wang, R., Xu, J., Juliette, L., Castilleja, A., Love, J., Sung, S. Y., et al. (2005). Three-dimensional co-culture models to study prostate cancer growth, progression, and metastasis to bone. *Seminars in Cancer Biology, 15*(5), 353–364.
6. Sung, S. Y., Hsieh, C. L., Law, A., Zhau, H. E., Pathak, S., Multani, A. S., et al. (2008). Coevolution of prostate cancer and bone stroma in three-dimensional coculture: Implications for cancer growth and metastasis. *Cancer Research, 68*(23), 9996–10003.
7. Parin, V. V., & Gazenko, O. G. (1963). Soviet experiments aimed at investigating the influence of space flight factors on the physiology of animals and man. *Life Sciences and Space Research, 1*, 113–127.
8. Antipov, V. V., Delone, N. L., Parfyonov, G. P., & Vysotsky, V. G. (1965). Results of biological experiments carried out under conditions of "Vostok" flights with the participation of cosmonauts A.G. Nikolajev, P.R. Popovich and V.F. Bykovsky. *Life Sciences and Space Research, 3*, 215–229.
9. Uri, J. J., & Haven, C. P. (2005). Accomplishments in bioastronautics research aboard International Space Station. *Acta Astronautica, 56*(9–12), 883–889.
10. Becker, J. L., & Souza, G. R. (2013). Using space-based investigations to inform cancer research on Earth. (Translated from eng). *Nature Reviews Cancer, 13*(5), 315–327 (in eng).
11. Taylor, G. R., Konstantinova, I., Sonnenfeld, G., & Jennings, R. (1997). Changes in the immune system during and after spaceflight. *Advances in Space Biology and Medicine, 6*, 1–32.

12. Herranz, R., Anken, R., Boonstra, J., Braun, M., Christianen, P. C., de Geest, M., et al. (2013). Ground-based facilities for simulation of microgravity: Organism-specific recommendations for their use, and recommended terminology. (Translated from eng). *Astrobiology, 13*(1), 1–17 (in eng).
13. King, J. A., & Miller, W. M. (2007). Bioreactor development for stem cell expansion and controlled differentiation. *Current Opinion in Chemical Biology, 11*(4), 394–398.
14. Twombly, R. (2003). Prostate modeling experiment success becomes part of legacy of shuttle astronauts. *Journal of the National Cancer Institute, 95*(7), 505–507.
15. Alva, A., & Hussain, M. (2013). The changing natural history of metastatic prostate cancer. *Cancer Journal, 19*(1), 19–24.
16. Eaton, C. L., & Coleman, R. E. (2003). Pathophysiology of bone metastases from prostate cancer and the role of bisphosphonates in treatment. *Cancer Treatment Reviews, 29*(3), 189–198.
17. Tu, S. M., Lopez, A., Leibovici, D., Bilen, M. A., Evliyaoglu, F., Aparicio, A., et al. (2009). Ductal adenocarcinoma of the prostate: Clinical features and implications after local therapy. *Cancer, 115*(13), 2872–2880.
18. Martinez-Outschoorn, U. E., Lisanti, M. P., & Sotgia, F. (2014). Catabolic cancer-associated fibroblasts (CAFs) transfer energy and biomass to anabolic cancer cells, fueling tumor growth. *Seminars in Cancer Biology, 25*, 47–60.
19. Chu, G. C., Zhau, H. E., Wang, R., Rogatko, A., Feng, X., Zayzafoon, M., et al. (2014). RANK- and c-Met-mediated signal network promotes prostate cancer metastatic colonization. *Endocrine-Related Cancer, 21*(2), 311–326.
20. Zhau, H. E., Goodwin, T. J., Chang, S. M., Baker, T. L., & Chung, L. W. (1997). Establishment of a three-dimensional human prostate organoid coculture under microgravity-simulated conditions: Evaluation of androgen-induced growth and PSA expression. *In Vitro Cellular & Developmental Biology. Animal, 33*(5), 375–380.
21. Cunha, G. R., & Chung, L. W. (1981). Stromal-epithelial interactions—I. Induction of prostatic phenotype in urothelium of testicular feminized (Tfm/y) mice. *Journal of Steroid Biochemistry, 14*(12), 1317–1324.
22. Cunha, G. R., Hayward, S. W., & Wang, Y. Z. (2002). Role of stroma in carcinogenesis of the prostate. *Differentiation, 70*(9–10), 473–485.
23. Cunha, G. R., & Hom, Y. K. (1996). Role of mesenchymal-epithelial interactions in mammary gland development. *Journal of Mammary Gland Biology and Neoplasia, 1*(1), 21–35.
24. Ingber, D. E. (2002). Cancer as a disease of epithelial-mesenchymal interactions and extracellular matrix regulation. *Differentiation, 70*(9–10), 547–560.
25. Taylor, M. A., Parvani, J. G., & Schiemann, W. P. (2010). The pathophysiology of epithelial-mesenchymal transition induced by transforming growth factor-beta in normal and malignant mammary epithelial cells. *Journal of Mammary Gland Biology and Neoplasia, 15*(2), 169–190.
26. Cunha, G. R., Shannon, J. M., Neubauer, B. L., Sawyer, L. M., Fujii, H., Taguchi, O., et al. (1981). Mesenchymal-epithelial interactions in sex differentiation. *Human Genetics, 58*(1), 68–77.
27. Neubauer, B. L., Chung, L. W., McCormick, K. A., Taguchi, O., Thompson, T. C., & Cunha, G. R. (1983). Epithelial-mesenchymal interactions in prostatic development. II. Biochemical observations of prostatic induction by urogenital sinus mesenchyme in epithelium of the adult rodent urinary bladder. *The Journal of Cell Biology, 96*(6), 1671–1676.
28. Chung, L. W., Zhau, H. E., & Ro, J. Y. (1990). Morphologic and biochemical alterations in rat prostatic tumors induced by fetal urogenital sinus mesenchyme. *The Prostate, 17*(2), 165–174.
29. Goldstein, A. S., Stoyanova, T., & Witte, O. N. (2010). Primitive origins of prostate cancer: In vivo evidence for prostate-regenerating cells and prostate cancer-initiating cells. *Molecular Oncology, 4*(5), 385–396.
30. Hayward, S. W., Haughney, P. C., Rosen, M. A., Greulich, K. M., Weier, H. U., Dahiya, R., et al. (1998). Interactions between adult human prostatic epithelium and rat urogenital sinus mesenchyme in a tissue recombination model. *Differentiation, 63*(3), 131–140.

31. Cunha, G. R., Chung, L. W., Shannon, J. M., Taguchi, O., & Fujii, H. (1983). Hormone-induced morphogenesis and growth: Role of mesenchymal-epithelial interactions. *Recent Progress in Hormone Research, 39*, 559–598.
32. Bhowmick, N. A., Chytil, A., Plieth, D., Gorska, A. E., Dumont, N., Shappell, S., et al. (2004). TGF-beta signaling in fibroblasts modulates the oncogenic potential of adjacent epithelia. *Science, 303*(5659), 848–851.
33. Ricke, E. A., Williams, K., Lee, Y. F., Couto, S., Wang, Y., Hayward, S. W., et al. (2012). Androgen hormone action in prostatic carcinogenesis: Stromal androgen receptors mediate prostate cancer progression, malignant transformation and metastasis. *Carcinogenesis, 33*(7), 1391–1398.
34. Camps, J. L., Chang, S. M., Hsu, T. C., Freeman, M. R., Hong, S. J., Zhau, H. E., et al. (1990). Fibroblast-mediated acceleration of human epithelial tumor growth in vivo. *Proceedings of the National Academy of Sciences of the United States of America, 87*(1), 75–79.
35. Gleave, M. E., Hsieh, J. T., von Eschenbach, A. C., & Chung, L. W. (1992). Prostate and bone fibroblasts induce human prostate cancer growth in vivo: Implications for bidirectional tumor-stromal cell interaction in prostate carcinoma growth and metastasis. *The Journal of Urology, 147*(4), 1151–1159.
36. Thalmann, G. N., Anezinis, P. E., Chang, S. M., Zhau, H. E., Kim, E. E., Hopwood, V. L., et al. (1994). Androgen-independent cancer progression and bone metastasis in the LNCaP model of human prostate cancer. *Cancer Research, 54*(10), 2577–2581.
37. Wu, H. C., Hsieh, J. T., Gleave, M. E., Brown, N. M., Pathak, S., & Chung, L. W. (1994). Derivation of androgen-independent human LNCaP prostatic cancer cell sublines: Role of bone stromal cells. *International Journal of Cancer, 57*(3), 406–412.
38. Kang, Y., & Pantel, K. (2013). Tumor cell dissemination: Emerging biological insights from animal models and cancer patients. *Cancer Cell, 23*(5), 573–581.
39. Cunha, G. R. (1994). Role of mesenchymal-epithelial interactions in normal and abnormal development of the mammary gland and prostate. *Cancer, 74*(3 Suppl), 1030–1044.
40. Rucki, A. A., & Zheng, L. (2014). Pancreatic cancer stroma: Understanding biology leads to new therapeutic strategies. *World Journal of Gastroenterology, 20*(9), 2237–2246.
41. Chambers, A. F., Groom, A. C., & MacDonald, I. C. (2002). Dissemination and growth of cancer cells in metastatic sites. *Nature Reviews Cancer, 2*(8), 563–572.
42. Fidler, I. J. (2001). Seed and soil revisited: Contribution of the organ microenvironment to cancer metastasis. *Surgical Oncology Clinics of North America, 10*(2), 257–269. vii–viiii.
43. Kerbel, R. S., Waghorne, C., Korczak, B., Lagarde, A., & Breitman, M. L. (1988). Clonal dominance of primary tumours by metastatic cells: Genetic analysis and biological implications. *Cancer Surveys, 7*(4), 597–629.
44. Talmadge, J. E. (2007). Clonal selection of metastasis within the life history of a tumor. *Cancer Research, 67*(24), 11471–11475.
45. Polyak, K. (2007). Breast cancer: Origins and evolution. *The Journal of Clinical Investigation, 117*(11), 3155–3163.
46. Ramel, C. (1992). Evolutionary aspects of human cancer. *Pharmacogenetics, 2*(6), 344–349.
47. Billaud, M., & Santoro, M. (2011). Is co-option a prevailing mechanism during cancer progression? *Cancer Research, 71*(21), 6572–6575.
48. Fidler, I. J. (2012). Biological heterogeneity of cancer: Implication to therapy. *Human Vaccines & Immunotherapeutics, 8*(8), 1141–1142.
49. Huang, S. (2013). Genetic and non-genetic instability in tumor progression: Link between the fitness landscape and the epigenetic landscape of cancer cells. *Cancer Metastasis Reviews, 32*(3–4), 423–448.
50. Nickerson, M. L., Im, K. M., Misner, K. J., Tan, W., Lou, H., Gold, B., et al. (2013). Somatic alterations contributing to metastasis of a castration-resistant prostate cancer. *Human Mutation, 34*(9), 1231–1241.
51. Kim, J. B., Stein, R., & O'Hare, M. J. (2004). Three-dimensional in vitro tissue culture models of breast cancer—A review. *Breast Cancer Research and Treatment, 85*(3), 281–291.

52. Gray, J. W. (2003). Evidence emerges for early metastasis and parallel evolution of primary and metastatic tumors. *Cancer Cell, 4*(1), 4–6.
53. Harbst, K., Staaf, J., Måsbäck, A., Olsson, H., Ingvar, C., Vallon-Christersson, J., et al. (2010). Multiple metastases from cutaneous malignant melanoma patients may display heterogeneous genomic and epigenomic patterns. *Melanoma Research, 20*(5), 381–391.
54. Langley, R. R., & Fidler, I. J. (2011). The seed and soil hypothesis revisited—The role of tumor-stroma interactions in metastasis to different organs. *International Journal of Cancer, 128*(11), 2527–2535.
55. Steinmetz, M. P., Mekhail, A., & Benzel, E. C. (2001). Management of metastatic tumors of the spine: Strategies and operative indications. *Neurosurgical Focus, 11*(6), e2.
56. Thalmann, G. N., Sikes, R. A., Wu, T. T., Degeorges, A., Chang, S. M., Ozen, M., et al. (2000). LNCaP progression model of human prostate cancer: Androgen-independence and osseous metastasis. *The Prostate, 44*(2), 91–103.
57. Xie, B. X., Zhang, H., Wang, J., Pang, B., Wu, R. Q., Qian, X. L., et al. (2011). Analysis of differentially expressed genes in LNCaP prostate cancer progression model. *Journal of Andrology, 32*(2), 170–182.
58. Spans, L., Helsen, C., Clinckemalie, L., Van den Broeck, T., Prekovic, S., Joniau, S., et al. (2014). Comparative genomic and transcriptomic analyses of LNCaP and C4-2B prostate cancer cell lines. *PLoS One, 9*(2), e90002.
59. Liu, A. Y., Brubaker, K. D., Goo, Y. A., Quinn, J. E., Kral, S., Sorensen, C. M., et al. (2004). Lineage relationship between LNCaP and LNCaP-derived prostate cancer cell lines. *The Prostate, 60*(2), 98–108.
60. Barron, D. A., & Rowley, D. R. (2012). The reactive stroma microenvironment and prostate cancer progression. *Endocrine-Related Cancer, 19*(6), R187–R204.
61. Lisanti, M. P., Martinez-Outschoorn, U. E., Chiavarina, B., Pavlides, S., Whitaker-Menezes, D., Tsirigos, A., et al. (2010). Understanding the "lethal" drivers of tumor-stroma co-evolution: Emerging role(s) for hypoxia, oxidative stress and autophagy/mitophagy in the tumor microenvironment. *Cancer Biology & Therapy, 10*(6), 537–542.
62. Josson, S., Matsuoka, Y., Chung, L. W., Zhau, H. E., & Wang, R. (2010). Tumor-stroma co-evolution in prostate cancer progression and metastasis. *Seminars in Cell & Developmental Biology, 21*(1), 26–32.
63. Alix-Panabieres, C., Riethdorf, S., & Pantel, K. (2008). Circulating tumor cells and bone marrow micrometastasis. *Clinical Cancer Research, 14*(16), 5013–5021.
64. Shao, C., Liao, C. P., Hu, P., Chu, C. Y., Zhang, L., Bui, M. H., et al. (2014). Detection of live circulating tumor cells by a class of near-infrared heptamethine carbocyanine dyes in patients with localized and metastatic prostate cancer. *PLoS One, 9*(2), e88967.
65. Huang, W. C., Xie, Z., Konaka, H., Sodek, J., Zhau, H. E., & Chung, L. W. (2005). Human osteocalcin and bone sialoprotein mediating osteomimicry of prostate cancer cells: Role of cAMP-dependent protein kinase A signaling pathway. *Cancer Research, 65*(6), 2303–2313.
66. Knerr, K., Ackermann, K., Neidhart, T., & Pyerin, W. (2004). Bone metastasis: Osteoblasts affect growth and adhesion regulons in prostate tumor cells and provoke osteomimicry. *International Journal of Cancer, 111*(1), 152–159.
67. Nomura, T., Huang, W.-C., Zhau, H. E., Josson, S., Mimata, H., & Kaur, M. (2014). Beta2-Microglobulin-mediated signaling as a target for cancer therapy. *Anti-Cancer Agents in Medicinal Chemistry, 14*(3), 343–352.
68. Rucci, N., & Teti, A. (2010). Osteomimicry: How tumor cells try to deceive the bone. *Frontiers in Bioscience, 2*, 907–915.
69. Hu, P., Chu, G. C., Zhu, G., Yang, H., Luthringer, D., Prins, G., et al. (2011). Multiplexed quantum dot labeling of activated c-Met signaling in castration-resistant human prostate cancer. *PLoS One, 6*(12), e28670.
70. Hu, P., Chung, L. W., Berel, D., Frierson, H. F., Yang, H., Liu, C., et al. (2013). Convergent RANK- and c-Met-mediated signaling components predict survival of patients with prostate cancer: An interracial comparative study. *PLoS One, 8*(9), e73081.

71. Zhau, H. E., Li, Q., & Chung, L. W. (2013). Interracial differences in prostate cancer progression among patients from the United States, China and Japan. *Asian Journal of Andrology, 15*(6), 705–707.
72. Huang, W. C., Havel, J. J., Zhau, H. E., Qian, W. P., Lue, H. W., Chu, C. Y., et al. (2008). Beta2-microglobulin signaling blockade inhibited androgen receptor axis and caused apoptosis in human prostate cancer cells. *Clinical Cancer Research, 14*(17), 5341–5347.
73. Huang, W. C., Wu, D., Xie, Z., Zhau, H. E., Nomura, T., Zayzafoon, M., et al. (2006). Beta2-microglobulin is a signaling and growth-promoting factor for human prostate cancer bone metastasis. *Cancer Research, 66*(18), 9108–9116.
74. Huang, W. C., Zhau, H. E., & Chung, L. W. (2010). Androgen receptor survival signaling is blocked by anti-beta2-microglobulin monoclonal antibody via a MAPK/lipogenic pathway in human prostate cancer cells. *The Journal of Biological Chemistry, 285*(11), 7947–7956.
75. Josson, S., Matsuoka, Y., Gururajan, M., Nomura, T., Huang, W. C., Yang, X., et al. (2013). Inhibition of beta2-microglobulin/hemochromatosis enhances radiation sensitivity by induction of iron overload in prostate cancer cells. *PLoS One, 8*(7), e68366.
76. Josson, S., Nomura, T., Lin, J. T., Huang, W. C., Wu, D., Zhau, H. E., et al. (2011). Beta2-microglobulin induces epithelial to mesenchymal transition and confers cancer lethality and bone metastasis in human cancer cells. *Cancer Research, 71*(7), 2600–2610.
77. Bernal, M., Ruiz-Cabello, F., Concha, A., Paschen, A., & Garrido, F. (2012). Implication of the beta2-microglobulin gene in the generation of tumor escape phenotypes. *Cancer Immunology, Immunotherapy, 61*(9), 1359–1371.
78. Cafri, G., Margalit, A., Tzehoval, E., Eisenbach, L., & Gross, G. (2013). Development of novel genetic cancer vaccines based on membrane-attached beta2 microglobulin. *Annals of the New York Academy of Sciences, 1283*, 87–90.
79. Yang, J., & Yi, Q. (2010). Killing tumor cells through their surface beta(2)-microglobulin or major histocompatibility complex class I molecules. *Cancer, 116*(7), 1638–1645.
80. Yang, J., & Yi, Q. (2011). Therapeutic monoclonal antibodies for multiple myeloma: An update and future perspectives. *American Journal of Blood Research, 1*(1), 22–33.
81. Yi, Q. (2009). Novel immunotherapies. *Cancer Journal, 15*(6), 502–510.
82. Chung, L. W., Li, W., Gleave, M. E., Hsieh, J. T., Wu, H. C., Sikes, R. A., et al. (1992). Human prostate cancer model: Roles of growth factors and extracellular matrices. *Journal of Cellular Biochemistry. Supplement, 16H*, 99–105.
83. Savorè, C., Zhang, C., Muir, C., Liu, R., Wyrwa, J., Shu, J., et al. (2005). Perlecan knockdown in metastatic prostate cancer cells reduces heparin-binding growth factor responses in vitro and tumor growth in vivo. *Clinical & Experimental Metastasis, 22*(5), 377–390.
84. Bisanz, K., Yu, J., Edlund, M., Spohn, B., Hung, M. C., Chung, L. W., et al. (2005). Targeting ECM-integrin interaction with liposome-encapsulated small interfering RNAs inhibits the growth of human prostate cancer in a bone xenograft imaging model. *Molecular Therapy, 12*(4), 634–643.
85. Edlund, M., Miyamoto, T., Sikes, R. A., Ogle, R., Laurie, G. W., Farach-Carson, M. C., et al. (2001). Integrin expression and usage by prostate cancer cell lines on laminin substrata. *Cell Growth & Differentiation, 12*(2), 99–107.
86. Josson, S., Anderson, C. S., Sung, S. Y., Johnstone, P. A., Kubo, H., Hsieh, C. L., et al. (2011). Inhibition of ADAM9 expression induces epithelial phenotypic alterations and sensitizes human prostate cancer cells to radiation and chemotherapy. *The Prostate, 71*(3), 232–240.
87. Liu, C. M., Hsieh, C. L., He, Y. C., Lo, S. J., Liang, J. A., Hsieh, T. F., et al. (2013). In vivo targeting of ADAM9 gene expression using lentivirus-delivered shRNA suppresses prostate cancer growth by regulating REG4 dependent cell cycle progression. *PLoS One, 8*(1), e53795.
88. Pen, C. C., Liu, C. M., Lin, C. C., Lin, C. C., Hsieh, T. F., Josson, S., et al. (2012). Combined dynamic alterations in urinary VEGF levels and tissue ADAM9 expression as markers for lethal phenotypic progression of prostate cancer. *The Chinese Journal of Physiology, 55*(6), 390–397.

89. Shigemura, K., Sung, S. Y., Kubo, H., Arnold, R. S., Fujisawa, M., Gotoh, A., et al. (2007). Reactive oxygen species mediate androgen receptor- and serum starvation-elicited downstream signaling of ADAM9 expression in human prostate cancer cells. *The Prostate, 67*(7), 722–731.
90. Sung, S. Y., Kubo, H., Shigemura, K., Arnold, R. S., Logani, S., Wang, R., et al. (2006). Oxidative stress induces ADAM9 protein expression in human prostate cancer cells. *Cancer Research, 66*(19), 9519–9526.
91. Wang, R., Sun, X., Wang, C. Y., Hu, P., Chu, C.-Y., Liu, S., et al. (2012). Spontaneous cancer-stromal cell fusion as a mechanism of prostate cancer androgen-independent progression. *PLoS One, 7*(8), e42653.
92. Chen, E. H., Grote, E., Mohler, W., & Vignery, A. (2007). Cell-cell fusion. (Translated from eng). *FEBS Letters, 581*(11), 2181–2193 (in eng).
93. Larsson, L. I., Bjerregaard, B., & Talts, J. F. (2008). Cell fusions in mammals. (Translated from eng). *Histochemistry and Cell Biology, 129*(5), 551–561 (in eng).
94. Oren-Suissa, M., & Podbilewicz, B. (2007). Cell fusion during development. (Translated from eng). *Trends in Cell Biology, 17*(11), 537–546 (in eng).
95. Pawelek, J. M. (2005). Tumour-cell fusion as a source of myeloid traits in cancer. (Translated from eng). *Lancet Oncology, 6*(12), 988–993 (in eng).
96. Cucinotta, F. A. (2014). Space radiation risks for astronauts on multiple International Space Station missions. (Translated from eng). *PLoS One, 9*(4), e96099 (in eng).
97. Buravkova, L., Romanov, Y., Rykova, M., Grigorieva, O., & Merzlikina, N. (2005). Cell-to-cell interactions in changed gravity: Ground-based and flight experiments. *Acta Astronautica, 57*(2–8), 67–74.
98. Kordyum, E. L., Shevchenko, G. V., Yemets, A. I., Nyporko, A. I., & Blume, Y. B. (2005). Application of GFP technique for cytoskeleton visualization onboard the International Space Station. *Acta Astronautica, 56*(6), 613–621.
99. Meloni, M. A., Galleri, G., Pani, G., Saba, A., Pippia, P., & Cogoli-Greuter, M. (2011). Space flight affects motility and cytoskeletal structures in human monocyte cell line J-111. *Cytoskeleton, 68*(2), 125–137.
100. Chang, T. T., Walther, I., Li, C. F., Boonyaratanakornkit, J., Galleri, G., Meloni, M. A., et al. (2012). The Rel/NF-kappaB pathway and transcription of immediate early genes in T cell activation are inhibited by microgravity. *Journal of Leukocyte Biology, 92*(6), 1133–1145.
101. Tauber, S., Hauschild, S., Crescio, C., Secchi, C., Paulsen, K., Pantaleo, A., et al. (2013). Signal transduction in primary human T lymphocytes in altered gravity – Results of the MASER-12 suborbital space flight mission. *Cell Communication and Signaling, 11*(1), 32.
102. Wang, C., Luo, H., Zhu, L., Yang, F., Chu, Z., Tian, H., et al. (2014). Microgravity inhibition of lipopolysaccharide-induced tumor necrosis factor-alpha expression in macrophage cells. *Inflammation Research, 63*(1), 91–98.
103. Guo, F., Li, Y., Liu, Y., Huang, J., Zhang, Z., Wang, J., et al. (2012). Identification of genes associated with tumor development in CaSki cells in the cosmic space. *Molecular Biology Reports, 39*(6), 6923–6931.

Chapter 8
Skeletal Muscle Culture Under Spaceflight Conditions

Mark S.F. Clarke

Historical Landmarks

1968—Yaffe describes establishment of a rat L6 myoblast cell line capable of generating spontaneously contractile myotubes in culture [1].

1975—Dietlein describes reduction in muscle volume and function associated with extended spaceflight during Apollo missions [2].

1977—Yaffe and Saxel describe the establishment of a mouse C2C12 myoblast cell line capable of generating myotubes in culture [3].

1982—Vandenburgh describes the ability of uni-axial stretching to promote the formation of contractile myotubes from primary avian myocytes within a collagen gel matrix [4].

1994—Kulesh et al. fly the first experiment aboard the Space Shuttle utilizing skeletal muscle cells (rat L8 myoblast cultures) resulting in the creation of a new, non-fusogenic myoblast variant known as the L8SF cell line which retained its altered phenotype even on return to Earth [5].

1997—Molnar et al. establish rat primary myoblast cultures in the rotating wall vessel (RWV) demonstrating myotube formation during microgravity analog conditions [6].

1999—Vanderburgh et al. fly the first experiment aboard the Space Shuttle to utilize cultured 3-D skeletal muscle organoids constructed from myoblasts to demonstrate that microgravity exposure can induce muscle atrophy even without the removal of extrinsic mechanical load [7].

(continued)

M.S.F. Clarke, PhD (✉)
Laboratory for Integrated Physiology, Department of Health and Human Performance, University of Houston, Houston, TX 77204-6015, USA
e-mail: mclarke@central.uh.edu

C.A. Nickerson et al. (eds.), *Effect of Spaceflight and Spaceflight Analogue Culture on Human and Microbial Cells*, DOI 10.1007/978-1-4939-3277-1_8

(continued)

2000—Dennis and Kosnik establish 3-D skeletal muscle organoids constructed from both myoblasts and fibroblasts that demonstrated superior contractile myotube properties than those constructed from myoblasts alone [8].

2005—Levenberg et al. establish 3-D skeletal muscle organoids created from myoblasts, fibroblasts, and endothelial cells capable of forming a functional microvascular network as well as contractile myotubes when transplanted into the whole animal [9].

2011—Van der Schaft et al. establish 3-D skeletal muscle organoids created from myotubes and endothelial cells in which microvascular formation and myotube alignment were driven by uni-axial load placed on the construct [10].

2013—Lee and Vandenburgh describe a contractile 3-D skeletal muscle organoid in which the removal of uni-axial load induced a dose-dependent myotube atrophy response and concomitant reduction in force production approximating that observed in intact muscle undergoing microgravity exposure [11].

8.1 Introduction

The physiological response of the human body to the microgravity environment encountered during spaceflight involves adaptations at numerous levels, ranging from those which affect the whole body, discrete physiological systems and/or tissues down to individual cell types. This chapter focuses primarily on the effect of microgravity exposure on skeletal muscle and how cell culture-based models have been used to investigate such effects. While the ultimate goal of such studies has been to understand the effects of spaceflight as it relates to maintaining skeletal muscle mass and function in crew members during and after spaceflight missions, such culture models also show potential for studying the underlying mechanisms at play in a variety of pathological conditions involving skeletal muscle in the terrestrial environment.

At the outset however, it must be clearly stated that while available tissue culture models provide an excellent means of studying the effects of microgravity exposure on the structure and function of skeletal muscle cells in isolation, information generated using these approaches must be viewed through a prism which recognizes the potential impact of various signals sensitive to spaceflight that emanate from other physiologic systems not represented within such skeletal muscle tissue culture analogs. For example, the complex, multi-cellular three-dimensional (3-D) architecture of skeletal muscle tissue (i.e., myofibers, satellite stem cells, neural cells, vascular cells) for the most part is not represented within skeletal muscle tissue culture analogs routinely available. In addition, in vivo biological signals which target skeletal muscle tissue that are derived from physiological systems known to be adversely affected by microgravity exposure, such as the endocrine, vascular, and neuronal systems, are absent in these culture models. Therefore, it is of the utmost importance that any conclusions drawn from skeletal muscle cell responses observed in such tissue microgravity culture analogs be interpreted relative to additional effects already known to impact the whole body during spaceflight. As such, where appropriate the author has included information relevant to the similarities and the differences inherent between various tissue culture models of skeletal muscle structure and function in vitro relative to the response of whole skeletal muscle tissue in vivo.

8.2 General Effects of Microgravity Exposure on Skeletal Muscle

In general, microgravity exposure during spaceflight of both crew members and animals results in a rapid loss of muscle mass and function [12–17]**, especially in those muscles associated with the maintenance of body posture and ambulation in terrestrial gravity known as the "anti-gravity" muscles.** It is the negative impact of such skeletal muscle adaptation/de-adaptation upon crew member health, performance, and ultimately mission success which drives the need to understand the deleterious physiological and biochemical effects of spaceflight upon skeletal muscle cell function. While numerous muscle group and species-specific effects have been reported, in general terms microgravity exposure of the musculoskeletal system in vivo results in: (1) disruption of protein balance within the myofiber [18, 19]; (2) an overall reduction in both whole muscle and individual myofiber cross-sectional area (CSA) [20–24]; (3) an enhanced Type II myofiber selective atrophy response [22, 25, 26]; (4) evidence in some situations of a Type I to Type II myofiber phenotypic shift [24, 26–32]; (5) a reduction in myonuclear to sarcoplasmic volume ratio [22, 33]; (6) a reduction in neuromuscular junction (NMJ) size/complexity; and (7) a reduction in the number of capillary vessel/myofiber contacts (i.e., capillary bed tortuosity) within the muscle [14, 15, 23, 34, 35]. These biochemical, cellular, and histological alterations are all paralleled with concomitant decreases in overall skeletal muscle physiological function such as muscle tone/stiffness, neuromuscular activation patterns, strength, and endurance at the whole body level [12, 16, 36–40].

While some of the negative effects of microgravity exposure on skeletal muscle function can be directly attributed to the loss of myofibrillar contractile proteins such as myosin heavy chain (MHC) protein [41], or the disruption of myofiber cytoskeletal proteins responsible for maintaining sarcomeric integrity such as titin and nebulin [42, 43], it is still unclear whether or not these myofiber effects are primarily mediated via intrinsic or extrinsic factors. For example, microgravity-induced reductions in circulating levels of muscle growth factors, such as pituitary-derived growth hormone (GH) or liver-derived insulin-like-growth factor-1 (IGF-1) [44–46], most likely have subsequent down-stream effects on intra-myofiber signaling pathways within skeletal muscle tissue [47]. These include down-regulation of the hypertrophic Akt/mTOR pathway [48, 49] and up-regulation of the atrophic myostatin/activin IIB receptor pathway [50–52], which acting in concert appear to be primarily responsible for myofiber protein balance and the overall maintenance of muscle mass in vivo [53–55]. In addition, the removal of mechanical load during spaceflight normally experienced by the musculoskeletal system under terrestrial conditions has already been shown to induce effects not only within the myofiber, such as disruption of normal gene transcription rates [56] and alterations in myonuclear-sarcoplasmic volume ratio [22], but also external to the myofiber but still within skeletal muscle tissue, such as reductions in the size and function of the NMJ [15] leading to possible myofiber type shifting [12, 23, 26, 57] and impaired neuromuscular function [35, 58]. Furthermore, the removal of mechanical load from skeletal muscle has also been shown to initiate intra-myofiber calcium-dependent proteolysis [59] and alterations in sarcoplasmic reticulum calcium pump (SERCA) expression [60, 61] due to an as yet undefined pathway potentially related to intrinsic membrane tension within the myofiber [62].

One additional level of complexity associated with the response of skeletal muscle specifically related to the mechanical unloading experienced during spaceflight is the apparent fiber type specific nature of these effects, linked not only to particular fiber types but also the species studied and the function of the specific muscle in which the fiber type resides. For example, during both short-term spaceflight [25] and bed rest [63], the human *m. vastus lateralis* muscle undergoes Type IIb myofiber selective atrophy whereas the human *soleus* and *gastrocnemius* muscle in the same leg undergo Type I myofiber selective atrophy following extended spaceflight [12] or bed rest [64]. Similar Type I myofiber selective atrophy has also been observed in the *soleus* muscle of the rat hind limbs following spaceflight [31] and terrestrial unloading resulting from hind limb suspension of the animals [65, 66]. It remains unclear why the same myofiber type in different muscles of the same individual crew member or animal exhibits different atrophy responses during spaceflight or terrestrial unloading. However, such myofiber selective responses appear related to the functional role of these particular muscles during standing/ambulation under terrestrial conditions (i.e., anti-gravity muscles), as well as the disruption of neural input to the myofibers due to reduced proprioceptive mechanoreceptor activation during spaceflight and terrestrial unloading in vivo [65, 67].

The complex response of whole skeletal muscle tissue to actual spaceflight or during terrestrial analogs of whole body microgravity exposure (i.e., bed rest, dry immersion, rodent hind limb suspension, etc.) clearly demonstrates that multiple cell types and a variety of signaling pathways are involved in the adaptive response of skeletal muscle tissue to microgravity. As such, in the context of this review it is important to recognize that most planar tissue culture experiments utilizing isolated skeletal muscle progenitor cells (i.e., satellite cells, primary myoblasts) or differentiated skeletal muscle cultures (i.e., myotubes), carried out under actual or terrestrial analog conditions of microgravity exposure, lack the complex multi-system, multi-cellular interplay required to truly mimic the overall response of intact skeletal muscle to spaceflight in vivo.

This is where tissue culture analogs in general and skeletal muscle tissue culture models in particular become extremely useful in investigating the cellular and biochemical responses of cells to a variety of environmental conditions such as microgravity exposure or mechanical unloading. By cataloging these responses while manipulating specific experimental conditions in a defined fashion, skeletal muscle tissue culture analogs can be used to develop causal relationships between particular stimuli and specific cellular responses. This approach provides a wealth of information on the response of individual muscle cells and/or more complex, 3-D organoid skeletal muscle tissue culture analogs to a variety of altered stimuli, which in turn can then be integrated into a larger model of how intact muscle tissue responds in vivo to altered environmental conditions such as microgravity exposure. In addition, the knowledge gained utilizing such tissue culture analogs relative to understanding the underlying cellular mechanisms involved in the response of skeletal muscle to spaceflight (including those primarily caused by mechanical unloading) has direct application in understanding various terrestrial disease states, such as age-related sarcopenia [68], cancer cachexia [69], muscle cell autophagy [70], and obesity-related muscle atrophy [71].

8.3 Skeletal Muscle Tissue Culture Models

8.3.1 General Considerations

As discussed above, the response of intact skeletal muscle to spaceflight involves a complex interplay between muscle function, myofiber type, neuronal innervations, and various circulating factors that modulate myofiber behavior and function. While it is generally accepted that the majority of effects initiated within skeletal muscle during spaceflight are as a consequence of the removal of the mechanical loading normally experienced by the musculoskeletal system under terrestrial conditions, the effects of microgravity exposure per se (i.e., removal or disruption of a defined gravitational vector) should not be ignored. However, it should be remembered that such relatively subtle effects are difficult to isolate even under well controlled tissue culture conditions when compared to the more potentially profound effects of removing the almost continuous mechanical stimulation experienced by skeletal muscle cells created as a consequence of voluntary muscle contraction combined with terrestrial gravity loading in vivo.

Regardless of the predominant affect of spaceflight (i.e., mechanical unloading vs. removal of gravitational vector) on skeletal muscle cells, the complex multifactorial response of myofibers to spaceflight or ground-based analogs of spaceflight in vivo raises an as yet unresolved issue relating to initial source of the skeletal muscle cells. For example, there is clear experimental evidence that the response to spaceflight is dependent on the specific muscle studied, its physiological function in vivo, and the myofiber type(s) present within the muscle [12, 23]. For example, cultured skeletal muscle cells isolated from an adult *soleus* muscle (predominantly made up of Type I myofibers) may respond differently to spaceflight or decreased mechanical loading under terrestrial conditions to those cells isolated from an adult *gastrocnemius* or *m. vastus lateralis* muscle (containing both Type I and Type II myofibers). Any responses observed may be further complicated when species variations (e.g., avian vs. mammalian, rodent vs. human) and/or muscle age differences (e.g., embryonic vs. adult tissue) are considered. As such, it is important to consider such "source" factors when interpreting observations made concerning the effects of microgravity exposure on cultured skeletal muscle cells in vitro. However, while skeletal muscle cells isolated from different muscles may initially retain some of their in vivo phenotypic characteristics, tissue culture studies which have specifically compared primary rodent myoblast cell populations derived from various muscles indicate that any such phenotypic programming appears to be lost after a relative short period of propagation under standard tissue culture conditions (i.e., three to four passages) resulting in myoblast cultures appear to have a uniform phenotype regardless of muscle origin [72].

8.3.2 Primary Skeletal Muscle Cell Cultures

In the case of skeletal muscle, the individual cell type most commonly used for tissue culture studies is the undifferentiated myoblast (derived from the muscle satellite cell population in vivo) or the differentiated myotube (consisting of multiple fused myoblasts forming a structure analogous to a myofiber in vivo) [73]. Primary myoblasts are isolated by mechanical and/or enzymatic disruption of fresh muscle tissue and can be isolated from a range of species and muscle types harvested from avian, amphibian, rodent, or human subjects. Skeletal muscle myoblasts are adherent cells requiring a culture substratum to attach and grow in culture. While primary myocytes can be directly cultured on a variety of tissue culture plastic surfaces, the use of extracellular matrix (ECM) coatings, such as collagen Type I or fibronectin, appear to be most suitable for ensuring cell attachment and proliferation in monolayer culture. Primary human myocytes become senescent within six to eight passages in culture, although addition of specific growth factors to the medium, such as fibroblast growth factor (FGF) or insulin-like growth factor-1 (IGF-1) [74, 75], have been reported to delay myoblast senescence while maintaining cell proliferation rates.

Since primary myoblast cultures are assumed to be derived from skeletal satellite cell populations, myoblast cultures have been used to study individual satellite cell responses to a variety of environmental and biochemical stimuli including microgravity exposure. In addition, under specific culture conditions myoblasts will fuse together to form a multi-nucleated structure known as a myotube, analogous to a myofiber in vivo. As such, primary myoblast cultures undergoing myotube formation have also been used to investigate the effects of various stimuli on the satellite cell–myofiber fusion process [76–78] shown to be central to the control of myofiber hypertrophy and the formation of new myofibers in response to injury in vivo [79]. In general, such primary myotube cultures will express markers of terminal differentiation, such as myosin heavy chain (MHC) [80] or creatine phosphokinase (CPK) [81], while organizing sarcomeric-like structures (i.e., striations) within the myotube [82]. While not usually spontaneously contractile, primary myotubes containing sarcomeres can be induced to contract after sarcolemma depolarization as a result of an externally applied electrical current or chemical depolarization [83].

8.3.3 Skeletal Muscle Cell Cultures from Established Myocyte Cell Lines

In order to overcome the limitations inherent in using primary myoblasts as a source for skeletal muscle cell tissue culture studies (such as a limited capacity for proliferation in culture and a reduction in myotube formation rate as culture age increases), investigators have turned to the use of a number of rodent myoblast cell lines that can be easily propagated in culture while still maintaining the ability to differentiate and fuse into myotubes (Table 8.1).

The three most commonly used, commercially available myoblast cell lines for studying skeletal muscle cell responses in vitro are the embryonic-derived rat omega (RMo) cell line [84], the adult mouse-derived C2C12 cell line [3], and the neonatal rat L6 cell line [1]. Each of these three cell lines has unique characteristics relevant to investigating the effects of various experimental conditions on specific skeletal muscle cell properties. For example, both rat-derived RMo and L6 myoblasts are capable of fusing to form multi-nucleated myotubes which are striated and exhibit spontaneous contraction even in the absence of neuronal cells [1, 84]. While contractility in RMo and L6 myotubes spontaneously arises, this contractile activity is capable of being "paced" by altering the myotube membrane potential either electrically [87] or chemically [88, 89]. Striated RMo myotubes also generate a complex intra-myofiber membrane system containing cholesterol (Fig. 8.1) similar to that observed in the sarcoplasmic reticulum system of intact human and rodent myofibers. Since the ability to regulate intra-myofiber calcium levels is a prerequisite for excitation–contraction (E–C) coupling within skeletal muscle myofibers in vivo [91], the formation of such intra-cellular membrane systems within cultured myotubes (Fig. 8.1) is almost certainly a requirement for spontaneous contractility to occur in vitro. Mouse-derived C2C12 myoblasts can also form striated, multinucleated myotubes but do not normally exhibit spontaneous contractile behavior. However, striated C2C12 myotubes can be induced to contract using an external

Table 8.1 Rodent myoblast cell lines commonly used to study muscle cell function in vitro

Myoblast cell line	Species	Tissue source	Myotube formation	Striations	Spontaneously contractile	References
RMo	Rat	Embryonic muscle	Yes	Yes	Yes	[84]
L6	Rat	Neonatal, quadriceps muscle	Yes	Yes	Yes	[1]
C2C12 (originally derived from C2 cell line)	Mouse	Adult, quadriceps muscle	Yes	Yes	No (inducible)	[3]
MM14	Mouse	Adult muscle	Yes	No	No	[85]
Sol8	Mouse	Adult, soleus muscle	Yes	No	No	[86]

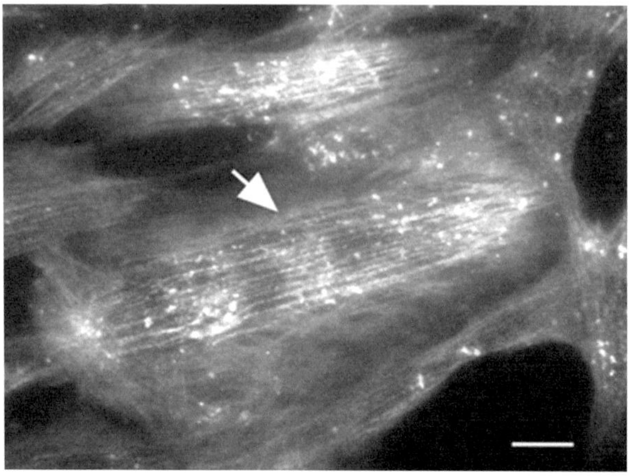

Fig. 8.1 Fluorescent micrograph of RMo myotubes immunostained for cholesterol. RMo myoblasts were differentiated into myotube structures for 7 days in culture medium containing 2 % horse serum on collagen Type I-coated Permanox plastic tissue culture chamber slides. Cells were fixed with freshly generated formaldehyde-PBS solution (pH 7.2), permeabilized using TX-100, and then stained with a monoclonal antibody directed against cholesterol as previously described [90]. Note the labeling of cholesterol-containing internal membrane components (*arrow*) aligned with myofibril direction within the RMo myotube (Bar—20 μm)

electrical current [92–94] to induce depolarization of the sarcolemma membrane, indicating that C2C12 myotubes possess the intra-cellular components required for E–C coupling and sarcomere contraction. One common issue encountered with the use of C2C12 cells to study muscle cell function in vitro is the loss of myoblast proliferative capacity and the ability to reproducibly form myotubes in culture. As such, it is very important to ensure that when utilizing C2C12 cells that the appropriate culture conditions are employed, such as high serum/low density seeding densities for myoblast propagation and low serum/high density seeding densities for myotube formation. By adhering to these conditions numerous multi-nucleated myotubes (Fig. 8.2) that express adult Type II MHC protein can be created in culture, thus providing for a well-defined and reproducible means of creating differentiated C2C12 myotube cultures for subsequent study.

8.3.4 Three-Dimensional Myoblast/Myotube Culture Approaches

There is wide recognition that conventional planar culture conditions which involve a single monolayer of adherent cells do not mimic the complex, three-dimensional (3-D) cellular architecture, intercellular interactions and biochemical signaling that occur within whole tissue in vivo. Recognition of this limitation has led many investigators to utilize 3-D tissue culture techniques to

Fig. 8.2 Light micrograph of C2C12 myotube cultures stained for myosin heavy chain (MHC) II protein using immune-peroxidase labeling. Proliferating C2C12 myoblasts were cultured in DMEM/F12 medium containing 10 % heat-inactivated fetal bovine serum (10%FBS.DMEM/F12) in 12 well culture plates coated with collagen Type I protein. Myoblasts were allowed to attach overnight in 10%FBS.DMEM/F12 after which time the medium was replaced with DMEM/F12 medium containing 2 % heat-inactivated horse serum (2%HS.DMEM/F12). Myoblasts were cultured for an additional 7 days with a single medium exchange at Day 3 of culture. Myotube formation begins at day 3–4 of culture under serum-deprived conditions. After 7 days, myotube cultures were fixed and stained using a rabbit anti-MHC II polyclonal antibody. Note the numerous large multi-nucleated myotubes staining for MHC II protein (*asterisks*) with little or no staining being observed in the remaining unfused myoblasts (*arrows*) (Bar—10 μm)

mimic such spatial interactions in a wide variety of cell types [95] including skeletal muscle [96, 97]. One of the simplest versions of this 3-D culture approach to study the effects of microgravity exposure on skeletal muscle is the formation of 3-D skeletal muscle organoids using the rotating wall vessel (RWV) tissue culture system [6, 98] originally developed by NASA as a means of modeling microgravity exposure (see Chap. 2). Both primary rat myoblast cells [6] and mouse C2C12 myoblasts [98] have been grown in the RWV system attached to collagen-coated Cytodex 1 micro-carrier beads to act as a culture substratum. 3-D skeletal muscle organoids generated in this fashion within the RWV consisted of myoblast cell/bead aggregates connected to each other by individual myoblast cells and/or myotubes. When 3-D organoids generated in the RWV were compared to identical 3-D organoids maintained under static culture conditions in Teflon culture bags, static 3-D organoids were much larger in size than those generated in rotational culture in the RWV [6]. These results were surprising considering the generally growth-promoting effects of RWV culture conditions on most other cell types related to mass transfer effects that result in better nutrient supply/waste product removal. Later studies utilizing mouse C2C12 myoblasts also grown on collagen-coated Cytodex-1 micro-carrier in the RWV demonstrated the formation of complex networks of multi-nucleated myotubes [98] that was paralleled by a reduction in overall C2C12 myoblast proliferation rate and reduction in 3-D organoid size relative to static control cultures. While similar to the effects previously observed in primary rat myoblasts grown in the RWV [6], these studies also indicated that C2C12-derived 3-D organoids grown under RWV conditions exhibited a less differentiated phenotype (i.e., reduced expression of myogenin, a-actinin, myosin, and tropomyosin) than 3-D organoids grown under static conditions [98]. The reduced differentiation observed in C2C12 3-D organoids grown in the RWV [98] was suggested to be a result of increased mechanical shear forces placed on the myoblasts during rotational culture as compared to static conditions. While RWV culture conditions are considered to produce relatively low levels of mechanical shear on individual cells during rotation, the shear effects created when larger 3-D organoids rotate around their own axis during RWV culture are less well understood. We have previously demonstrated that mechanical loading induces skeletal muscle cell membrane damage both in vivo [99, 100] and in vitro [81], resulting in the release of both CPK and fibroblast growth factor (FGF) from the sarcoplasm, the latter being a potent proliferative stimulus for myoblasts. As such, it is entirely possible that mechanical shear stress induced by RWV culture may result in a similar release of FGF that may account for the reduced level of differentiation observed when 3-D C2C12 organoids are grown under RWV conditions [98].

An alternative approach to creating modeled microgravity in myoblast/myotube cultures is the use of clinostat rotation of adherent monolayers grown in planar culture, rather than utilizing 3-D organoids maintained in suspension culture using the RWV system. Studies utilizing monolayers of rat L6 myoblasts grown under clinostat rotation demonstrate an inhibition of myoblast differentiation and myoblast fusion/myotube formation in modeled microgravity [101], although without a reduction in myoblast proliferation rate as observed in RWV culture [98]. These results suggest that the use of either 3-D myoblast organoids directly formed using the RWV system [6, 98] or monolayer myoblast cultures maintained under clinostat rotation [101] may not be the most appropriate model for investigating the effects of spaceflight on fully differentiated skeletal muscle cells (i.e., myotubes). However, these types of tissue culture analogs may be more appropriate for probing the potentially more subtle disruptive effects of microgravity exposure on myoblast/myotube fusion events [5, 102]. Since satellite cell (SC) activation and SC–myofiber fusion are important events in the maintenance and repair of muscle tissue in the terrestrial environment [79], understanding the underlying mechanisms which control these responses may also shed light on the etiology of a variety of terrestrial skeletal muscle diseases involving significant myofiber atrophy, such as age-related sarcopenia [68], muscular dystrophy [103], and cancer cachexia [70].

Skeletal muscle 3-D organoids can also be created by culturing myoblast cells embedded within a variety of polymerized extracellular matrix materials, such as collagen Type I gels or hydrogels containing hyaluronic acid (HA), fibrinogen, and synthetic peptides [11, 73, 96, 104]. Primary myoblast cells of avian, rodent, and human origin, as well as rodent myoblast cell lines have all been utilized to create such 3-D skeletal muscle organoids containing myotubes demonstrating that embedding myoblasts within a biologically compatible 3-D matrix does not, in of itself, inhibit myoblast proliferation, fusion, and/or myotube formation. Generation of such 3-D skeletal muscle organoids allows application of defined amounts of external mechanical force (e.g., stretch) to be applied to the 3-D organoid [105, 106], or to create internal mechanical loads within the 3-D organoid by initiating myotube contraction using external electrical stimulation [11]. The ability to combine both defined external loading and/or intrinsic loading [11] allows the 3-D skeletal muscle organoid to be exposed to mechanical load profiles nearly identical to that experienced by skeletal muscle tissue in vivo. Indeed, intrinsic mechanical tension (created within the polymerized extracellular material as a consequence of the formation of myotubes in 3-D organoids tethered at both ends to prevent matrix contraction) [107, 108], or exogenous mechanical load by active stretching of the 3-D organoid has been shown to enhance myotube formation and promote the parallel alignment of myotubes within the 3-D organoid [11, 109, 110]. The importance of mechanical load in inducing a phenotype in culture resembling that of skeletal muscle tissue in vivo is highlighted by the up-regulation of a variety of protein markers associated with myoblast differentiation, myotube formation, and skeletal muscle tissue remodeling in vivo (i.e., sarcomeric tropomyosin, myogenin, creatine phosphokinase, IGF-1, and metalloproteinase 2) [11, 73, 105, 107, 109, 110] in these mechanically active tissue culture models.

Skeletal muscle 3-D organoids constructed from myoblasts (or a mixture of myoblasts and other relevant cell types such as fibroblasts and endothelial cells) exhibit many of the histological, biochemical, and mechanical load reactive properties of skeletal muscle in vivo [8]. As such, 3-D skeletal muscle organoids of this type exhibit the greatest physiologically relevant similarities to whole muscle in vivo of any skeletal muscle tissue culture analog developed to date. Hence, if available, such 3-D skeletal muscle organoids are the most appropriate in vitro analog model in which to investigate the in vivo effects of spaceflight or mechanical unloading on the structure and function of skeletal muscle.

8.4 Spaceflight Experiments Utilizing Skeletal Muscle Tissue Culture

To date, very few actual spaceflight experiments have been conducted on-orbit utilizing skeletal muscle cell cultures. In repeat experiments flown aboard two Space Shuttle missions, STS-45 and STS-63, exposure of a variant of the rat L6 myoblast cell line (known as Clone L8) to spaceflight resulted in both the inhibition of myoblast fusion and myotube formation [5]. During these flight experiments, L8 myoblasts were cultured under static, three-dimensional perfusion culture conditions [111]. In addition to the reduction in myoblast fusion and myotube formation rates observed in the original L8 myoblast cultures as a consequence of spaceflight, microgravity exposure also resulted in the conversion of the L8 cell line to a non-fusogenic variant cell line known as the L8SF clone. Interestingly, the non-fusogenic phenotype exhibited by the L8SF variant became permanent when these cells were returned to Earth and propagated in long-term culture [5]. If confirmed by additional spaceflight experiments, the possibility that microgravity exposure may induce permanent phenotypic and/or genetic alterations associated with reduced myoblast fusogenic capacity may have significant negative implications for SC function as it relates to the repair or regeneration of damaged skeletal muscle in crew members. Interestingly, a reduced ability to induce SC activation and subsequent fusion of SCs with existing myofibers appears to be a major cellular "road block" in reversing the skeletal myofiber atrophy observed in age-related sarcopenia [68], adding credence to the concept that exposure to microgravity conditions during spaceflight may induce effects similar to premature aging.

The development of 3-D organoids as a tissue culture analog of skeletal muscle was in part driven by the need for an appropriate in vitro model to study the effects of spaceflight on skeletal muscle structure and function. The pioneering work of Vandenburgh and colleagues in developing such 3-D skeletal muscle organoids [4, 106, 108, 112, 113], also known as bioartificial muscles (BAMs) [96], led to a series of experiments being carried out on two Space Shuttle missions, STS-66 and STS-77, designed to assess the effects of spaceflight on these analogs [7]. The BAMs utilized in these spaceflight experiments were generated pre-flight using embryonic avian myoblasts embedded in a collagen/Matrigel ECM initially supported by a flat culture surface consisting of a wire mesh. After several days in culture, myotubes form within the collagen/Matrigel ECM causing the center of gel/cell sheet containing the myotubes to detach itself from the wire mesh. The central part of the cell sheet proceeds to roll itself into a long tubular structure while the ends of the "tube" remains anchored to the wire mesh (see reference [96] for images). Replicate BAMs were then transferred to a perfusion tissue culture system for subsequent flight aboard the Space Shuttle. Prior to spaceflight, BAMs contained numerous parallel aligned, multi-nucleated myotubes exhibiting striations and expressing sarcomeric tropomyosin. BAMs also expressed MHC protein and fibronectin consistent with a differentiated skeletal muscle tissue phenotype. BAMs were fixed during flight on Mission Day 9, or upon return to Earth. Ground synchronous controls were also processed in an identical fashion to the flight samples. Microgravity exposure of BAMs for 9 days during spaceflight induced a significant reduction in myotube cross-sectional area (CSA) of approximately 12 %, a reduction in the synthesis of non-collagenous, myofibrillar-associated protein (i.e., MHC and fibronectin), but did not appear to impact protein degradation rates [7]. These experiments were the first to demonstrate that microgravity exposure alone, absent the removal of mechanical load (produced either by active contraction or passive stretch), resulted in a muscle atrophy response analogous to that seen in intact muscle.

Interestingly, the rate of myotube atrophy observed in these space flown BAMs was very similar to that observed in myofibers from both intact human and rodent muscle after spaceflight [13–15, 23, 25], as well as in ground-based analogs such as human bed rest [100, 114] or rodent hind limb suspension [65, 115]. However, one significant difference observed between the effects of spaceflight on BAMs as compared to intact skeletal muscle is the distinct lack of increased myofibrillar protein degradation characteristically observed in whole skeletal muscle as a consequence of spaceflight or terrestrial unloading [18, 116, 117]. This difference may be related to effects associated with the removal of the extrinsic mechanical loading (i.e., either passive or dynamic stretch) during spaceflight normally experienced under terrestrial conditions by intact muscle in both crew members and rodents as compared to the constant static loading experienced by the BAMs used in these experiments. Conversely, the maintenance of constant GH/IGF-1 levels in the tissue culture medium of the space flown BAMs as compared to a reduction in circulating levels of these same hormones observed in crew members [44] or space flown rodents [118] may explain the difference in protein degradation rates. Both the removal of extrinsic mechanical load or a reduction in pituitary-derived GH/IGF-1 signaling within intact skeletal muscle under terrestrial conditions has been shown to negatively impact myofiber protein balance [48], specifically leading to an increase in myofibrillar protein degradation via the ubiquitin-proteasome system [47, 59]. As such, it is very possible that the culture conditions experienced by these BAMs during spaceflight (no change in either extrinsic mechanical loading or culture levels of GH/IGF-1) may explain why no increases in myofibrillar protein degradation rates were observed in this space flown culture model as compared to that observed in intact skeletal muscle tissue after spaceflight or terrestrial unloading in vivo.

8.5 Modeling Spaceflight Exposure by Removal of Mechanical Load

One of the recurring themes in this chapter is the possibility that skeletal muscle adaptation to spaceflight is a consequence of two distinct yet related environmental factors experienced by the tissue, namely true microgravity exposure and/or the removal of the normal mechanical loading experienced by the intact musculoskeletal system while living under terrestrial gravity conditions. Actual spaceflight experiments utilizing both cultured monolayer [5] and 3-D skeletal muscle organoids [7] have shown that microgravity exposure alone, without alteration in the mechanical load profile experienced by skeletal muscle cells during spaceflight, causes individual cellular responses consistent with the spaceflight-induced atrophy response seen in intact skeletal muscle tissue in vivo. For example, myofiber atrophy is correlated with a reduction in mechanically induced sarcolemma damage during both spaceflight and 6°-head-down-tilt bed rest in human subjects [100], which in turn has been linked to a reduction in the release of FGF from myofibers via these sarcolemma wounds or “micro-tears.” However, acute exposure of myotubes to brief periods of microgravity during parabolic flight inhibits the membrane–membrane fusion events required for membrane resealing and repair of such mechanically induced sarcolemma damage in vitro [102]. Such divergent responses suggest that the effects of microgravity exposure at the cellular level are not solely associated with removal of mechanical load from the muscle tissue as a whole.

In order to investigate the specific effects of mechanical unloading on skeletal muscle, a variety of mechanically active tissue culture models have been developed which allow the application of defined extrinsic mechanical load to myoblasts and/or myotube cultures grown both in monolayer and 3-D organoid cultures. Extrinsic mechanical load stimulus can be applied directly by stretching the 3-D skeletal muscle organoid (as in the case of BAMs) [11], indirectly by stretching of the 3-D support matrix in which the myotubes are embedded (as is the case for collagen gels or fibrin hydrogels) [119], or stretching of a flexible (i.e., silicon) culture surface on which the myotubes are growing [81, 106, 120]. One of the most interesting and exciting applications of mechanically active 3-D skeletal muscle organoids relative to understanding the effects of spaceflight is the use of such models to first induce and then study the muscle atrophy response by mimicking the mechanical loading profiles experienced by intact, skeletal muscle when it undergoes a "terrestrial to spaceflight" transition in vivo. This can be achieved by first applying mechanical load to 3-D skeletal muscle organoids in the form of both external loading (i.e., stretch) combined with intrinsic loading (i.e., electrically elicited myotube contraction) [11], and then removing part or all of these load stimuli in order to mimic spaceflight conditions.

For example, Vanderburgh and colleagues [11] have used a 3-D skeletal muscle organoid culture (*a.k.a.* BAM) generated using primary mouse myoblasts which included regular periods of electrically induced myotube contraction over a period of 80 days, followed by a reduction in the amount of extrinsic mechanical load (i.e., stretch) placed on the BAM. The reduction in stretch of the BAM under terrestrial conditions resulted in a rapid myotube atrophy response characterized by reductions in myotube CSA content, tetanic force (Po), total protein synthesis rate, and non-collagenous protein content [11]. Similar atrophic responses were observed in BAMs flown aboard the Space Shuttle [7], even though the atrophy response aboard the space craft was induced solely by microgravity exposure and not the removal of either extrinsic or intrinsic mechanical loading as was observed in the case of terrestrial BAMs [11]. Interestingly, no significant increase in muscle protein degradation rates (a hallmark of spaceflight-induced muscle atrophy and terrestrial unloading in vivo) was observed in atrophied BAMs from either spaceflight or terrestrial conditions. Taken together, these data indicate that the underlying mechanism(s) involved in the initiation of the "hallmark" increase in myofiber protein degradation observed during myofiber atrophy in vivo as a consequence of either spaceflight or mechanical unloading in a terrestrial setting [16, 48, 59, 121] remains to be fully elucidated.

8.6 Knowledge Gained Applicable to Terrestrial Disease States

While the use of skeletal muscle tissue culture models for studying a variety of disease states in terrestrial populations has a relatively long history, most of these studies have utilized myotube cultures grown as conventional monolayers rather than 3-D skeletal muscle organoids. The more traditional myotube model has been used to investigate myofiber-induced cancer cachexia [121–123], screening pharmaceutical agents [124], and nutritional supplements [125] potentially useful in combating muscle loss, understanding insulin signaling and glucose transport mechanisms as they relate to the etiology of Type II diabetes [126] and obesity [127], as well as investigating the underlying signaling pathways operating during muscle atrophy [121, 128] or hypertrophy [129]. However, the development of 3-D skeletal muscle organoids, driven in part by the spaceflight research community space, has significantly increased the ability of investigators to study a variety of other pathological conditions using tissue culture analogs much closer to the skeletal muscle tissue milieu existing in vivo.

For example, 3-D skeletal muscle organoids have been used to mimic myoblast behavior in aged skeletal muscle. Constructed from "high passage" myoblasts, these organoids exhibit many of the histological and biochemical characteristics of aged sarcopenic muscle [130], including myotube atrophy, decreased expression of metalloproteinase and IGF-1/IGF receptor mRNA, increased expression of myostatin mRNA, and an overall reduction in peak force production. All of these changes are consistent with the myofiber atrophy and impaired ability for muscle differentiation and regeneration observed in age-related sarcopenia [62, 68]. Interestingly many of the same histological and biochemical features reported in both actual and ground-based analogs of spaceflight exposure [13, 15, 23, 37] appear in this model of sarcopenic muscle, lending credence to the concept that spaceflight may induce "premature" aging of the skeletal muscle cells in vivo.

The use of 3-D skeletal muscle organoid cultures that exhibit histological, biochemical, and contractile properties nearly identical to that of intact skeletal muscle will provide many new opportunities for truly mechanistic study of skeletal muscle function as it relates to a wide variety of pathological conditions. The added benefit of being able to simultaneously modulate the levels of extrinsic (i.e., stretch) and intrinsic (i.e., contractile) mechanical load placed on 3-D skeletal muscle organoids, as well as allowing control of various circulating factors (i.e., GH, IGF-1, FGF, myostatin, follastatin, reactive oxygen species, etc.) in the tissue culture medium will allow the development of a range of models capable of explaining a wide number of skeletal muscle pathologies in a causal fashion.

8.7 Questions for Future Research

This review chapter has primarily focused on the use of conventional monolayer and 3-D organoid cultures of skeletal muscle cells as tissue culture analogs of skeletal muscle structure and function related to effects of spaceflight. It should also be obvious to the reader that such tissue culture analogs can also be utilized to understand the effects of a myriad of environmental, genetic, and epigenetic effects on skeletal muscle cells of interest in the terrestrial environment. By being able to manipulate mechanical loading levels and a wide range of culture conditions, 3-D skeletal muscle organoids generated from either normal or diseased muscle can be used to investigate the underlying etiology and signaling mechanisms involved in almost any disease state which involves skeletal muscle tissue. However, while the highly specialized, tissue culture hardware required for achieving this level of experimental control are currently under development in a number of terrestrial laboratories, no such culture systems capable of operating in the space craft environment presently exist.

Specific research questions which remain unanswered relating to the effects of spaceflight on skeletal muscle structure and function center not only on the effects of microgravity exposure and unloading, but also the potentially negative combinatory effects of other environmental risk factors encountered by crew members during spaceflight or exploration class missions. For example, the potential effects of space radiation on satellite stem cell populations [131–133], elevated oxidative stress levels [134, 135], disrupted nutritional status [136, 137], and/or genetic risk factors [138] related to altered one-carbon metabolism (1-CM) [139] as it relates to skeletal muscle have yet to be investigated. Similarly, the effects of reloading on skeletal muscle after extended periods of inactivity/atrophy relative to the ability recover muscle mass [140, 141], as well as the capacity of atrophied muscle for tissue repair and regeneration after injury [77] remain unknown. The use of 3-D skeletal muscle organoid cultures where such combinatory effects can be mimicked will in part allow the investigation of many of these spaceflight-related issues, while furthering our understanding of how to combat these challenges in crew members.

While in the early stages of development, the creation of even more complex 3-D skeletal muscle organoids containing additional cell types such as endothelial, neuronal, and/or fibroblast cells [8–10] has demonstrated that it may be possible in the future to develop truly bioengineered skeletal muscle tissue nearly identical to intact skeletal muscle in vivo. Such bioengineered tissue will serve not only as an in vitro tissue culture analog of skeletal muscle structure and function, but potentially as means of repairing or replacing damaged or dysfunctional skeletal muscle tissue in vivo. The rapid and truly paradigm shifting advances being made in the area of 3-D tissue "printing" [142] using a variety of ECM compounds and primary cell types promises to accelerate all areas of tissue bioengineering, including that involving skeletal muscle tissue. For example, the ability to "print" a 3-D skeletal muscle organoid in three dimensions which contains myoblasts arranged for optimal formation of aligned myotubes while also having a network of capillary endothelial cells simulating a functional vascular system with strategically located motor-neuron cells has become a realistic and achievable goal [142]. Combining such "printed" 3-D skeletal muscle organoids with existing systems capable of providing defined contractile stimulation and manipulation of external mechanical loads only furthers our ability to mimic the tissue milieu/microenvironment within intact skeletal muscle. Analysis of these 3-D models will be advanced by the use of cutting edge technologies, like spatially resolved molecular pathology (SRMP) techniques [60], allowing the study of individual cellular components within skeletal muscle using a combination of laser capture microscopy and "state-of-the-art" genomic and proteomic analysis [60, 143]. Development of such bioengineered models of skeletal muscle will provide many opportunities to further our understanding of the effects of spaceflight exposure on skeletal muscle, while providing a broad-based platform technology in which to probe the causes and potential cures for a wide variety of human muscle diseases.

References

1. Yaffe, D. (1968). Retention of differentiation potentialities during prolonged cultivation of myogenic cells. *Proceedings of the National Academy of Sciences of the United States of America, 61*(2), 477–483.
2. Dietlein, L. F. (1975). Summary and Conclusions. In: R.S. Johnson, L.F. Dietlein & C.A. Berry (Eds.), *Biomedical results of Apollo, NASA SP-368*. Washington, D. C.: National Aeronautics and Space Administration.
3. Yaffe, D., & Saxel, O. (1977). Serial passaging and differentiation of myogenic cells isolated from dystrophic mouse muscle. *Nature, 270*(5639), 725–727.
4. Vandenburgh, H. H. (1982). Dynamic mechanical orientation of skeletal myofibers in vitro. *Developmental Biology, 93*(2), 438–443.
5. Kulesh, D. A., Anderson, L. H., Wilson, B., Otis, E. J., Elgin, D. M., Barker, M. J., et al. (1994). Space shuttle flight (STS-45) of L8 myoblast cells results in the isolation of a nonfusing cell line variant. *Journal of Cellular Biochemistry, 55*(4), 530–544.
6. Molnar, G., Schroedl, N. A., Gonda, S. R., & Hartzell, C. R. (1997). Skeletal muscle satellite cells cultured in simulated microgravity. *In Vitro Cellular & Developmental Biology. Animal, 33*(5), 386–391.
7. Vandenburgh, H., Chromiak, J., Shansky, J., Del Tatto, M., & Lemaire, J. (1999). Space travel directly induces skeletal muscle atrophy. *FASEB Journal, 13*(9), 1031–1038.
8. Dennis, R. G., & Kosnik, P. E., 2nd. (2000). Excitability and isometric contractile properties of mammalian skeletal muscle constructs engineered in vitro. *In Vitro Cellular & Developmental Biology. Animal, 36*(5), 327–335.
9. Levenberg, S., Rouwkema, J., Macdonald, M., Garfein, E. S., Kohane, D. S., Darland, D. C., et al. (2005). Engineering vascularized skeletal muscle tissue. *Nature Biotechnology, 23*(7), 879–884.
10. van der Schaft, D. W., van Spreeuwel, A. C., van Assen, H. C., & Baaijens, F. P. (2011). Mechanoregulation of vascularization in aligned tissue-engineered muscle: a role for vascular endothelial growth factor. *Tissue Engineering. Part A, 17*(21–22), 2857–2865.
11. Lee, P. H., & Vandenburgh, H. H. (2013). Skeletal muscle atrophy in bioengineered skeletal muscle: A new model system. *Tissue Engineering. Part A, 19*(19–20), 2147–2155.
12. Fitts, R. H., Trappe, S. W., Costill, D. L., Gallagher, P. M., Creer, A. C., Colloton, P. A., et al. (2010). Prolonged space flight-induced alterations in the structure and function of human skeletal muscle fibres. *Journal of Physiology, 588*(Pt 18), 3567–3592.
13. Baldwin, K. M. (1996). Effect of spaceflight on the functional, biochemical, and metabolic properties of skeletal muscle. *Medicine and Science in Sports and Exercise, 28*(8), 983–987.
14. Riley, D. A., Ellis, S., Slocum, G. R., Sedlak, F. R., Bain, J. L., Krippendorf, B. B., et al. (1996). In-flight and postflight changes in skeletal muscles of SLS-1 and SLS-2 spaceflown rats. *Journal of Applied Physiology, 81*(1), 133–144.
15. Riley, D. A., Ilyina-Kakueva, E. I., Ellis, S., Bain, J. L., Slocum, G. R., & Sedlak, F. R. (1990). Skeletal muscle fiber, nerve, and blood vessel breakdown in space-flown rats. *FASEB Journal, 4*(1), 84–91.
16. Adams, G. R., Caiozzo, V. J., & Baldwin, K. M. (2003). Skeletal muscle unweighting: Spaceflight and ground-based models. *Journal of Applied Physiology, 95*(6), 2185–2201.
17. Michel, E. L., Johnston, R. S., & Dietlein, L. F. (1976). Biomedical results of the Skylab Program. *Life Sciences and Space Research, 14*, 3–18.

18. Stein, T. P., Leskiw, M. J., & Schluter, M. D. (1996). Diet and nitrogen metabolism during spaceflight on the shuttle. *Journal of Applied Physiology, 81*(1), 82–97.
19. Grigoriev, A. I., Bugrov, S. A., Bogomolov, V. V., Egorov, A. D., Kozlovskaya, I. B., Pestov, I. D., et al. (1991). Medical results of the Mir year-long mission. *Physiologist, 34*(1 Suppl), S44–S48.
20. LeBlanc, A., Lin, C., Shackelford, L., Sinitsyn, V., Evans, H., Belichenko, O., et al. (2000). Muscle volume, MRI relaxation times (T2), and body composition after spaceflight. *Journal of Applied Physiology, 89*(6), 2158–2164.
21. Akima, H., Kawakami, Y., Kubo, K., Sekiguchi, C., Ohshima, H., Miyamoto, A., et al. (2000). Effect of short-duration spaceflight on thigh and leg muscle volume. *Medicine and Science in Sports and Exercise, 32*(10), 1743–1747.
22. Day, M. K., Allen, D. L., Mohajerani, L., Greenisen, M. C., Roy, R. R., & Edgerton, V. R. (1995). Adaptations of human skeletal muscle fibers to spaceflight. *Journal of Gravitational Physiology, 2*(1), P47–P50.
23. Edgerton, V. R., & Roy, R. R. (1994). Neuromuscular adaptation to actual and simulated weightlessness. *Advances in Space Biology and Medicine, 4*, 33–67.
24. Shenkman, B. S., Belozerova, I. N., Lee, P., Nemirovskaya, T. L., & Kozlovskaya, I. B. (2003). Effects of weightlessness and movement restriction on the structure and metabolism of the soleus muscle in monkeys after space flight. *Neuroscience and Behavioral Physiology, 33*(7), 717–722.
25. Edgerton, V. R., Zhou, M. Y., Ohira, Y., Klitgaard, H., Jiang, B., Bell, G., et al. (1995). Human fiber size and enzymatic properties after 5 and 11 days of spaceflight. *Journal of Applied Physiology, 78*(5), 1733–1739.
26. Zhou, M. Y., Klitgaard, H., Saltin, B., Roy, R. R., Edgerton, V. R., & Gollnick, P. D. (1995). Myosin heavy chain isoforms of human muscle after short-term spaceflight. *Journal of Applied Physiology, 78*(5), 1740–1744.
27. Staron, R. S., Kraemer, W. J., Hikida, R. S., Reed, D. W., Murray, J. D., Campos, G. E., et al. (1998). Comparison of soleus muscles from rats exposed to microgravity for 10 versus 14 days. *Histochemistry and Cell Biology, 110*(1), 73–80.
28. Martin, T. P., Edgerton, V. R., & Grindeland, R. E. (1988). Influence of spaceflight on rat skeletal muscle. *Journal of Applied Physiology, 65*(5), 2318–2325.
29. Ohira, Y., Jiang, B., Roy, R. R., Oganov, V., Ilyina-Kakueva, E., Marini, J. F., et al. (1992). Rat soleus muscle fiber responses to 14 days of spaceflight and hindlimb suspension. *Journal of Applied Physiology, 73*(2 Suppl), 51S–57S.
30. Caiozzo, V. J., Baker, M. J., Herrick, R. E., Tao, M., & Baldwin, K. M. (1994). Effect of spaceflight on skeletal muscle: Mechanical properties and myosin isoform content of a slow muscle. *Journal of Applied Physiology, 76*(4), 1764–1773.
31. Harrison, B. C., Allen, D. L., Girten, B., Stodieck, L. S., Kostenuik, P. J., Bateman, T. A., et al. (2003). Skeletal muscle adaptations to microgravity exposure in the mouse. *Journal of Applied Physiology, 95*(6), 2462–2470.
32. Shenkman, B. S., Kozlovskaya, I. B., Kuznetsov, S. L., Nemirovskaya, T. L., & Desplanches, D. (1994). Plasticity of skeletal muscle fibres in space-flown primates. *Journal of Gravitational Physiology, 1*(1), P64–P66.
33. Roy, R. R., Zhong, H., Talmadge, R. J., Bodine, S. C., Fanton, J. W., Koslovskaya, I., et al. (2001). Size and myonuclear domains in Rhesus soleus muscle fibers: Short-term spaceflight. *Journal of Gravitational Physiology, 8*(2), 49–56.
34. Tidball, J. G., & Quan, D. M. (1992). Reduction in myotendinous junction surface area of rats subjected to 4-day spaceflight. *Journal of Applied Physiology, 73*(1), 59–64.
35. Deschenes, M. R., Britt, A. A., Gomes, R. R., Booth, F. W., & Gordon, S. E. (2001). Recovery of neuromuscular junction morphology following 16 days of spaceflight. *Synapse, 42*(3), 177–184.
36. Tesch, P. A., Berg, H. E., Bring, D., Evans, H. J., & LeBlanc, A. D. (2005). Effects of 17-day spaceflight on knee extensor muscle function and size. *European Journal of Applied Physiology, 93*(4), 463–468.

37. Trappe, S., Costill, D., Gallagher, P., Creer, A., Peters, J. R., Evans, H., et al. (2009). Exercise in space: Human skeletal muscle after 6 months aboard the International Space Station. *Journal of Applied Physiology, 106*(4), 1159–1168.
38. Kozlovskaia, I. B., Grigor'eva, L. S., & Gevlich, G. I. (1984). Comparative analysis of the effect of weightlessness and its model on the velocity-strength properties and tonus of human skeletal muscles. *Kosmicheskaia Biologiia i Aviakosmicheskaia Meditsina, 18*(6), 22–26.
39. Gallasch, E., Moser, M., Kozlovskaya, I. B., Kenner, T., & Noordergraaf, A. (1997). Effects of an eight-day space flight on microvibration and physiological tremor. *American Journal of Physiology, 273*(1 Pt 2), R86–R92.
40. Layne, C. S., & Spooner, B. S. (1994). Microgravity effects on "postural" muscle activity patterns. *Advances in Space Research, 14*(8), 381–384.
41. Widrick, J. J., Knuth, S. T., Norenberg, K. M., Romatowski, J. G., Bain, J. L., Riley, D. A., et al. (1999). Effect of a 17 day spaceflight on contractile properties of human soleus muscle fibres. *The Journal of Physiology, 516*(Pt 3), 915–930.
42. Shenkman, B. S., Litvinova, K. S., Nemirovskaya, T. L., Podlubnaya, Z. A., Vikhlyantsev, I. M., & Kozlovskaya, I. B. (2004). Afferent and peripheral control of muscle fiber properties during gravitational unloading. *Journal of Gravitational Physiology, 11*(2), P111–P114.
43. Shenkman, B. S., & Nemirovskaya, T. L. (2008). Calcium-dependent signaling mechanisms and soleus fiber remodeling under gravitational unloading. *Journal of Muscle Research and Cell Motility, 29*(6–8), 221–230.
44. McCall, G. E., Goulet, C., Roy, R. R., Grindeland, R. E., Boorman, G. I., Bigbee, A. J., et al. (1999). Spaceflight suppresses exercise-induced release of bioassayable growth hormone. *Journal of Applied Physiology, 87*(3), 1207–1212.
45. Adams, G. R., McCue, S. A., Bodell, P. W., Zeng, M., & Baldwin, K. M. (2000). Effects of spaceflight and thyroid deficiency on hindlimb development. I. Muscle mass and IGF-I expression. *Journal of Applied Physiology, 88*(3), 894–903.
46. Lalani, R., Bhasin, S., Byhower, F., Tarnuzzer, R., Grant, M., Shen, R., et al. (2000). Myostatin and insulin-like growth factor-I and -II expression in the muscle of rats exposed to the microgravity environment of the NeuroLab space shuttle flight. *Journal of Endocrinology, 167*(3), 417–428.
47. Kandarian, S. C., & Jackman, R. W. (2006). Intracellular signaling during skeletal muscle atrophy. *Muscle and Nerve, 33*(2), 155–165.
48. Bodine, S. C., Stitt, T. N., Gonzalez, M., Kline, W. O., Stover, G. L., Bauerlein, R., et al. (2001). Akt/mTOR pathway is a crucial regulator of skeletal muscle hypertrophy and can prevent muscle atrophy in vivo. *Nature Cell Biology, 3*(11), 1014–1019.
49. Lipina, C., Kendall, H., McPherron, A. C., Taylor, P. M., & Hundal, H. S. (2010). Mechanisms involved in the enhancement of mammalian target of rapamycin signalling and hypertrophy in skeletal muscle of myostatin-deficient mice. *FEBS Letters, 584*(11), 2403–2408.
50. Kim, J. S., Cross, J. M., & Bamman, M. M. (2005). Impact of resistance loading on myostatin expression and cell cycle regulation in young and older men and women. *American Journal of Physiology. Endocrinology and Metabolism, 288*(6), E1110–E1119.
51. Reardon, K. A., Davis, J., Kapsa, R. M., Choong, P., & Byrne, E. (2001). Myostatin, insulin-like growth factor-1, and leukemia inhibitory factor mRNAs are upregulated in chronic human disuse muscle atrophy. *Muscle and Nerve, 24*(7), 893–899.
52. Elkina, Y., von Haehling, S., Anker, S. D., & Springer, J. (2011). The role of myostatin in muscle wasting: an overview. *Journal of Cachexia, Sarcopenia and Muscle, 2*(3), 143–151.
53. Zhang, P., Chen, X., & Fan, M. (2007). Signaling mechanisms involved in disuse muscle atrophy. *Medical Hypotheses, 69*(2), 310–321.
54. Tidball, J. G. (2005). Mechanical signal transduction in skeletal muscle growth and adaptation. *Journal of Applied Physiology, 98*(5), 1900–1908.
55. Kraemer, W. J., & Ratamess, N. A. (2005). Hormonal responses and adaptations to resistance exercise and training. *Sports Medicine, 35*(4), 339–361.
56. Allen, D. L., Bandstra, E. R., Harrison, B. C., Thorng, S., Stodieck, L. S., Kostenuik, P. J., et al. (2009). Effects of spaceflight on murine skeletal muscle gene expression. *Journal of Applied Physiology, 106*(2), 582–595.

57. Maggs, A. M., Huxley, C., & Hughes, S. M. (2008). Nerve-dependent changes in skeletal muscle myosin heavy chain after experimental denervation and cross-reinnervation and in a demyelinating mouse model of Charcot-Marie-Tooth disease type 1A. *Muscle and Nerve, 38*(6), 1572–1584.
58. Butikofer, L., Zurlinden, A., Bolliger, M. F., Kunz, B., & Sonderegger, P. (2011). Destabilization of the neuromuscular junction by proteolytic cleavage of agrin results in precocious sarcopenia. *FASEB Journal, 25*(12), 4378–4393.
59. Bassel-Duby, R., & Olson, E. N. (2006). Signaling pathways in skeletal muscle remodeling. *Annual Review of Biochemistry, 75*, 19–37.
60. Vanderburg, C. R., & Clarke, M. S. (2012). Laser capture microdissection of metachromatically stained skeletal muscle allows quantification of fiber type specific gene expression. *Molecular and Cellular Biochemistry, 375*(1–2), 159–170.
61. Schulte, L. M., Navarro, J., & Kandarian, S. C. (1993). Regulation of sarcoplasmic reticulum calcium pump gene expression by hindlimb unweighting. *American Journal of Physiology, 264*(5 Pt 1), C1308–C1315.
62. Clarke, M. S. (2004). The effects of exercise on skeletal muscle in the aged. *Journal of Musculoskeletal & Neuronal Interactions, 4*(2), 175–178.
63. Bamman, M. M., Clarke, M. S., Feeback, D. L., Talmadge, R. J., Stevens, B. R., Lieberman, S. A., et al. (1998). Impact of resistance exercise during bed rest on skeletal muscle sarcopenia and myosin isoform distribution. *Journal of Applied Physiology, 84*(1), 157–163.
64. Trappe, S., Creer, A., Slivka, D., Minchev, K., & Trappe, T. (2007). Single muscle fiber function with concurrent exercise or nutrition countermeasures during 60 days of bed rest in women. *Journal of Applied Physiology, 103*(4), 1242–1250.
65. Kyparos, A., Feeback, D. L., Layne, C. S., Martinez, D. A., & Clarke, M. S. (2005). Mechanical stimulation of the plantar foot surface attenuates soleus muscle atrophy induced by hindlimb unloading in rats. *Journal of Applied Physiology, 99*(2), 739–746.
66. Stelzer, J. E., & Widrick, J. J. (2003). Effect of hindlimb suspension on the functional properties of slow and fast soleus fibers from three strains of mice. *Journal of Applied Physiology, 95*(6), 2425–2433.
67. Layne, C. S., & Forth, K. E. (2008). Plantar stimulation as a possible countermeasure to microgravity-induced neuromotor degradation. *Aviation, Space, and Environmental Medicine, 79*(8), 787–794.
68. Narici, M. V., & Maffulli, N. (2010). Sarcopenia: Characteristics, mechanisms and functional significance. *British Medical Bulletin, 95*, 139–159.
69. Han, H. Q., & Mitch, W. E. (2011). Targeting the myostatin signaling pathway to treat muscle wasting diseases. *Current Opinion in Supportive and Palliative Care, 5*(4), 334–341.
70. Penna, F., Costamagna, D., Pin, F., Camperi, A., Fanzani, A., Chiarpotto, E. M., et al. (2013). Autophagic degradation contributes to muscle wasting in cancer cachexia. *American Journal of Pathology, 182*(4), 1367–1378.
71. Knoblauch, M. A., O'Connor, D. P., & Clarke, M. S. (2013). Obese mice incur greater myofiber membrane disruption in response to mechanical load compared with lean mice. *Obesity (Silver Spring), 21*(1), 135–143.
72. LaFramboise, W. A., Guthrie, R. D., Scalise, D., Elborne, V., Bombach, K. L., Armanious, C. S., et al. (2003). Effect of muscle origin and phenotype on satellite cell muscle-specific gene expression. *Journal of Molecular and Cellular Cardiology, 35*(10), 1307–1318.
73. Sharples, A. P., & Stewart, C. E. (2011). Myoblast models of skeletal muscle hypertrophy and atrophy. *Current Opinion in Clinical Nutrition and Metabolic Care, 14*(3), 230–236.
74. Sumino, Y., Hirata, Y., Sato, F., & Mimata, H. (2007). Growth mechanism of satellite cells in human urethral rhabdosphincter. *Neurourology and Urodynamics, 26*(4), 552–561.
75. Johnson, S. E., & Allen, R. E. (1990). The effects of bFGF, IGF-I, and TGF-beta on RMo skeletal muscle cell proliferation and differentiation. *Experimental Cell Research, 187*(2), 250–254.
76. Cornelison, D. D. (2008). Context matters: In vivo and in vitro influences on muscle satellite cell activity. *Journal of Cellular Biochemistry, 105*(3), 663–669.

77. Karalaki, M., Fili, S., Philippou, A., & Koutsilieris, M. (2009). Muscle regeneration: Cellular and molecular events. *In Vivo, 23*(5), 779–796.
78. Hindi, S. M., Tajrishi, M. M., & Kumar, A. (2013). Signaling mechanisms in mammalian myoblast fusion. *Science Signaling, 6*(272), re2.
79. Harridge, S. D. (2007). Plasticity of human skeletal muscle: Gene expression to in vivo function. *Experimental Physiology, 92*(5), 783–797.
80. Bader, D., Masaki, T., & Fischman, D. A. (1982). Immunochemical analysis of myosin heavy chain during avian myogenesis in vivo and in vitro. *Journal of Cell Biology, 95*(3), 763–770.
81. Clarke, M. S., & Feeback, D. L. (1996). Mechanical load induces sarcoplasmic wounding and FGF release in differentiated human skeletal muscle cultures. *FASEB Journal, 10*(4), 502–509.
82. Abdul-Hussein, S., van der Ven, P. F., & Tajsharghi, H. (2013). Expression profiles of muscle disease-associated genes and their isoforms during differentiation of cultured human skeletal muscle cells. *BMC Musculoskeletal Disorders, 13*, 262.
83. Cherednichenko, G., Zhang, R., Bannister, R. A., Timofeyev, V., Li, N., Fritsch, E. B., et al. (2012). Triclosan impairs excitation-contraction coupling and Ca2+ dynamics in striated muscle. *Proceedings of the National Academy of Sciences of the United States of America, 109*(35), 14158–14163.
84. Merrill, G. F. (1989). Clonal derivation of a rat muscle cell strain that forms contraction-competent myotubes. *In Vitro Cellular & Developmental Biology, 25*(5), 471–476.
85. Linkhart, T. A., Clegg, C. H., & Hauschika, S. D. (1981). Myogenic differentiation in permanent clonal mouse myoblast cell lines: Regulation by macromolecular growth factors in the culture medium. *Developmental Biology, 86*(1), 19–30.
86. Daubas, P., Klarsfeld, A., Garner, I., Pinset, C., Cox, R., & Buckingham, M. (1988). Functional activity of the two promoters of the myosin alkali light chain gene in primary muscle cell cultures: Comparison with other muscle gene promoters and other culture systems. *Nucleic Acids Research, 16*(4), 1251–1271.
87. Kawahara, Y., Yamaoka, K., Iwata, M., Fujimura, M., Kajiume, T., Magaki, T., et al. (2006). Novel electrical stimulation sets the cultured myoblast contractile function to 'on'. *Pathobiology, 73*(6), 288–294.
88. Krebs, J. M., & Denney, R. M. (1997). Effect of atrophy and contractions on myogenin mRNA concentration in chick and rat myoblast omega muscle cells. *In Vitro Cellular & Developmental Biology. Animal, 33*(3), 187–194.
89. Vali, S., Carlsen, R., Pessah, I., & Gorin, F. (2000). Role of the sarcoplasmic reticulum in regulating the activity-dependent expression of the glycogen phosphorylase gene in contractile skeletal muscle cells. *Journal of Cellular Physiology, 185*(2), 184–199.
90. Clarke, M. S., Vanderburg, C. R., Bamman, M. M., Caldwell, R. W., & Feeback, D. L. (2000). In situ localization of cholesterol in skeletal muscle by use of a monoclonal antibody. *Journal of Applied Physiology, 89*(2), 731–741.
91. Lamb, G. D. (2009). Mechanisms of excitation-contraction uncoupling relevant to activity-induced muscle fatigue. *Applied Physiology, Nutrition, and Metabolism, 34*(3), 368–372.
92. Nedachi, T., Fujita, H., & Kanzaki, M. (2008). Contractile C2C12 myotube model for studying exercise-inducible responses in skeletal muscle. *American Journal of Physiology. Endocrinology and Metabolism, 295*(5), E1191–E1204.
93. Cooper, S. T., Maxwell, A. L., Kizana, E., Ghoddusi, M., Hardeman, E. C., Alexander, I. E., et al. (2004). C2C12 co-culture on a fibroblast substratum enables sustained survival of contractile, highly differentiated myotubes with peripheral nuclei and adult fast myosin expression. *Cell Motility and the Cytoskeleton, 58*(3), 200–211.
94. Connor, M. K., Irrcher, I., & Hood, D. A. (2001). Contractile activity-induced transcriptional activation of cytochrome C involves Sp1 and is proportional to mitochondrial ATP synthesis in C2C12 muscle cells. *Journal of Biological Chemistry, 276*(19), 15898–15904.
95. Baker, B. M., & Chen, C. S. (2012). Deconstructing the third dimension: How 3D culture microenvironments alter cellular cues. *Journal of Cell Science, 125*(Pt 13), 3015–3024.

96. Vandenburgh, H., Shansky, J., Del Tatto, M., & Chromiak, J. (1999). Organogenesis of skeletal muscle in tissue culture. *Methods in Molecular Medicine, 18*, 217–225.
97. Chromiak, J. A., Shansky, J., Perrone, C., & Vandenburgh, H. H. (1998). Bioreactor perfusion system for the long-term maintenance of tissue-engineered skeletal muscle organoids. *In Vitro Cellular & Developmental Biology. Animal, 34*(9), 694–703.
98. Torgan, C. E., Burge, S. S., Collinsworth, A. M., Truskey, G. A., & Kraus, W. E. (2000). Differentiation of mammalian skeletal muscle cells cultured on microcarrier beads in a rotating cell culture system. *Medical and Biological Engineering and Computing, 38*(5), 583–590.
99. Knoblauch, M. A., O'Connor, D. P., & Clarke, M. S. (2010). Capillary and venous samples of total creatine kinase are similar after eccentric exercise. *Journal of Strength and Conditioning Research, 24*(12), 3471–3475.
100. Clarke, M. S., Bamman, M. M., & Feeback, D. L. (1998). Bed rest decreases mechanically induced myofiber wounding and consequent wound-mediated FGF release. *Journal of Applied Physiology, 85*(2), 593–600.
101. Hirasaka, K., Nikawa, T., Yuge, L., Ishihara, I., Higashibata, A., Ishioka, N., et al. (2005). Clinorotation prevents differentiation of rat myoblastic L6 cells in association with reduced NF-kappa B signaling. *Biochimica et Biophysica Acta, 1743*(1–2), 130–140.
102. Clarke, M. S., Vanderburg, C. R., & Feeback, D. L. (2001). The effect of acute microgravity on mechanically-induced membrane damage and membrane-membrane fusion events. *Journal of Gravitational Physiology, 8*(2), 37–47.
103. Clarke, M. S., Khakee, R., & McNeil, P. L. (1993). Loss of cytoplasmic basic fibroblast growth factor from physiologically wounded myofibers of normal and dystrophic muscle. *Journal of Cell Science, 106*(Pt 1), 121–133.
104. Wang, W., Fan, M., Zhang, L., Liu, S. H., Sun, L., & Wang, C. Y. (2009). Compatibility of hyaluronic acid hydrogel and skeletal muscle myoblasts. *Biomedical Materials, 4*(2), 025011.
105. Powell, C. A., Smiley, B. L., Mills, J., & Vandenburgh, H. H. (2002). Mechanical stimulation improves tissue-engineered human skeletal muscle. *American Journal of Physiology. Cell Physiology, 283*(5), C1557–C1565.
106. Vandenburgh, H. H. (1988). A computerized mechanical cell stimulator for tissue culture: Effects on skeletal muscle organogenesis. *In Vitro Cellular & Developmental Biology, 24*(7), 609–619.
107. Shansky, J., Del Tatto, M., Chromiak, J., & Vandenburgh, H. (1997). A simplified method for tissue engineering skeletal muscle organoids in vitro. *In Vitro Cellular & Developmental Biology. Animal, 33*(9), 659–661.
108. Vandenburgh, H., & Kaufman, S. (1979). In vitro model for stretch-induced hypertrophy of skeletal muscle. *Science, 203*(4377), 265–268.
109. Cheema, U., Yang, S. Y., Mudera, V., Goldspink, G. G., & Brown, R. A. (2003). 3-D in vitro model of early skeletal muscle development. *Cell Motility and the Cytoskeleton, 54*(3), 226–236.
110. Cheema, U., Brown, R., Mudera, V., Yang, S. Y., McGrouther, G., & Goldspink, G. (2005). Mechanical signals and IGF-I gene splicing in vitro in relation to development of skeletal muscle. *Journal of Cellular Physiology, 202*(1), 67–75.
111. http://lsda.jsc.nasa.gov/scripts/hardware/hardw.aspx?hardware_id=920.
112. Hatfaludy, S., Shansky, J., & Vandenburgh, H. H. (1989). Metabolic alterations induced in cultured skeletal muscle by stretch-relaxation activity. *American Journal of Physiology, 256*(1 Pt 1), C175–C181.
113. Vandenburgh, H. H., Karlisch, P., & Farr, L. (1988). Maintenance of highly contractile tissue-cultured avian skeletal myotubes in collagen gel. *In Vitro Cellular & Developmental Biology, 24*(3), 166–174.
114. Bamman, M. M., Clarke, M. S., Feeback, D. L., Talmadge, R. J., Stevens, B. R., Lieberman, S. A., et al. (1998). Impact of resistance exercise during bed rest on skeletal muscle sarcopenia and myosin isoform distribution. *Journal of Applied Physiology, 84*(1), 157–163.
115. Booth, F. W. (1982). Effect of limb immobilization on skeletal muscle. *Journal of Applied Physiology, 52*(5), 1113–1118.

116. Jaspers, S. R., & Tischler, M. E. (1984). Atrophy and growth failure of rat hindlimb muscles in tail-cast suspension. *Journal of Applied Physiology: Respiratory, Environmental and Exercise Physiology, 57*(5), 1472–1479.
117. Morris, C. A., Morris, L. D., Kennedy, A. R., & Sweeney, H. L. (2005). Attenuation of skeletal muscle atrophy via protease inhibition. *Journal of Applied Physiology, 99*(5), 1719–1727.
118. Hymer, W. C., Grindeland, R., Krasnov, I., Victorov, I., Motter, K., Mukherjee, P., et al. (1992). Effects of spaceflight on rat pituitary cell function. *Journal of Applied Physiology, 73*(2 Suppl), 151S–157S.
119. Auluck, A., Mudera, V., Hunt, N. P., & Lewis, M. P. (2005). A three-dimensional in vitro model system to study the adaptation of craniofacial skeletal muscle following mechano-stimulation. *European Journal of Oral Sciences, 113*(3), 218–224.
120. Banes, A. J., Tsuzaki, M., Hu, P., Brigman, B., Brown, T., Almekinders, L., et al. (1995). PDGF-BB, IGF-I and mechanical load stimulate DNA synthesis in avian tendon fibroblasts in vitro. *Journal of Biomechanics, 28*(12), 1505–1513.
121. Lokireddy, S., Mouly, V., Butler-Browne, G., Gluckman, P. D., Sharma, M., Kambadur, R., et al. (2011). Myostatin promotes the wasting of human myoblast cultures through promoting ubiquitin-proteasome pathway-mediated loss of sarcomeric proteins. *American Journal of Physiology. Cell Physiology, 301*(6), C1316–C1324.
122. Sanders, P. M., Russell, S. T., & Tisdale, M. J. (2005). Angiotensin II directly induces muscle protein catabolism through the ubiquitin-proteasome proteolytic pathway and may play a role in cancer cachexia. *British Journal of Cancer, 93*(4), 425–434.
123. Bonetto, A., Aydogdu, T., Kunzevitzky, N., Guttridge, D. C., Khuri, S., Koniaris, L. G., et al. (2011). STAT3 activation in skeletal muscle links muscle wasting and the acute phase response in cancer cachexia. *PloS One, 6*(7), e22538.
124. Russell, S. T., Siren, P. M., Siren, M. J., & Tisdale, M. J. (2009). Attenuation of skeletal muscle atrophy in cancer cachexia by D-myo-inositol 1,2,6-triphosphate. *Cancer Chemotherapy and Pharmacology, 64*(3), 517–527.
125. Olivan, M., Busquets, S., Figueras, M., de Oliveira, C. F., Toledo, M., Sette, A., et al. (2010). Nutraceutical inhibition of muscle proteolysis: a role of diallyl sulphide in the treatment of muscle wasting. *Clinical Nutrition, 30*(1), 33–37.
126. Lauritzen, H. P., & Schertzer, J. D. (2010). Measuring GLUT4 translocation in mature muscle fibers. *American Journal of Physiology. Endocrinology and Metabolism, 299*(2), E169–E179.
127. Mao, X., Zeng, X., Huang, Z., Wang, J., & Qiao, S. (2012). Leptin and leucine synergistically regulate protein metabolism in C2C12 myotubes and mouse skeletal muscles. *British Journal of Nutrition, 110*(2), 256–264.
128. Polge, C., Heng, A. E., Jarzaguet, M., Ventadour, S., Claustre, A., Combaret, L., et al. (2011). Muscle actin is polyubiquitinylated in vitro and in vivo and targeted for breakdown by the E3 ligase MuRF1. *FASEB Journal, 25*(11), 3790–3802.
129. Morissette, M. R., Cook, S. A., Buranasombati, C., Rosenberg, M. A., & Rosenzweig, A. (2009). Myostatin inhibits IGF-I-induced myotube hypertrophy through Akt. *American Journal of Physiology. Cell Physiology, 297*(5), C1124–C1132.
130. Sharples, A. P., Player, D. J., Martin, N. R., Mudera, V., Stewart, C. E., & Lewis, M. P. (2012). Modelling in vivo skeletal muscle ageing in vitro using three-dimensional bioengineered constructs. *Aging Cell, 11*(6), 986–995.
131. Bandstra, E. R., Thompson, R. W., Nelson, G. A., Willey, J. S., Judex, S., Cairns, M. A., et al. (2009). Musculoskeletal changes in mice from 20–50 cGy of simulated galactic cosmic rays. *Radiation Research, 172*(1), 21–29.
132. Harfouche, G., & Martin, M. T. (2010). Response of normal stem cells to ionizing radiation: A balance between homeostasis and genomic stability. *Mutation Research, 704*(1–3), 167–174.
133. Trosko, J. E. (2005). The role of stem cells and cell-cell communication in radiation carcinogenesis: Ignored concepts. *BJR Supplement, 27*, 132–138.
134. Stein, T. P. (2002). Space flight and oxidative stress. *Nutrition, 18*(10), 867–871.

135. Lawler, J. M., Song, W., & Demaree, S. R. (2003). Hindlimb unloading increases oxidative stress and disrupts antioxidant capacity in skeletal muscle. *Free Radical Biology and Medicine, 35*(1), 9–16.
136. Smith, S. M., Zwart, S. R., Block, G., Rice, B. L., & Davis-Street, J. E. (2005). The nutritional status of astronauts is altered after long-term space flight aboard the International Space Station. *Journal of Nutrition, 135*(3), 437–443.
137. Smith, S. M., & Zwart, S. R. (2008). Nutritional biochemistry of spaceflight. *Advances in Clinical Chemistry, 46*, 87–130.
138. Yang, Q. H., Botto, L. D., Gallagher, M., Friedman, J. M., Sanders, C. L., Koontz, D., et al. (2008). Prevalence and effects of gene-gene and gene-nutrient interactions on serum folate and serum total homocysteine concentrations in the United States: Findings from the third National Health and Nutrition Examination Survey DNA Bank. *American Journal of Clinical Nutrition, 88*(1), 232–246.
139. Zwart, S. R., Gibson, C. R., Mader, T. H., Ericson, K., Ploutz-Snyder, R., Heer, M., et al. (2012). Vision changes after spaceflight are related to alterations in folate- and vitamin B-12-dependent one-carbon metabolism. *Journal of Nutrition, 142*(3), 427–431.
140. Oishi, Y., Ogata, T., Yamamoto, K. I., Terada, M., Ohira, T., Ohira, Y., et al. (2008). Cellular adaptations in soleus muscle during recovery after hindlimb unloading. *Acta Physiologica (Oxford, England), 192*(3), 381–395.
141. Seene, T., Kaasik, P., & Riso, E. M. (2011). Review on aging, unloading and reloading: Changes in skeletal muscle quantity and quality. *Archives of Gerontology and Geriatrics, 54*, 374–380.
142. Ozbolat, I. T., & Yu, Y. (2013). Bioprinting toward organ fabrication: Challenges and future trends. *IEEE Transactions on Biomedical Engineering, 60*(3), 691–699.
143. Clarke, M. S. (2011). Proteomic analysis of skeletal muscle tissue using SELDI-TOF MS: Application to disuse atrophy. *Methods in Molecular Biology, 818*, 131–141.

Chapter 9
Microgravity and Microgravity Analogue Studies of Cartilage and Cardiac Tissue Engineering

Kacey Ronaldson and Gordana Vunjak-Novakovic

Historical Landmarks

1995—The first report on engineering cartilage tissue constructs using primary cells, biomaterial scaffolds, and rotating bioreactors [1].

1996—Freed and Vunjak-Novakovic (MIT) engineered cartilage tissues in rotating bioreactors on ground and in space, demonstrating the feasibility of long-term spaceflight studies, and the utility of tissue engineering in understanding the effects of the space environment on human physiology. The study was reported in PNAS [2] as a cover article [3] and highlighted in JAMA [4].

1997—Cardiac tissues engineered in rotating bioreactors [2].

1998–2002—Use of rotating bioreactors in quantitative studies of cartilage development [5–7].

2000—Mathematical model of cartilage development in rotating bioreactors [7, 8].

2001—Use of rotating bioreactors in quantitative studies of cartilage healing [9].

(continued)

K. Ronaldson
Department of Biomedical Engineering, Columbia University, New York, NY USA

G. Vunjak-Novakovic (✉)
Department of Biomedical Engineering, Columbia University, New York, NY USA

Mikati Foundation Professor of Biomedical Engineering and Medical Sciences, Laboratory for Stem Cells and Tissue Engineering, Columbia University, 622 West 168th Street, VC12-234, New York, NY 10032, USA
e-mail: gv2131@columbia.edu

C.A. Nickerson et al. (eds.), *Effect of Spaceflight and Spaceflight Analogue Culture on Human and Microbial Cells*, DOI 10.1007/978-1-4939-3277-1_9

(continued)

2001—Ground and rocket flight studies of cartilage formation from pig chondrocytes [8, 10, 11].

2003—Fluid-mechanic analysis of cartilage development in rotating bioreactors [12].

2006–2013—Cartilage formation from bone marrow cells [13–15].

2011—Cardiac differentiation of mouse embryonic cells in rotating bioreactors [16].

9.1 Introduction

Outer space is seen as the ultimate frontier, and a focus of keen interest for as long as humankind has existed. It is conceivable that experimentation in space, a "gravity knock-out," will similarly advance our understanding of the largely unknown roles of gravity in physical and biological processes. It is also possible that such simplified experimental settings, that lack the force of gravity, will result in more accurate measurements of parameters that normally interfere with gravity. Besides serving as a platform for creative science, the International Space Station (ISS) can also be viewed as a "microgravity garage" for developing technologies, and testing of materials, instruments, and processes under extreme conditions.

Tissue engineering has been one of the prime areas of space research. Interestingly, tissue engineering and space research rapidly developed over the last 3 decades, contributing mutually beneficial advances to each field. By definition, tissue engineering is "the application of principles and methods of engineering and life sciences toward fundamental understanding of structure–function relationships in normal and pathological tissues and the development of biological substitutes to restore, maintain or improve tissue function" [17].

The ability of engineered tissue constructs to recapitulate some of the critical developmental events and physiological functions not mimicked by conventional cell cultures opened many exciting avenues of research. NIH describes the role of tissue engineering as: "a rapidly growing multidisciplinary field involving the life, physical and engineering sciences that seeks to develop functional, tissue and organ substitutes to repair, replace or enhance biological function that has been lost due to congenital abnormalities, injury, disease, or aging." Engineered tissues can also serve as enabling platforms for scientific inquiry, as evidenced in the study of human stem cells for their ability to differentiate and functionally assemble into specific tissue structures. Bioengineered systems that combine microenvironmental control with tissue-specific transport and signaling are critical for our efforts to study tissue regeneration and disease under conditions that predict the human context [18].

Finally, engineered tissues are opening new frontiers in modeling the initiation and progression of human diseases, with etiologies associated with microbial pathogens, environmental toxins, and genetic predisposition, thereby allowing identification of new therapeutic targets and determining drug safety and efficacy. The mere prospect of recapitulating human physiology and pathology in vitro using "organ-on-a-chip" platforms that combine silicon chip technology with patient-derived cell samples are but one example of bioengineered human surrogates that are having a major impact on drug development and personalized medicine [19].

Each tissue within our body is composed of specific cell types interacting with each other towards providing functional roles, and supported by a vascular system responsible for the exchange of oxygen, nutrients, regulatory factors, and metabolic waste. The cells reside within extracellular matrix (ECM) that, in combination with other microenvironmental components including biochemical and physical signals, ultimately forms well-organized and fully functional tissues. In order to harness the potential to engineer such complex tissues, we must first understand how to guide the individual cells by closely mimicking native conditions of the microenvironment in which that tissue functions [20].

Conventional cell culture techniques utilize environments constrained to two dimensions (2-D), such as cells cultured on flat impermeable surfaces in flasks and petri dishes, which do not recapitulate the environments found in vivo. The shift to more physiologically relevant three-dimensional (3-D) culture systems over the last few decades has been largely driven by methods offered by tissue engineering (Fig. 9.1).

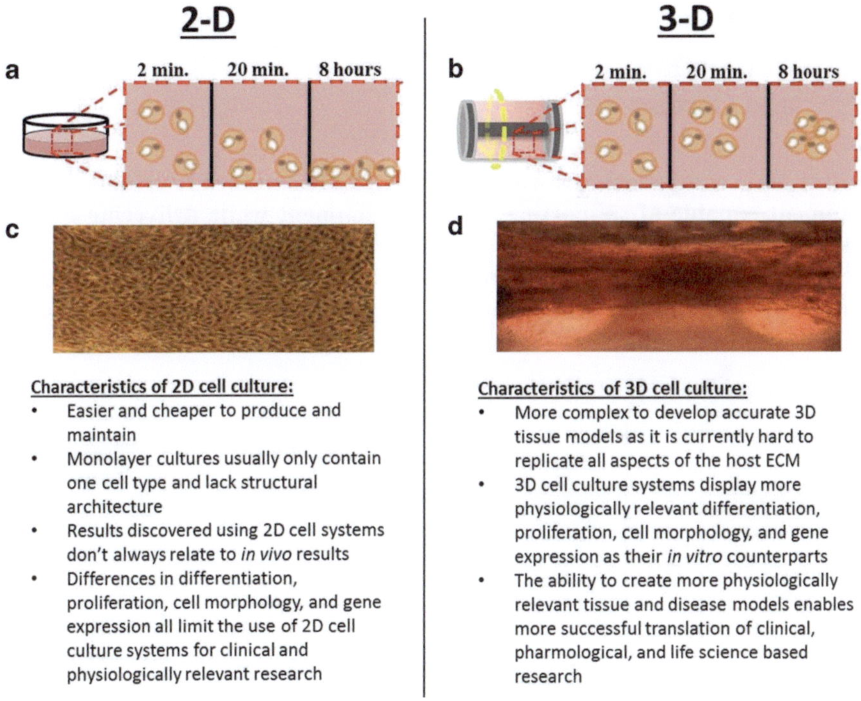

Fig. 9.1 Cell culture in 2-D vs 3-D systems. (**a**) Schematic of cell culture over time in 2-D monolayers grown on plastic. (**b**) Schematic of cell culture over time in 3-D. (**c**) Phase contrast image of rat neonatal cardiomyocytes cultured in 2-D on a tissue culture treated polystyrene petri dish. (**d**) Phase contrast image of rat neonatal cardiomyocytes cultured in a 3-D system where the cells are suspended within a 0.8 mg/mL Collagen type I hydrogel

Since the 1970s, human exploration of space has included cell culture studies that have yielded valuable data describing the effects of spaceflight on cell morphology, locomotion, growth, and function (reviewed in [8, 11]). Since the 1990s, 120–160 papers on various aspects of microgravity research were published every year, according to the ISI Web of Science. The number of citations has increased, from ~500 per year in early 1990s to a steady 2500–3000 per year in the last 10 years. Among the top-10 microgravity studies that were most frequently cited, roughly one half are in biology and medicine: orthostatic intolerance [21], tissue engineering [2], bone loss [22, 23], and the other half are in materials and physics: crystals [24], colloids [25], plasma [26], materials processing [27, 28].

From the early 1990s, NASA started to offer opportunities to the scientific community to conduct short- and long-duration tissue engineering research in true microgravity (spaceflight) or in microgravity analogue environments on Earth (microgravity analogue suspension culture using the rotating wall vessel/RWV bioreactor) [29]. Tissue engineering of cartilage and heart—two tissues known to be affected in humans during spaceflight as well as in spaceflight tissue culture experiments were extensively studied [1, 2, 11, 21, 30–35] are the main focus of this chapter that reviews microgravity studies of tissue engineering.

9.2 Culture Environments for Engineering Cartilage and Cardiac Tissues

Three main components are often optimized and combined to create an environment capable of supporting tissue development while delivering biomimetic biochemical and physical stimuli: (1) the cell type and source, (2) a scaffold for the cells, and (3) a bioreactor to provide necessary mass transport regimens and physical stimulation (Fig. 9.2). In this system, the cells are

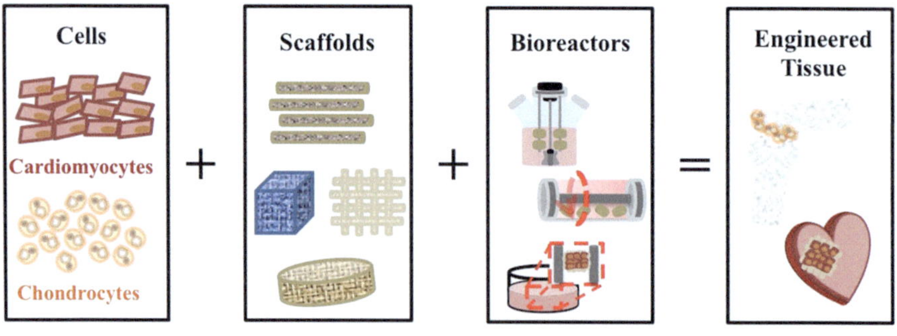

Fig. 9.2 Tissue engineering system. The three major components (cells, scaffolds, and bioreactors) are combined to form engineered tissue constructs with ability to replace damaged tissues as depicted schematically

the actual tissue engineers, the scaffold serves as an instructive structural and logistic template for tissue formation, and the bioreactor plays a role of a "substitute body," providing environmental control, molecular, and physical regulatory factors [18, 36].

When determining what cell type and source to use, it is imperative to consider the capability of the cells to predictably proliferate and differentiate in vitro and to form the desired tissue, and to consider the eventual application of the engineered tissue. With respect to cartilage tissue engineering, there are multiple cell types to choose from including articular chondrocytes and bone marrow stromal cells (BMSC), which have both shown capabilities for repairing large cartilage defects in animal models [37, 38]. For cardiac cells, most current methods involve the use of cardiomyocytes (CMs) from neonatal rats hearts [35, 39]. The development of advanced protocols for differentiating stem cells, particularly human derived induced pluripotent stem cells (hiPS) into these desired tissues is of great importance [40] and thus, we expect future work to include these sources more relevant to human applications.

Cells typically require attachment to a surface for survival and thus, recreating these surfaces so that they mimic native tissue ECM can help direct cellular functions and enable successful tissue formation. One method for implementing 3-D culture systems involves the use of polymeric scaffolds that can be reproducibly fabricated, are highly porous (to increase the surface area for cell attachment and facilitate nutrient transport), and are composed of biocompatible materials that can be controllably degraded at similar rates of in vitro tissue formation. Scaffolds generally exhibit physical and biochemical properties similar to those of the tissue being engineered. Thus, scaffolds for engineering articular cartilage should be mechanically stable to withstand the loads associated with cartilage function in vivo. Since the mechanical properties of the tissue greatly depend on the composition of the ECM, an ideal scaffold would promote glycosaminoglycans (GAG) and collagen deposition by the seeded chondrocytes [41]. Cardiac scaffolds should enhance the formation on aligned cell distributions and thus, highly aligned scaffolds—synthetic or derived from native tissues—result in more aligned and electrically synchronous cardiac tissue [42, 43].

Finally, bioreactors enable the introduction of convective-diffusive transport regimens, electrical, and mechanical forces that allow for a more precise control over cell distribution within a scaffold, adequate mass transfer of nutrients, gases, and desired soluble factors, and ultimately support the formation of larger, more complex tissues [34]. Each bioreactor should be designed to provide cues specific to the desired tissue. A few of the many types of bioreactors used in tissue engineering are listed in Fig. 9.3. For cartilage tissue engineering bioreactors, relevant parameters include low-friction, articulating scaffolds that can withstand and transmit compressive loading and shear forces [44]. On the other hand, cardiac tissue engineering bioreactors should incorporate electrical signals that enable cardiac constructs to contract synchronously, and provide low fluid shear environments [33].

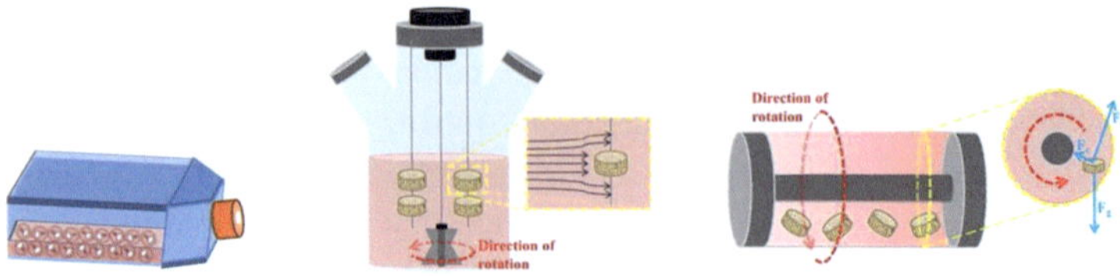

	Tissue Culture Flask	Spinner Flask Bioreactor	Rotating Wall Vessel (RWV)
Microgravity?	No	No	Microgravity Analogue
Mechanism	Cells either attach to a tissue-culture treated surface where they are then covered in static media or are grown in suspension in static media.	A spinner is used to mix oxygen and nutrients while scaffolds are physically suspended in media.	Horizontally rotating cylindrical, fluid filled, low shear vessel containing a concentric inner cylinder (used for gas exchange) that maintains cells in a suspended state thereby reducing the effects of gravity.
Advantages for Tissue Engineering	Relatively easy and cheap to use.	Tissues can be grown larger than their 2D counterparts as mass transport is much more efficient.	Low shear, and high mass transport of enables cellular differentiation, larger 3D tissue formation, and ECM production.

Fig. 9.3 Common platforms for tissue engineering

9.3 Investigating the Role of Microgravity in Tissue Engineering

Surprisingly, microgravity tissue engineering resulted from the lack of our understanding of the complex and multiple effects of spaceflight on human physiology. Already at the beginning of the era of space exploration, scientists realized that their previous hypotheses that the lack of gravitational effects would cause only negligible effects on the astronauts' tissues and organs were actually false. One direction of work was to try to mimic the conditions of microgravity on Earth, by employing systems that randomize the direction of the gravitational vector, such that the net sum of gravitational forces, averaged over time, equals zero. Clearly, these conditions do not correspond to zero gravity or reduced gravity, as the gravitational force acts on any point in the system at any point in time. Nevertheless, randomization of gravitational forces by rotational flow maintaining the tissue constructs in a state of optimized suspension in the fluid, have made major contributions to cell culture and tissue engineering [29, 31, 32, 45, 46]. The first mechanically functional cartilage was in fact grown under these conditions [2], as were many other tissues.

9.3.1 Engineering Microgravity Cell Culture Environments on Earth

As the goal of tissue engineering is to build tissues in vitro, requirements for successful cell culture conditions can be established by mimicking the in vivo environment. Cell-scaffold-bioreactor culture systems more closely mimic the native environment of the parent tissues and thus are desired over traditional flat 2-D culture systems. 3-D environments also enable the cells to migrate and remodel in a manner dictated by biomimetic cues, replicating the tissue developmental processes characterized by in vivo morphogenesis [1]. Another desired condition involves the ability of cells to co-localize and form connections with one another. Initiating the formation of cell-to-cell contacts allows the cells to establish necessary intercellular pathways responsible for transmitting cell signals and soluble molecules typical of native tissue [8].

A major requirement for generating large 3-D cell aggregates involves sufficient mass transport. Cells require adequate oxygen and nutrient supplies and the removal of metabolic waste. Without vasculature, cells internally located within the cell aggregates cannot be adequately supported by sufficient mass transfer of Efforts to provide adequate mass transport often introduce high levels of fluid shear, which negatively affect cells. An ideal tissue culture system should be associated with low fluid shear transport.

Role of Bioreactors in Tissue Engineering

Traditional 3-D cell culture techniques are limited in their ability to exchange nutrients and metabolites between the cells and the culture media. Mechanical stirring and fluid flow enhanced the transport rates but at the expense of exposing the cells to shear stress. To enhance transport while limiting hydrodynamic shear, scientists turned to specialized bioreactors. While bioreactors have been used in many disciplines to create controlled environment's for biological and biochemical processes, they have subsequently been adapted for tissue engineering purposes to create controlled environments that deliver mechanical stimuli to enhance cell differentiation, homogenous cell populations, and stimulate the production of ECM necessary for tissue formation [18, 44].

Bioreactors can be designed to recapitulate several key aspects of the in vivo milieu and thereby guide cellular maturation and organization into the specific cellular phenotypes and assemble functional tissues. Bioreactor designs enable the creation of a cellular environment that can be controlled and monitored for parameters such as nutrient supply, metabolic waste removal, temperature, gases, pH, chemical gradients, electrical stimulation, and fluid shear forces. By controlling mass transport within the system, bioreactors can support the culture of large 3-D tissues that require increased nutrient transport and subsequent waste removal [18, 31, 33, 34].

Development of Microgravity Bioreactors for Tissue Engineering

To actually experience zero gravity, an object would need to be infinitely far from any gravitational force. As this is impossible to recreate in systems on Earth, and not characteristic of most regions of space chartered by humans, simulating conditions of microgravity is an attractive option. Microgravity, defined as an environment with greatly reduced gravitational forces, is typical of conditions found in space, including the low Earth orbit.

Investigations into the effects of microgravity are appealing to tissue engineering studies. Low-shear fluid environment promotes high-density cell interactions while minimizing the disruptions typical of 3-D in vitro environments such as gravity-induced sedimentation, buoyancy, inadequate nutrient transfer, and density-driven convection. By reducing the effects of gravity, cells can grow more uniformly as they divide and build an ECM evenly in all directions. It is thought that this setting creates tissue-like aggregates that are more biomimetic in nature, and provides a platform for creating larger 3-D tissue constructs [29, 32, 37, 47] (Fig. 9.4).

Larger constructs require increased nutrient transport and effective removal of metabolic wastes. Bioreactors with flow and mixing are able to provide this, and were shown to increase the growth rates of chondrocytes [48]. Tissue constructs

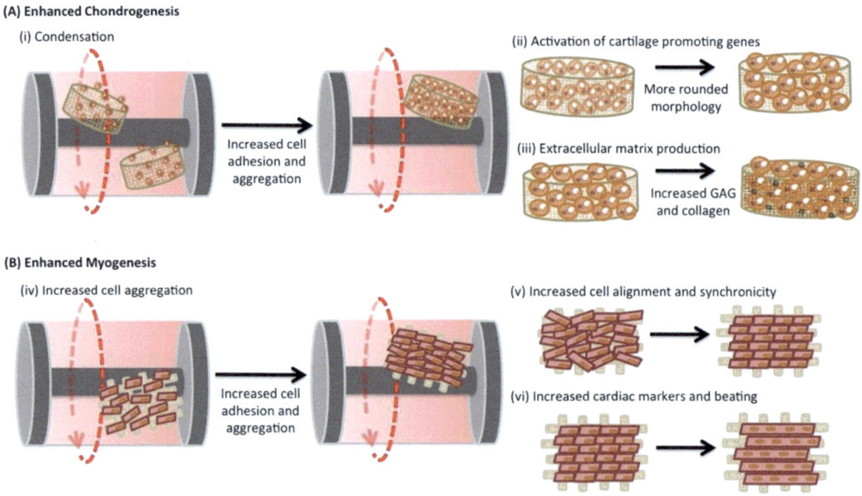

Fig. 9.4 Bioreactors for cartilage and cardiac tissue formation. Bioreactors can be utilized to (**a**) increase chondrogenesis by (*i*) increasing condensation through cell aggregation, (*ii*) activating genes that promote cartilage formation, and (*iii*) enhancing extracellular matrix production and (**b**) increasing myogenesis by (*iv*) enhancing cell aggregation to create larger tissue-like constructs, (*v*) increasing cardiac cell alignment and cell synchronicity, and (vi) increasing genetic up-regulation of cardiac specific genes which when combined with more aligned cardiomyocytes, result in enhanced beating of cardiac tissues

cultured with mixing, such as in spinner flasks were larger and produced the ECM at higher rates when compared to static conditions [49]. The beneficial effects of mixing provide a platform for researchers to manipulate the fluid dynamics of cell culture systems in order to maximize transport of nutrients. So-called simulated/modeled microgravity systems can be seen as a natural progression of these results, ultimately leading to bioreactors providing the necessary environments to direct, enhance, and support cell differentiation and growth of 3-D tissue constructs similar to native tissues [9].

Rotating Wall Vessel Bioreactors

NASA developed a 3-D cell culture system able to meet several requirements essential for establishing organotypic tissue models that are discussed in the previous section. Initially, NASA intended to use this system to culture mammalian cells during transport from Earth to space so that the effects of microgravity on an astronaut's health could be investigated. Basing the culture system on clinorotation principles, the forces of gravity could be randomized by slowly rotating the culture chamber around a central axis. Rotation established the condition of "simulated/modeled microgravity" during culture of the cells on the ground, prior to actual microgravity conditions that the cells would experience once in space. Scientists noticed that cells cultured in this system during ground-based experiments actually started to form larger 3-D aggregates and noted the applicability of this technology to further tissue engineering applications.

The development of this original NASA 3-D cell culture bioreactor, the Rotating Wall Vessel (RWV) [12, 47], established a microgravity-analogue environment termed low shear modeled microgravity [50]. Details of the operational and design principles of the RWV are discussed in Chap. 2. In brief, the design of the RWV enables an optimized form of suspension culture, which mimics several aspects of the microgravity environment. Thus, within the RWV, the cultured cells, cell aggregates or macroscopic tissue constructs can freely move within the media so that they remain suspended in the media as the wall of the vessel rotates [2, 51]. This condition is established as the downward force of gravity is offset by the upward hydrodynamic drag force imposed as the wall of the vessel rotates. Consequently, the RWV design allows for fluid transport, similar to that in spinner flasks, gas exchange through a central cylinder containing a membrane, and a more uniform cell suspension that ultimately leads to more homogenously seeded scaffolds. As tissues begin to grow within the RWV, the rotational speed can be adjusted accordingly to maintain "simulated/modeled microgravity" conditions [7, 9, 12, 31].

9.4 Microgravity Studies of Cartilage and Cardiac Tissue Engineering

Microgravity-based bioreactor designs promote in vivo-like tissue growth by having adequate mass transfer while limiting hydrodynamic shear stress, thus allowing cells to aggregate based on natural cellular affinities and differentiate into 3-D organotypic tissue constructs [32]. These functional constructs are of great significance in the search for cartilage and cardiac tissue engineered replacements and cell-based treatments. Thus, research into the effects of microgravity on both cartilage and cardiac tissue formation has proven to be useful to further progress in the field to reach such potential [5, 8, 14, 15, 31, 32, 35, 49, 52].

9.4.1 How Does Simulated/Modeled Microgravity Support Cartilage Tissue Formation?

Adult cartilage is a connective tissue that contains relatively low numbers of chondrocytes within an ECM consisting of proteoglycans, collagens, and elastic fibers [5]. While avascular, cartilage still requires nutrient exchange via diffusion through its ECM. Mechanical stimuli, imposed in vivo through joint movement, induce chondrocytes to produce more ECM and thus should be mimicked in vitro through the use of bioreactors. Chondrogenesis involves a series of steps starting from mesenchymal condensation, activation of genes that promote cartilage formation resulting in a more rounded cellular morphology and increased differentiation, and the secretion of ECM proteins typical of cartilage [5, 31, 32, 37, 49]. Under simulated/modeled microgravity conditions, the RWV bioreactor creates an environment that enables the initial cellular aggregation necessary for chondrogenesis which was previously hard to obtain in typical cell culture systems (Fig. 9.4).

Several previous studies have demonstrated that RWV-cultured chondrocytes were able to form viable cartilage constructs [22]. The functionality of these tissue engineered cartilage constructs revealed mechanical properties that were characteristic of native cartilage (Table 9.1). The ability to grow larger 3-D tissues, while adequately providing media and nutrient exchange within the constructs to meet their already low metabolic demands, allows the incorporation of microgravity analogue culture conditions to develop physiologically relevant cartilage tissues.

In a set of experiments carried out by Freed and Vunjak-Novakovic, the effects of various cell culture regimens including static 2-D culture, RWV culture, and spinner flask culture to enhance media perfusion of scaffolds were directly compared. Engineered cartilage was best supported by the simulated/modeled microgravity conditions in the RWV, as evidenced by the highest extent of enhanced cartilage growth, rounded cell morphology, and an ECM high in both GAG and collagen. The authors demonstrated that exposing the cartilage tissues to hydrodynamically active environments markedly enhanced chondrogenesis [35, 49].

Table 9.1 Comparison of tissue engineered cartilage constructs grown on Earth and in spaceflight with native cartilage

Parameter	Tissue engineered constructs			Natural cartilage[b]
	At launch[a]	Mir-grown	Earth-grown	
Culture time (months)	3	7	7	0
Chondrocyte function				
Sulfate incorporation (ng/μg DNA/day)	80.8 ± 37.2 (3)	88.9 ± 8.0 (3)	83.4 ± 13.4 (3)	**332 ± 36.3 (3)**
Proline incorporation (ng/μg DNA/day)	82.0 ± 20.6 (3)	93.0 ± 3.0 (3)	87.9 ± 22.7 (3)	**205 ± 46.8 (3)**
Construct composition				
Weight (mg wet)	250 ± 38 (20)	*330 ± 25 (5)*	**429 ± 14 (5)**	**461 ± 72.8 (7)**
Water (%)	**89.2 ± 0.89 (3)**	**89.2 ± 0.31 (3)**	*86.3 ± 0.53 (3)*	82.8 ± 0.93 (7)
Cells (% wet weight)	**0.64 ± 0.03 (3)**	0.4 ± 0.01 (3)	0.46 ± 0.02 (3)	0.66 ± 0.09 (7)
GAG (% wet weight)	6.03 ± 0.84 (3)	3.59 ± 0.22 (3)	**8.83 ± 0.93 (3)**	*7.05 ± 0.56 (6)*
Collagen (% wet weight)	2.70 ± 0.75 (3)	3.42 ± 0.17 (3)	3.68 ± 0.27 (3)	**10.7 ± 0.91 (6)**
Total of above components (% wet weight)	*98.6 ± 0.07 (3)*	96.6 ± 0.68 (3)	*99.3 ± 0.60 (3)*	**101 ± 0.46 (6)**
Type II collagen (% total collagen)	91.6 ± 19.1 (2)	78.0 ± 4.1 (4)	75.3 ± 7.8 (4)	90.3 ± 17.9 (5)
Mechanical behavior				
Aggregate modulus (MPa)	0.108 ± 0.047 (2)	*0.313 ± 0.045 (4)*	**0.932 ± 0.049 (3)**	**0.949 ± 0.021 (3)**
Hydraulic permeability ($\times 10^{-15}$ m^4/Ns)	**8.25 ± 1.94 (2)**	6.73 ± 3.02 (4)	3.72 ± 0.167 (3)	2.72 ± 0.641 (3)
Dynamic stiffness (MPa, 1 Hz)	2.23 ± 0.12 (2)	3.80 ± 0.39 (4)	*3.80 ± 0.39 (4)*	**16.8 ± 1.14 (3)**

Data represent average ± standard deviation. Parentheses indicate the number of samples analyzed per group. Statistical significance was assessed using ANOVA ($\alpha = 0.05$) in conjunction with Tukey's studentized range test. For each parameter, statistically significant differences between groups ($P < 0.05$) are denoted by bold and italic values

[a]Wet weights at 3 months were those of the actual constructs used in 7-month Mir space station and Earth cultivations; all other parameters in this group were assessed using additional 3-month constructs from the same batch

[b]Wet weights for natural cartilage were scaled by a factor of 5 for better comparison with corresponding values measured for Earth-grown tissue engineered constructs. Reproduced with permission from Freed et al. [2]

Studies of primary chondrocytes cultured on scaffolds in the RWV revealed that hydrodynamic forces enhanced GAG and collagen synthesis through compressive forces on the chondrocytes, thereby increasing chondrogenesis [1, 2, 5, 14, 49, 51, 53–55]. The culture conditions in the RWV resulted in enhanced cartilage tissue formation by effectively providing mass transfer of nutrients, simulating chondrocytes to synthesis ECM comparable to native tissue, and dynamic laminar fluid flow patterns that served to promote desired cell differentiation and resulting matrix formation [49]. Studies of chondrogenesis and cartilage formation starting from mesenchymal stem cells also demonstrated the benefits of dynamic flow environment in rotating bioreactor vessels [13, 14].

Functional cartilaginous constructs were grown in rotating bioreactors starting from chondrocytes immobilized on polymeric scaffolds. To facilitate data interpretation and optimize cultivation conditions, a mathematical model was developed which yields the concentrations of oxygen and GAG as functions of time and position in growing tissue. Calculated GAG concentrations were qualitatively and quantitatively consistent with profiles measured via high-resolution image processing of tissue samples cultured at two different oxygen tensions for various periods of time [6]. This study built upon previous modeling work [1] to establish theoretical foundation for optimizing the conditions and duration of cartilage tissue engineering in rotating bioreactors.

The dynamic flow conditions around freely suspended tissue constructs, with oscillating fluid velocity, shear, and pressure are considered a key factor in stimulating cartilage tissue growth. In vivo, cartilage health is also associated with loading-induced fluctuations in fluid flow and mechanical stress. Notably, the shape of cartilage tissue constructs reflected the levels of local hydrodynamic shear, with more rapid tissue formation in areas of higher dynamic shear [2, 12]. Because of their favorable fluid-dynamic and mass transport properties [12], rotating bioreactors were successfully used to optimize the gas exchange conditions for cartilage tissue engineering [6], to modulate mechanical properties of tissue-engineered cartilage [7] and to investigate integration between engineered cartilage and adjacent native articular cartilage [9].

9.4.2 How Does Simulated/Modeled Microgravity Support Cardiac Tissue Formation?

Cardiac muscle is composed of cardiac myocytes arranged longitudinally so that they connect with one another to form contractile units known as muscle fibers. For such a highly metabolically active tissue, dense vasculature is essential for maintaining proper supply of nutrients and regulatory factors and removal of metabolites. The development of functional cardiac tissue relies on adequate oxygen delivery and metabolic waste removal from cells, a uniform and physiologically dense cell distribution, and the delivery of stimuli to promote electrically responsive and mechanically strong tissue formation. In addition, cardiomyocytes couple with fibroblasts and exchange paracrine signals with endothelial cells.

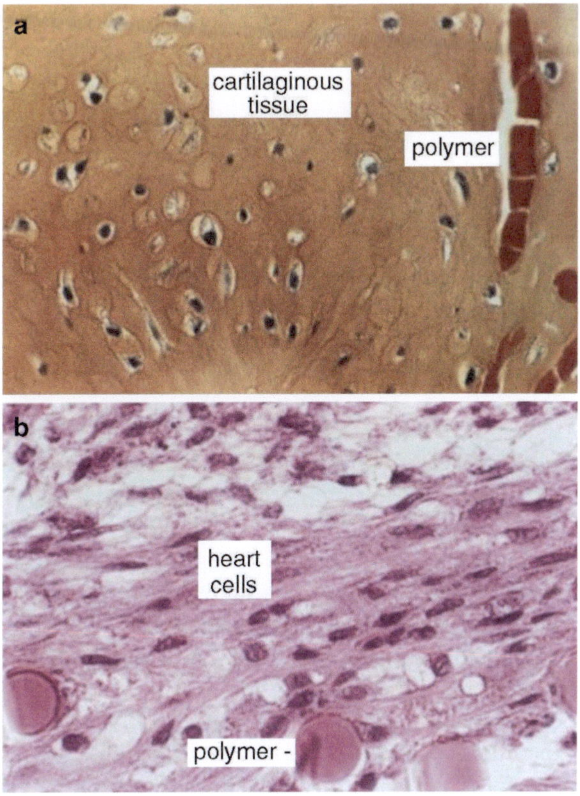

Fig. 9.5 Tissue constructs grown in simulated/modeled microgravity. (**a**) Bovine articular chondrocytes grown for 2 weeks formed cartilage tissue under simulated/modeled microgravity conditions (safranin-O stain, original magnification ×400); (**b**) Embryonic chick cells grown for 3 weeks formed cardiac tissue under simulated/modeled microgravity conditions (hematoxylin and eosin stain, original magnification ×1000). Image reprinted with permission (Freed et al. 1997)

While 3-D aggregates can be successfully formed in bioreactors with mixing and/or perfusion, high-fluid shear environments can be detrimental to the formation of cardiac tissues [33]. Thus, low-fluid shear environments, such as the microgravity analogue conditions in RWV bioreactors, would be beneficial to successfully create physiologically functional cardiac tissues (Fig. 9.5). Similar to in vivo-like cartilage formation in the RWV, physiologically relevant 3-D cardiac constructs have also been engineered under these culture conditions [35, 56, 57].

Cardiac constructs based on rat heart cells and cultured for up to 3 weeks were observed to contract spontaneously and synchronously at a rate of 30–130 beats per min as assessed by videomicroscopy. Specifically, as compared to conventional 2-D cell culture, RWV-derived 3-D cardiac constructs displayed more physiologically relevant cellular morphology and organization, including higher fractions of elongated cells that were better aligned and spontaneously contracted at enhanced rates of 30–130 beats per min [35, 56]. In addition, RWV culture enhanced the presence of cardiac-specific proteins such as Cx43, a gap junction protein critical for cell–cell communication in the heart, and CK-MM, a primary creatine kinase present in the sarcomeres. The RWV cultures also resulted in improved electromechanical function of engineered heart tissues, as evidenced by higher maximum capture rates, which is the maximum frequency at which a tissue can beat synchronously in response to electrical stimulation [58].

Collectively, these findings demonstrate that RWV culture resulted in enhanced molecular, structural, and electrophysiological properties of engineered cardiac tissues. These cardiac tissue constructs could potentially be used in vivo, as implants, or in vitro, to study the effects of hypoxia, drugs, or microgravity [19, 56]. Dynamic flow at low hydrodynamic shear was the likely key contributing factor to these favorable cardiac tissue engineering outcomes, as suggested from the comparison of RWV microgravity settings with the RWV solid body rotation. Simulated microgravity promoted the growth of cardiac tissue constructs with elongated cells that contracted spontaneously and synchronously at a rate of 30–130 beats per min, whereas the solid body rotation did not [35].

9.5 Investigating the Role of True Microgravity in Spaceflight

In order to determine the effects of the actual microgravity on tissue engineering, cell culture must be conducted during spaceflight. Aboard the ISS National Laboratory, microgravity conditions are routinely used to explore the effects and contributions of this environment to growing engineered tissues. Previous evidence from space shuttle experiments during the NASA Mir program has revealed the enhanced ability to culture larger, more functional tissues in true microgravity conditions as compared to ground-based controls [4, 8]. It is hoped that recent completion of the ISS National Laboratory will enable longer duration microgravity experiments that will advance fundamental tissue engineering knowledge with potential for downstream applications (Fig. 9.6).

9.5.1 Cartilage Tissue Engineering in Space

The effects of microgravity during spaceflight on engineered cartilage were investigated during a 4-month long study aboard the Mir Space Station [2]. Since cartilage is extremely resilient to the harsh conditions that might be expected during spaceflight, as well as being clinically relevant in investigating the effects of spaceflight on skeletal tissues, it was chosen as a model tissue. Engineered cartilage constructs were grown by cultivation of primary bovine chondrocytes on biodegradable scaffolds in form of a mesh made of polyglycolic acid fibers. The constructs were grown in the RWV on Earth for 3 months, and then transferred to a pair of identical rotating perfused bioreactors designed for spaceflight to enable parallel studies under conditions of simulated/modeled microgravity on Earth and true microgravity of space onboard Mir (Fig. 9.7). After 4 months, these constructs were returned to Earth by another space shuttle and comparisons were made between the flight and ground cultures. The rotational speed of the ground-based perfused bioreactors was adjusted to have the engineered constructs remain in suspension, thus balancing gravity, centrifugal force, and the fluid shear force.

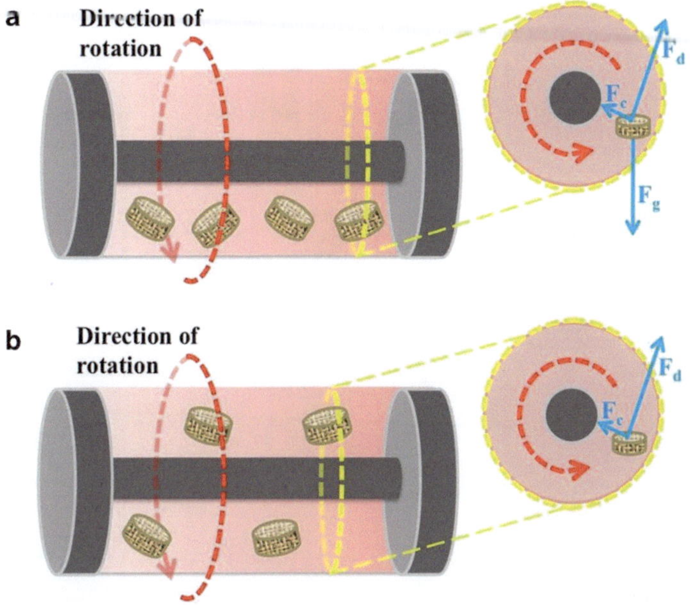

Fig. 9.6 Simulated/modeled versus true microgravity. (**a**) A free-body diagram representative of "simulated/modeled microgravity conditions" and (**b**) a free-body diagram representative of true microgravity conditions in space

In this study, the longest tissue engineering experiment carried out in space, the cartilage constructs cultured in the microgravity environment of spaceflight underwent morphological changes in shape from an initial discoid construct to a more spherical construct shape, an indicator of randomized, spatially uniform hydrodynamic shear. Tissue constructs cultured under control ground-based conditions, however, took shapes reflecting the levels of hydrodynamic shear (high at the construct edges and at the bottom face of the disc, and low at the top face of the disc [12] (Fig. 9.8). Further characterization of the samples from the two different culture conditions revealed similar construct cellularity and similar increases in wet weight. However, the equilibrium compressive modulus of the ground-based constructs, the standard parameter measured to determine biomechanical properties under compressive loading, was three times higher than that of their counterparts cultured in space. Notably, engineered cartilage grown I this study in RWVs based on ground was the first ever to match the mechanical properties of native articular cartilage [2].

Thus, while both culture methods were successful in maintaining viable cartilage constructs [2], the constructs grown in ground-based conditions were larger and mechanically superior to those cultured in the actual microgravity conditions of space. This finding was accredited to a reduction in physical forces found in microgravity conditions of spaceflight, as those constructs experienced conditions of free-fall aboard the space station whereas those cultured in the ground-based perfused reactors still encountered the forces of gravity settling on Earth.

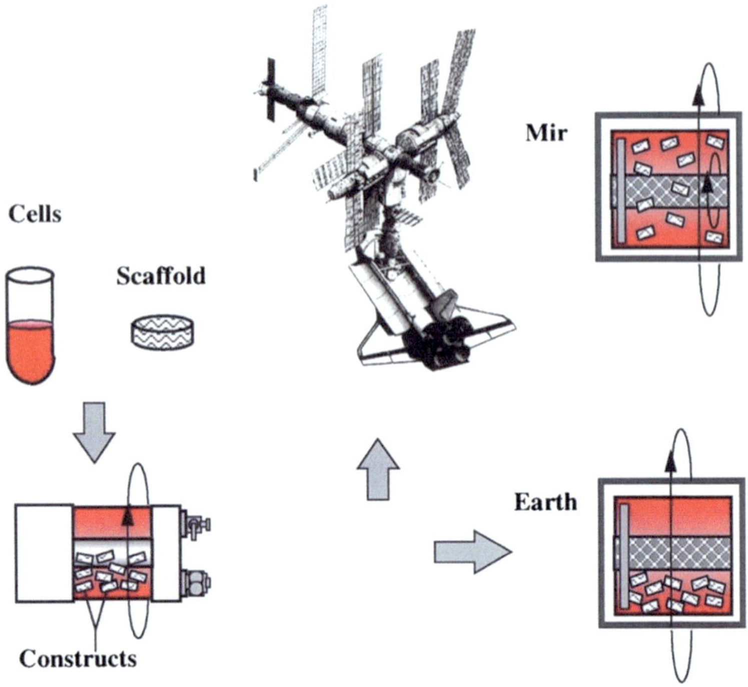

Fig. 9.7 Experimental design used to determine the effects of microgravity on engineered cartilage. Cartilage cells were seeded onto polymer scaffolds, and the resulting constructs were cultivated in RWV bioreactors first for 3 months on Earth and then for 4 additional months in perfused rotating bioreactors either on the Mir Space Station or on Earth. Reproduced with permission from (Freed et al. 1996)

The important results of this study are: (1) demonstrated ability of rotating perfused bioreactors and associated components to support the cultivation of viable cartilage tissues over a prolonged period of time (4 months), (2) striking differences in the compositions, morphologies, and mechanical properties of cartilage constructs cultured in simulated/modeled and true microgravity, and (3) consistency between the space-induced changes of cartilage construct properties with the weakening of skeletal tissues observed in astronauts. These results highlight the ability of conducting microgravity tissue engineering studies in space, and using bioreactor cultured engineered tissues as models for studying human physiology.

However, the lack of isolating and controlling the experimental conditions in this and other spaceflight studies prevented deeper insights into the direct effects of microgravity as compared to other contributing factors of spaceflight, such as the changes in gravity during the launch and landing, and the increased levels of cosmic radiation. The results of this study emphasized the need for a unit-gravity control in

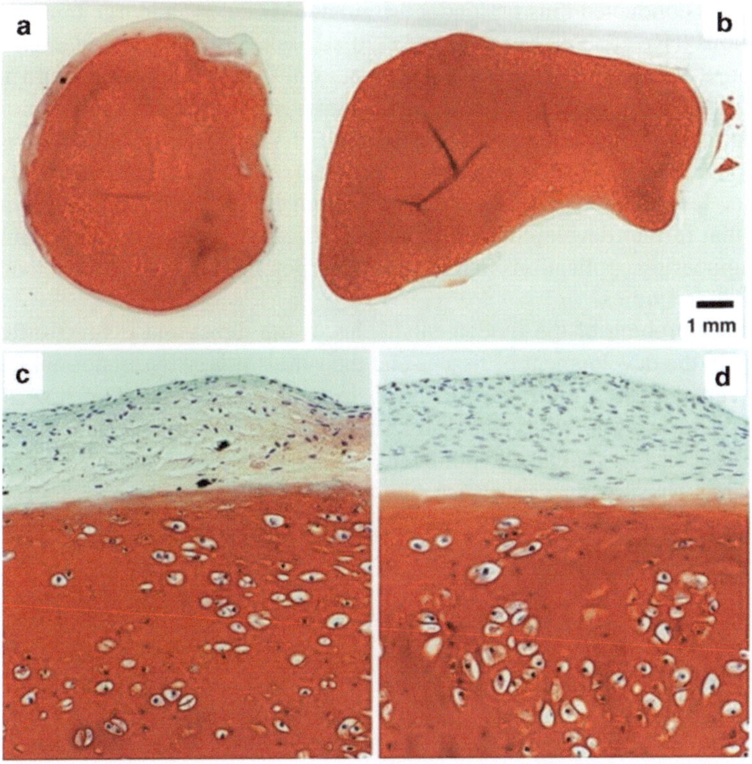

Fig. 9.8 Structure of engineered cartilage in true versus stimulated/modeled microgravity regimens. (**a**, **b**) Full cross sections of constructs from space and ground-based culture. (**c**, **d**) Representative areas at the construct surfaces are shown. GAG is stained red with Safranin-O. Reproduced with permission from Freed et al. [2]

space, such as a centrifuge onboard the Space Station, that would add beneficial insight to the mechanistic effects of microgravity space conditions versus the characteristic unit-gravity environment found on Earth.

9.6 Conclusions

It is clear that culturing cells under the conditions of microgravity can help us understand the effect of spaceflight on human health, as well as decipher some of the effects of gravity on the development and function of our tissues and organs. As our current tissue engineering systems are getting closer to recapitulating the actual human physiology, these systems can provide controllable platforms of high fidelity meeting the clinical and basic research needs.

Studies conducted thus far, under "simulated/modeled" and true microgravity conditions have enabled cartilage and cardiac tissue growth via mechanisms that promote 3-D cell aggregation, adequate mass transfer, in conjunction with appropriate types and levels of physical stimuli that have resulted in organotypic models that mimic key aspects of the parental tissue. These studies have also contributed insight toward advancing our understanding of mechanisms leading to tissue loss and deconditioning during prolonged stay in space [2]. It is anticipated that further development of dynamic bioreactor conditions will enable tissue engineering applications to live up to their potential in regenerative medicine and disease studies.

Further probing of the fundamental roles of the direct and indirect effects of gravity in the development of tissues and their transition between normal homeostasis and disease are critical for advancing our understanding of how tissues develop and function. With constantly improving control of molecular and physical factors in the cellular environment, further studies are expected to lead to increasing physiological relevance of the collected data. Spaceflight makes us think creatively, catalyzing the development of products and technologies that would not be available otherwise.

References

1. Freed, L. E., & Vunjak-Novakovic, G. (1995). Cultivation of cell-polymer tissue constructs in simulated microgravity. *Biotechnology and Bioengineering, 46*(4), 306–313.
2. Freed, L. E., Langer, R., Martin, I., Pellis, N. R., & Vunjak-Novakovic, G. (1997). Tissue engineering of cartilage in space. *Proceedings of the National Academy of Sciences of the United States of America, 94*(25), 13885–13890.
3. Saltzman, W. M. (1997). Weaving cartilage at zero g: The reality of tissue engineering in space. *Proceedings of the National Academy of Sciences of the United States of America, 94*(25), 13380–13382.
4. Vastag, B. (2001). Cell biology update: A decade of simulating space on earth. *JAMA, 285*(17), 2181–2182.
5. Freed, L. E., Hollander, A. P., Martin, I., Barry, J. R., Langer, R., & Vunjak-Novakovic, G. (1998). Chondrogenesis in a cell-polymer-bioreactor system. *Experimental Cell Research, 240*(1), 58–65.
6. Obradovic, B., Carrier, R. L., Vunjak-Novakovic, G., & Freed, L. E. (1999). Gas exchange is essential for bioreactor cultivation of tissue engineered cartilage. *Biotechnology and Bioengineering, 63*(2), 197–205.
7. Martin, I., Obradovic, B., Treppo, S., Grodzinsky, A. J., Langer, R., Freed, L. E., et al. (2000). Modulation of the mechanical properties of tissue engineered cartilage. *Biorheology, 37*(1–2), 141–147.
8. Barzegari, A., & Saei, A. A. (2012). An update to space biomedical research: Tissue engineering in microgravity bioreactors. *Bioimpacts, 2*(1), 23–32.
9. Obradovic, B., Martin, I., Padera, R. F., Treppo, S., Freed, L. E., & Vunjak-Novakovic, G. (2001). Integration of engineered cartilage. *Journal of Orthopaedic Research, 19*(6), 1089–1097.
10. Conza, N., Mainil-Varlet, P., Rieser, F., Kraemer, J., Bittmann, P., Huijser, R., et al. (2001). Tissue engineering in space. *Journal of Gravitational Physiology, 8*(1), 17–20.
11. Lewis, M. L., & Hughes-Fulford, M. (1997). Response of living systems to spaceflight. In S. Churchill (Ed.), *Fundamentals of space life sciences* (pp. 21–39). Malabar: Krieger Publishing Company.

12. Lappa, M. (2003). Organic tissues in rotating bioreactors: Fluid-mechanical aspects, dynamic growth models, and morphological evolution. *Biotechnology and Bioengineering, 84*(5), 518–532.
13. Ohyabu, Y., Kida, N., Kojima, H., Taguchi, T., Tanaka, J., & Uemura, T. (2006). Cartilaginous tissue formation from bone marrow cells using rotating wall vessel (RWV) bioreactor. *Biotechnology and Bioengineering, 95*(5), 1003–1008.
14. Sakai, S., Mishima, H., Ishii, T., Akaogi, H., Yoshioka, T., Ohyabu, Y., et al. (2009). Rotating three-dimensional dynamic culture of adult human bone marrow-derived cells for tissue engineering of hyaline cartilage. *Journal of Orthopaedic Research, 27*(4), 517–521.
15. Wu, X., Li, S. H., Lou, L. M., & Chen, Z. R. (2013). The effect of the microgravity rotating culture system on the chondrogenic differentiation of bone marrow mesenchymal stem cells. *Molecular Biotechnology, 54*(2), 331–336.
16. Rungarunlert, S., Klincumhom, N., Bock, I., Nemes, C., Techakumphu, M., Pirity, M. K., et al. (2011). Enhanced cardiac differentiation of mouse embryonic stem cells by use of the slow-turning, lateral vessel (STLV) bioreactor. *Biotechnology Letters, 33*(8), 1565–1573.
17. Nerem, R. M. (1992). Tissue engineering in the USA. *Medical and Biological Engineering and Computing, 30*(4), CE8–CE12.
18. Vunjak-Novakovic, G., & Scadden, D. T. (2011). Biomimetic platforms for human stem cell research. *Cell Stem Cell, 8*(3), 252–261.
19. Vunjak-Novakovic, G., Bhatia, S., Chen, C., & Hirschi, K. (2013). HeLiVa platform: Integrated heart-liver-vascular systems for drug testing in human health and disease. *Stem Cell Research & Therapy, 4*(1 Suppl), S8. doi:10.1186/scrt369.
20. Langer, R., & Vacanti, J. P. (1993). Tissue engineering. *Science, 260*(5110), 920–926.
21. Buckey, J. C., Jr., Lane, L. D., Levine, B. D., Watenpaugh, D. E., Wright, S. J., Moore, W. E., et al. (1996). Orthostatic intolerance after spaceflight. *Journal of Applied Physiology, 81*(1), 7–18.
22. Lang, T., LeBlanc, A., Evans, H., Lu, Y., Genant, H., & Yu, A. (2004). Cortical and trabecular bone mineral loss from the spine and hip in long-duration spaceflight. *Journal of Bone and Mineral Research, 19*(6), 1006–1012.
23. Vico, L., Collet, P., Guignandon, A., Lafage-Proust, M. H., Thomas, T., Rehaillia, M., et al. (2000). Effects of long-term microgravity exposure on cancellous and cortical weight-bearing bones of cosmonauts. *Lancet, 355*(9215), 1607–1611.
24. Vaney, M. C., Maignan, S., Riès-Kautt, M., & Ducruix, A. (1996). High-Resolution Structure (1.33 Å) of a HEW Lysozyme Tetragonal Crystal Grown in the APCF Apparatus. Data and Structural Comparison with a Crystal Grown under Microgravity from SpaceHab-01 Mission. *Acta Crystallographica. Section D, 52*(3), 505–517.
25. Jixiang, Z., Min, L., Rogers, R., Meyer, W., Ottewill, R. H., Crew, S.-S. S., et al. (1997). Crystallization of hard-sphere colloids in microgravity. *Nature, 387*, 883–885.
26. Goree, J., Morfill, G. E., Tsytovich, V. N., & Vladimirov, S. V. (1999). Theory of dust voids in plasmas. *Physical Review. E, Statistical Physics, Plasmas, Fluids, and Related Interdisciplinary Topics, 59*(6), 7055–7067.
27. Herlach, D. M., Cohrane, R. F., Egry, I., Fecht, H. J., & Greer, A. L. (1993). Containerless processing in the study of metallic melts and their solidification. *International Materials Review, 38*, 273–347.
28. Rhim, W.-K., Chung, S. K., Barber, D., Man, K. F., Gutt, G., Rulison, A., et al. (1993). An electrostatic levitator for high-temperature containerless materials processing in 1g. *Review of Scientific Instruments, 64*, 7055–7067.
29. Unsworth, B. R., & Lelkes, P. I. (1998). Growing tissues in microgravity. *Nature Medicine, 4*(8), 901–907.
30. Pellis, N. R., Goodwin, T. J., Risin, D., McIntyre, B. W., Pizzini, R. P., Cooper, D., et al. (1997). Changes in gravity inhibit lymphocyte locomotion through type I collagen. *In Vitro Cellular & Developmental Biology. Animal, 33*(5), 398–405.
31. Freed, L. E., Rupnick, M. A., Schaefer, D., & Vunjak-Novakovic, G. (2003). Engineering functional cartilage and cardiac tissue: In vitro culture parameters. In F. Guilak, D. Butler, D. Mooney, & S. Goldstein (Eds.), *Functional tissue engineering: The role of biomechanics* (pp. 360–376). New York: Springer.
32. Freed, L. E., & Vunjak-Novakovic, G. (2002). Spaceflight bioreactor studies of cells and tissues. In *Advances in space biology and medicine* (pp. 177–95). Elsevier Science.

33. Vunjak Novakovic, G., Eschenhagen, T., & Mummery, C. (2014). Myocardial Tissue Engineering: in vitro models. In *The Biology of the Heart (textbook)*. Cold Spring Harbor Laboratories 4(3): a014076.
34. Grayson, W., Chao, P. G., Marolt, D., Radisic, M., Cannizzaro, C., Figallo, E., et al. (2007). Bioreactors for tissue engineering and regenerative medicine. In J. J. Mao, G. Vunjak-Novakovic, A. Mikos, & A. Atala (Eds.), *Translational approaches in tissue engineering and regenerative medicine* (pp. 353–374). Norwood: Artech House.
35. Freed, L. E., & Vunjak-Novakovic, G. (1997). Microgravity tissue engineering. *In Vitro Cellular & Developmental Biology, 33*, 381–385.
36. Burdick, J. A., & Vunjak-Novakovic, G. (2009). Engineered microenvironments for controlled stem cell differentiation. *Tissue Engineering. Part A, 15*(2), 205–219.
37. Freed, L. E., Martin, I., & Vunjak-Novakovic, G. (1999). Frontiers in tissue engineering. In vitro modulation of chondrogenesis. In *Clinical Orthopaedics and Related Research*, (367 Suppl), S46–S58.
38. Kawamura, S., Wakitani, S., Kimura, T., Maeda, A., Caplan, A. I., Shino, K., et al. (1998). Articular cartilage repair. Rabbit experiments with a collagen gel-biomatrix and chondrocytes cultured in it. *Acta Orthopaedica Scandinavica, 69*(1), 56–62.
39. Radisic, M., Park, H., Shing, H., Consi, T., Schoen, F. J., Langer, R., et al. (2004). Functional assembly of engineered myocardium by electrical stimulation of cardiac myocytes cultured on scaffolds. *Proceedings of the National Academy of Sciences of the United States of America, 101*(52), 18129–18134.
40. Zhang, J., Wilson, G. F., Soerens, A. G., Koonce, C. H., Yu, J., Palecek, S. P., et al. (2009). Functional cardiomyocytes derived from human induced pluripotent stem cells. *Circulation Research, 104*(4), e30–e41.
41. Moutos, F. T., Freed, L. E., & Guilak, F. (2007). A biomimetic three-dimensional woven composite scaffold for functional tissue engineering of cartilage. *Nature Materials, 6*(2), 162–167.
42. Engelmayr, G. C., Jr., Cheng, M., Bettinger, C. J., Borenstein, J. T., Langer, R., & Freed, L. E. (2008). Accordion-like honeycombs for tissue engineering of cardiac anisotropy. *Nature Materials, 7*(12), 1003–1010.
43. Godier-Furnemont, A. F., Martens, T. P., Koeckert, M. S., Wan, L., Parks, J., Arai, K., et al. (2011). Composite scaffold provides a cell delivery platform for cardiovascular repair. *Proceedings of the National Academy of Sciences of the United States of America, 108*(19), 7974–7979.
44. Bhumiratana, S., Bernhard, J., Cimetta, E., Vunjak-Novakovic, G. (2013). Principles of bioreactor design for tissue engineering. In: R. Lanza, R. Langer, & J. P. Vacanti (Eds.), *Principles of tissue engineering* (4th ed., pp. 261–78). Academic.
45. Jessup, J. M., Frantz, M., Sonmez-Alpan, E., Locker, J., Skena, K., Waller, H., et al. (2000). Microgravity culture reduces apoptosis and increases the differentiation of a human colorectal carcinoma cell line. *In Vitro Cellular & Developmental Biology. Animal, 36*(6), 367–373.
46. Hammond, T. G., Benes, E., O'reilly, K. C., Wolf, D. A., Linnehan, R. M., Taher, A., et al. (2000). Mechanical culture conditions effect gene expression: Gravity-induced changes on the space shuttle. *Physiological Genomics, 3*(3), 163–173.
47. Schwarz, R. P., Goodwin, T. J., & Wolf, D. A. (1992). Cell culture for three-dimensional modeling in rotating-wall vessels: An application of simulated microgravity. *Journal of Tissue Culture Methods, 14*(2), 51–57.
48. Freed, L. E., Marquis, J. C., Langer, R., & Vunjak-Novakovic, G. (1994). Kinetics of chondrocyte growth in cell-polymer implants. *Biotechnology and Bioengineering, 43*(7), 597–604.
49. Vunjak-Novakovic, G., Obradovic, B., Martin, I., & Freed, L. E. (2002). Bioreactor studies of native and tissue engineered cartilage. *Biorheology, 39*(1–2), 259–268.
50. Wilson, J. W., Ramamurthy, R., Porwollik, S., McClelland, M., Hammond, T., Allen, P., et al. (2002). Microarray analysis identifies Salmonella genes belonging to the low-shear modeled microgravity regulon. *Proceedings of the National Academy of Sciences of the United States of America, 99*(21), 13807–13812.

51. Duke, J., Daane, E., Arizpe, J., & Montufar-Solis, D. (1996). Chondrogenesis in aggregates of embryonic limb cells grown in a rotating wall vessel. *Advances in Space Research, 17*(6–7), 289–293.
52. Stamenkovic, V., Keller, G., Nesic, D., Cogoli, A., & Grogan, S. P. (2010). Neocartilage formation in 1 g, simulated, and microgravity environments: Implications for tissue engineering. *Tissue Engineering. Part A, 16*(5), 1729–1736.
53. Emin, N., Koc, A., Durkut, S., Elcin, A. E., & Elcin, Y. M. (2008). Engineering of rat articular cartilage on porous sponges: Effects of TGF-beta 1 and microgravity bioreactor culture. *Artificial Cells, Blood Substitutes, and Immobilization Biotechnology, 36*(2), 123–137.
54. Ulbrich, C., Westphal, K., Pietsch, J., Winkler, H. D., Leder, A., Bauer, J., et al. (2010). Characterization of human chondrocytes exposed to simulated microgravity. *Cellular Physiology and Biochemistry, 25*(4–5), 551–560.
55. Aleshcheva, G., Sahana, J., Ma, X., Hauslage, J., Hemmersbach, R., Egli, M., et al. (2013). Changes in morphology, gene expression and protein content in chondrocytes cultured on a random positioning machine. *PLoS One, 8*(11), e79057.
56. Akins, R. E., Schroedl, N. A., Gonda, S. R., & Hartzell, C. R. (1997). Neonatal rat heart cells cultured in simulated microgravity. *In Vitro Cellular & Developmental Biology. Animal, 33*(5), 337–343.
57. Akins, R. E., Boyce, R. A., Madonna, M. L., Schroedl, N. A., Gonda, S. R., McLaughlin, T. A., et al. (1999). Cardiac organogenesis in vitro: Reestablishment of three-dimensional tissue architecture by dissociated neonatal rat ventricular cells. *Tissue Engineering, 5*(2), 103–118.
58. Papadaki, M., Bursac, N., Langer, R., Merok, J., Vunjak-Novakovic, G., & Freed, L. E. (2001). Tissue engineering of functional cardiac muscle: Molecular, structural, and electrophysiological studies. *American Journal of Physiology. Heart and Circulatory Physiology, 280*(1), H168–H178.

Part III
Microbial Investigations

Chapter 10
Microbial Investigations: Overview

Duane L. Pierson and C. Mark Ott

Historical Landmarks

1970—A crewmember aboard Apollo 13 suffers from a severe pseudomonas urinary tract infection demonstrating the potential risk to crew health and need to understand infectious disease during spaceflight [1, 2].

1982—*Staphylococcus aureus* and *Escherichia coli* that were cultured during the Cytos 2 experiment aboard Salyut 7 displayed an increased resistance to antibiotics compared to ground controls [3].

1998—Long et al. report enhanced rhinovirus replication in HeLa cells cultured in the Rotating Wall Vessel (RWV) bioreactor as compared to static cultures or cells grown in roller bottles [4].

1999—Based upon Space Shuttle investigations of *Bacillus subtilis* and *E. coli* using liquid and semi-solid growth media, Kacena et al. report that the differences in microbial response are likely the result of external physical forces, such as fluid dynamics or extracellular transport [5].

2000—Nickerson et al. report increased virulence and stress resistance in *Salmonella enterica* serovar Typhimurium cultured in the RWV bioreactor compared to identical cultures grown in a reoriented control [6].

2002—In the first use of whole genome microarray analysis of a spaceflight analogue culture, Wilson et al. identify 163 differentially expressed genes in *S.* Typhimurium cultured in the RWV compared to identical cultures grown in a reoriented control [7].

(continued)

D.L. Pierson • C. Mark Ott (✉)
Biomedical Research and Environmental Sciences Division, NASA/Johnson Space Center, 2101 NASA Parkway, Mail: SK4, Houston, TX 77058, USA
e-mail: c.m.ott@nasa.gov

C.A. Nickerson et al. (eds.), *Effect of Spaceflight and Spaceflight Analogue Culture on Human and Microbial Cells*, DOI 10.1007/978-1-4939-3277-1_10

(continued)

2007—Nauman et al. report a correlation between fluid shear levels experienced by pathogens in the RWV and those naturally encountered in the infected host [8]. Different microbial responses in *Salmonella* were observed as a function of incremental changes in fluid shear.

2007—The increased virulence and key changes in gene expression in *S.* Typhimurium observed in the RWV are confirmed in a spaceflight experiment aboard the Space Shuttle. This was the first study to examine the effect of spaceflight on microbial virulence, obtain the entire gene expression profile of a bacterium to spaceflight, and identify a global regulatory protein, Hfq, as having a role in this response [9].

2008—Independent validation of increased *S.* Typhimurium virulence when cultured during spaceflight was reproduced aboard Space Shuttle mission, STS-123. This mission also demonstrated that the altered virulence was dependent on media ion concentration [10].

2013—Brinley et al. reported culture in the RWV decreased DNA repair in cells infected with Epstein Barr virus (EBV) and greater DNA damage from radiation compared to EBV negative cells [11].

10.1 Introduction

Microorganisms, including bacteria, fungi, viruses, and protozoa, are exceedingly small organisms that generally cannot be seen by the unaided eye, and are ubiquitous on Earth. They are found in and adapted to every ecological niche of the Earth including hot, dry deserts, the extreme low temperatures of Antarctica, the depths and pressures of the ocean, in the atmosphere, and subterranean as well as the more favorable temperate and tropical zones. Thus, the concept of microbial response in and adaptation to the spaceflight environment is a logical assumption.

Microbes have been present on every human occupied spacecraft or space habitat and will play essential roles in long term habitation of humans in space. In addition to their role in crew health, microorganisms may also be useful in the remediation of solid and liquid wastes, water and air purification, and food sources. Their powerful biosynthetic and biodegradative metabolic capabilities will likely be harnessed for long term human habitation in space.

From a crew health perspective, microorganisms have often been described as the enemy when in fact most microbes are beneficial and do not threaten our health; indeed, relatively few species adversely affect our well-being. The relationship between the human host and the accompanying microbiome is one of remarkable symbiosis. Our microbiota assists in food digestion, absorption, vitamin production, and nutrient release along with many other functions [12]. It is now understood that the microbes are essential to development of an effective immune system [13]. Olszak et al. [14] have identified the value of microbes of the gut to stimulate natural killer cells to develop tolerance without which some immune-related diseases such as irritable bowel syndrome and asthma can result. Knowledge of the human microbiome is expanding rapidly due to efforts, such as the NIH Human Microbiome Project (http://commonfund.nih.gov/hmp/index) and the European program, Metagenomics of the Human Intestinal Tract—MetaHIT (http://www.metahit.eu/). It is now understood that the human organism consists of approximately one trillion human cells and approximately 10 trillion microorganisms. This translates into several pounds of microorganisms. Perhaps more important is that the human microbiome consists of 5–8 million genes compared to about 20,000 genes in the human genome [15]. NASA has joined in the research to understand alterations in the astronaut microbiome that result from spaceflight. As on Earth, the interactions between the human during spaceflight and their microbes ultimately determine health or disease for the host. Thus, understanding how microorganisms respond to spaceflight and how this alters their interaction with the crew during a mission is of critical importance for mission success.

How microorganisms function in the microgravity environment associated with spaceflight had been an open question for 50 years [16]. A variety of unique microbial responses have been observed in experiments throughout the spaceflight program [16–18]; however, the impact of these responses on medical operations was never thoroughly investigated. Answers were slow coming for several reasons, including a low priority for microbial studies because of the perception that microbes have relatively small impact on short duration spaceflights on the Space Shuttle. However, as the goals of the National Aeronautics and Space Administration (NASA) have shifted to much longer exploration missions, the potential impact of microbes to human health and ultimately long term habitation in space became evident. Longer stays in space resulted in the realization of emerging problems associated with human occupation of the relatively small, closed environments of spacecraft/habitats.

10.2 The Rotating Wall Vessel Ground-Based Spaceflight Analogue

Another hurdle in our efforts to understand the cause and impact of spaceflight associated changes in microbial responses was the lack of suitable ground-based models to enhance preflight development and post-flight validation for relatively rare spaceflight opportunities. This limitation was overcome with the successful culture and analysis of microorganisms with the Rotating Wall Vessel (RWV) bioreactor (Fig. 10.1). As mentioned in Chap. 2, the RWV is a rotating bioreactor in which cells are maintained in suspension in a gentle fluid orbit that creates a sustained low fluid shear, low turbulence environment for cell growth—thus enabling the evaluation of cellular responses under conditions that simulate aspects of the microgravity environment [8, 17, 19–21]. The low-shear environment produced in the RWV has been coined low-shear modeled microgravity (LSMMG) [17]. This environment, designed to reproduce aspects of the spaceflight environment, also mimics the fluid shear levels encountered by pathogens in the host, including the intestinal, respiratory, and urogenital tracts, which are the most commonly infected sites in the body [8]. While originally designed for studies of mammalian cells in a

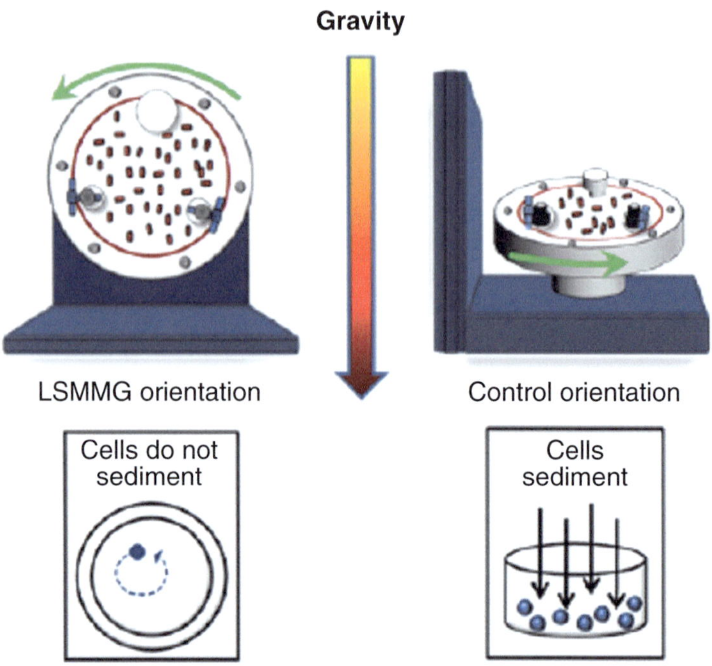

Fig. 10.1 LSMMG and control orientation of the RWV

space-like environment [20], the RWV quickly proved to also be an excellent tool for determining trends in microbial responses that are expected from microorganisms cultured in space [9]. The first RWV studies with microorganisms were conducted by Fang et al. [22] and demonstrated the value of the RWV for microbial studies. These early pioneering studies of *Escherichia coli* demonstrated that bacteria respond differently in response to LSMMG culture. Several additional studies by these researchers demonstrated the flexibility of this system in the investigation of various microorganisms [23–26], thus opening a path for future microbial studies with the RWV.

Proper operation of the RWV bioreactor requires the rotation of an aerated cylinder that is completely filled with liquid resulting in a solid-body rotation of the media within. This design allows a constant rotation normal to the gravitational field, resulting in an environmental culture condition that mimics several aspects of the spaceflight environment, e.g., low fluid shear across the microbial cell surface [17, 27]. For a meaningful comparative control, the LSMMG environment should be disrupted (e.g., create higher fluid shear) while maintaining other environmental conditions. To accomplish this type of control, many investigators reorient the RWV such that the axis of rotation is parallel to the gravity vector as a control (Fig. 10.1). An alternative method that has been used to disrupt the RWV culture environment is the addition of beads into the reactor vessel to modulate fluid shear [8, 22]. Previous mathematical modeling has demonstrated that the low fluid shear environ-

The Physiological Relevance of RWV Studies

Previous mathematical and computational modeling indicated that the fluid shear experienced by cells in the LSMMG environment is less than 0.01 dynes/cm^2 [8]. Thus, the environment of the RWV and the ability to incrementally increase shear levels in the RWV have particular physiological relevance. During the course of the infection process, pathogenic bacteria can encounter a wide range of fluctuations in fluid shear levels during the natural course of infection, ranging from 4 to 50 dynes/cm^2 along blood vessel walls [28] to less than 1 dyne/cm^2 in utero and between the brush border microvilli of epithelial cells [29–33]. As survival in these conditions is essential for pathogen proliferation, it is logical that these organisms would develop response mechanisms to adapt to their changing environment. Indeed, this type of response has been demonstrated by microorganisms in response to increased fluid shear [34, 35] and thus, the quiescent environment of the RWV and true spaceflight should be considered as potential stimuli for unique responses in microbial pathogens.

ment of the RWV can be incrementally increased by increasing the diameter of the bead in the vessel [8]. These findings are especially interesting as the authors showed that when exposed to incrementally higher physiological fluid shear levels in the RWV, cultures of *S.* Typhimurium displayed corresponding progressive changes in both phenotypic (acid and thermal stresses) and targeted gene expression responses [8]. These changes suggest the intriguing possibility that different fluid shear areas in the body encountered by pathogens during infection might serve as biophysical signals to trigger virulence responses not previously identified during growth under conventional culture conditions.

10.3 Development of Microbial Virulence Investigations

One additional factor that has driven interest in the microbial spaceflight studies is the investigation of the impact of spaceflight and spaceflight analogue culture on virulence, which was first addressed by Nickerson et al. in 2000 [6]. This landmark study demonstrated increased virulence in *Salmonella enterica* serotype Typhimurium when cultivated in the LSMMG environment and opened new avenues of microbial research in the RWV. Subsequently, Wilson et al. [9, 10] demonstrated global alterations in transcriptomic and proteomic expression profiles and higher levels of virulence in *S.* Typhimurium when cultured aboard the Space Shuttle in microgravity as compared to otherwise identical ground controls. A thorough description of these studies can be found in Chap. 11. This quantum leap in knowledge of bacterial function in the microgravity environment of space spawned a frenzy of microbial studies in the RWV and actual spaceflight by investigators using a variety of medically significant microorganisms, including *Pseudomonas aeruginosa* [36–40], *Candida albicans* [41–44], adherent-invasive *E. coli* [45], *Streptococcus pneumoniae* [46], and *Staphylococcus aureus* [47–50]. The *S.* Typhimurium virulence studies introduced entire classes of microbial genes/proteins involved in host interactions not previously identified during growth under conventional culture conditions on Earth, where the force of gravity can mask key cellular responses. As a result, multiple spaceflight investigations have followed to determine the medical relevance of how the unique microbial responses observed during spaceflight could increase infectious disease risk to the crew. These studies have implications beyond acute infections during spaceflight. In an intriguing study comparing EBV negative cells and EBV positive cells (derived from Burkitt's lymphoma containing an abortive EBV replicative system), Brinley et al. reported that culture of the EBV positive cells in the RWV decreased their DNA repair and increased resulting DNA damage and production of reactive oxygen species from radiation exposure compared to uninfected cells [11]. As EBV is a latent virus carried by the majority of the population, this study suggests those infected with EBV may have greater risk from radiation during spaceflight missions.

10.4 Conclusion

In summary, studies of microorganisms in the RWV and true spaceflight began as simple investigations to understand the basic phenotypic characteristics of microorganisms when cultured in the spaceflight environment. However, the combination of intriguing discoveries and advances in technology has rapidly stimulated the field of spaceflight and spaceflight analogue research, especially those studies investigating novel responses relevant to how microbes maintain the balance between homeostasis and disease causing potential. The application of this research is extensive, including Gram positive and Gram negative bacteria (including gastrointestinal, respiratory tract, and skin pathogens) and eukaryotic yeast. Taken together, the study of microorganisms in the spaceflight environment should continue to provide unique, exciting findings that are important to both our space exploration efforts and scientific advances for the general public on Earth.

Expanding Microbial Studies to Eukaryotic Opportunistic Pathogens
Isolation of *Candida albicans* from astronauts is not uncommon [51, 52]. Thus, the impact of spaceflight and spaceflight analogue culture on this opportunistic pathogen holds great interest when considering risk assessment on long duration missions. In early studies using the RWV, DNA microarray analysis indicated that culture in LSMMG induced differential gene expression in 1372 genes when compared to a reoriented control [41]. These genes

(continued)

(continued)

were globally distributed and were associated with budding, cell polarity establishment, and cell separation. In a separate study, growth in the RWV produced a random budding pattern, as opposed to the expected bipolar budding, and an increase in the filamentous form compared to reoriented controls [42]. The increase in filamentous morphology is notable as transition to this form has been hypothesized to play a role in adhesion to host tissue and virulence of this organism. When cultured during spaceflight, *C. albicans* displayed 452 differentially regulated genes compared to otherwise identical ground controls. Among these were genes involved in cell aggregation, induction of ABC transporters, downregulation of ergosterol-encoding genes, downregulation of genes involved in actin cytoskeleton, and upregulation of genes involved in oxidative stress resistance [43]. As had been observed in the RWV, the spaceflight-cultured organisms displayed random budding pattern. While no differences were observed between mice infected by spaceflight or ground-based control cultures, these molecular genetic changes are indicative of a general stress response and may provide novel insight into the way this organism causes disease on Earth.

References

1. Taylor, G. R. (1974). Space microbiology. *Annual Review of Microbiology, 28*, 121–137.
2. Ferguson, J. K., Taylor, G. R., & Mieszkuc, B. J. (1975). Microbiological investigations. In R. S. Johnson, L. F. Dietlein, & C. A. Berry (Eds.), *Biomedical results of Apollo* (pp. 83–103). Washington: Scientific and Technical Information Office, NASA.
3. Tixador, R., Richoilley, G., Gasset, G., Templier, J., Bes, J. C., Moatti, N., et al. (1985). Study of minimal inhibitory concentration of antibiotics on bacteria cultivated in vitro in space (Cytos 2 experiment). *Aviation, Space, and Environmental Medicine, 56*(8), 748–751.
4. Long, J. P., Pierson, S. S., & Hughes, J. H. (1998). Rhinovirus replication in HeLa cells cultured under conditions of simulated microgravity. *Aviation, Space, and Environmental Medicine, 69*, 851–856.
5. Kacena, M. A., Leonard, P. E., Todd, P., & Luttges, M. W. (1997). Low gravity and inertial effects on the growth of *E. coli* and *B. subtilis* in semi-solid media. *Aviation, Space, and Environmental Medicine, 68*(12), 1104–1108.
6. Nickerson, C. A., Ott, C. M., Mister, S. J., Morrow, B. J., Burns-Keliher, L., & Pierson, D. L. (2000). Microgravity as a novel environmental signal affecting *Salmonella enterica* serovar Typhimurium virulence. *Infection and Immunity, 68*(6), 3147–3152.
7. Wilson, J. W., Ramamurthy, R., Porwollik, S., McClelland, M., Hammond, T., Allen, P., et al. (2002). Microarray analysis identifies *Salmonella* genes belonging to the low-shear modeled microgravity regulon. *Proceedings of the National Academy of Sciences of the United States of America, 99*(21), 13807–13812.
8. Nauman, E. A., Ott, C. M., Sander, E., Tucker, D. L., Pierson, D., Wilson, J. W., et al. (2007). Novel quantitative biosystem for modeling physiological fluid shear stress on cells. *Applied and Environmental Microbiology, 73*(3), 699–705.
9. Wilson, J. W., Ott, C. M., Honer zu Bentrup, K., Ramamurthy, R., Quick, L., Porwollik, S., et al. (2007). Space flight alters bacterial gene expression and virulence and reveals a role for global regulator Hfq. *Proceedings of the National Academy of Sciences of the United States of America, 104*(41), 16299–16304.

10. Wilson, J. W., Ott, C. M., Quick, L., Davis, R., Honer zu Bentrup, K., Crabbe, A., et al. (2008). Media ion composition controls regulatory and virulence response of Salmonella in spaceflight. *PLoS One, 3*(12), e3923.
11. Brinley, A. A., Theriot, C. A., Nelman-Gonzalez, M., Crucian, B., Stowe, R. P., Barrett, A. D., et al. (2013). Characterization of Epstein-Barr virus reactivation in a modeled spaceflight system. *Journal of Cellular Biochemistry, 114*(3), 616–624.
12. Nicholson, J. K., Holmes, E., Kinross, J., Burcelin, R., Gibson, G., Jia, W., et al. (2012). Host-gut microbiota metabolic interactions. *Science, 336*(6086), 1262–1267.
13. Cerf-Bensussan, N., & Gaboriau-Routhiau, V. (2010). The immune system and the gut microbiota: Friends or foes? *Nature Reviews Immunology, 10*(10), 735–744.
14. Olszak, T., An, D., Zeissig, S., Vera, M. P., Richter, J., Franke, A., et al. (2012). Microbial exposure during early life has persistent effects on natural killer T cell function. *Science, 336*(6080), 489–493.
15. Balter, M. (2012). Taking stock of the human microbiome and disease. *Science, 336*(6086), 1246–1247.
16. Dickson, K. J. (1991). Summary of biological spaceflight experiments with cells. *ASGSB Bulletin, 4*(2), 151–260.
17. Nickerson, C. A., Ott, C. M., Wilson, J. W., Ramamurthy, R., & Pierson, D. L. (2004). Microbial responses to microgravity and other low-shear environments. *Microbiology and Molecular Biology Reviews, 68*(2), 345–361.
18. Horneck, G., Klaus, D. M., & Mancinelli, R. L. (2010). Space microbiology. *Microbiology and Molecular Biology Reviews, 74*(1), 121–156.
19. Nickerson, C. A., Ott, C. M., Wilson, J. W., Ramamurthy, R., LeBlanc, C. L., Honer zu Bentrup, K., et al. (2003). Low-shear modeled microgravity: A global environmental regulatory signal affecting bacterial gene expression, physiology, and pathogenesis. *Journal of Microbiological Methods, 54*(1), 1–11.
20. Wolf, D. A., & Schwarz, R. P. (1991). *Analysis of gravity-induced particle motion and fluid perfusion flow in the NASA-designed rotating zero-head space tissue culture vessel.*
21. Unsworth, B. R., & Lelkes, P. I. (1998). Growing tissues in microgravity. *Nature Medicine, 4*(8), 901–907.
22. Fang, A., Pierson, D. L., Koenig, D. W., Mishra, S. K., & Demain, A. L. (1997). Effect of simulated microgravity and shear stress on microcin B17 production by *Escherichia coli* and on its excretion into the medium. *Applied and Environmental Microbiology, 63*(10), 4090–4092.
23. Fang, A., Pierson, D. L., Mishra, S. K., Koenig, D. W., & Demain, A. L. (1997). Secondary metabolism in simulated microgravity: Beta-lactam production by *Streptomyces clavuligerus. Journal of Industrial Microbiology and Biotechnology, 18*(1), 22–25.
24. Fang, A., Pierson, D. L., Mishra, S. K., Koenig, D. W., & Demain, A. L. (1997). Gramicidin S production by *Bacillus brevis* in simulated microgravity. *Current Microbiology, 34*(4), 199–204.
25. Fang, A., Pierson, D. L., Mishra, S. K., & Demain, A. L. (2000). Relief from glucose interference in microcin B-17 biosynthesis by growth in a rotating-wall bioreactor. *Letters in Applied Microbiology, 31*, 39–41.
26. Gao, Q., Fang, A., Pierson, D. L., Mishra, S. K., & Demain, A. L. (2001). Shear stress enhances microcin B17 production in a rotating wall bioreactor, but ethanol stress does not. *Applied Microbiology and Biotechnology, 56*(3–4), 384–387.
27. Klaus, D. (2001). Clinostats and bioreactors. *Gravitational and Space Biology Bulletin, 14*(2), 55–64.
28. Davies, D. G., & Geesey, G. G. (1995). Regulation of the alginate biosynthesis gene algC in Pseudomonas aeruginosa during biofilm development in continuous culture. *Applied and Environmental Microbiology, 61*(3), 860–867.
29. Beeson, J. G., Rogerson, S. J., Cooke, B. M., Reeder, J. C., Chai, W., Lawson, A. M., et al. (2000). Adhesion of Plasmodium falciparum-infected erythrocytes to hyaluronic acid in placental malaria. *Nature Medicine, 6*(1), 86–90.

30. Cai, Z., Xin, J., Pollock, D. M., & Pollock, J. S. (2000). Shear stress-mediated NO production in inner medullary collecting duct cells. *American Journal of Physiology Renal Physiology, 279*(2), F270–F274.
31. Creasy, R. K., & Reznik, R. (1984). *Maternal-fetal medicine: Principles and practice*. Philadelphia: WB Saunders Company.
32. Guo, P., Weinstein, A. M., & Weinbaum, S. (2000). A hydrodynamic mechanosensory hypothesis for brush border microvilli. *American Journal of Physiology Renal Physiology, 279*(4), F698–F712.
33. Stock, U. A., & Vacanti, J. P. (2001). Cardiovascular physiology during fetal development and implications for tissue engineering. *Tissue Engineering, 7*(1), 1–7.
34. Thomas, W. E., Nilsson, L. M., Forero, M., Sokurenko, E. V., & Vogel, V. (2004). Shear-dependent 'stick-and-roll' adhesion of type 1 fimbriated *Escherichia coli. Molecular Microbiology, 53*(5), 1545–1557.
35. Thomas, W., Forero, M., Yakovenko, O., Nilsson, L., Vicini, P., Sokurenko, E., et al. (2006). Catch-bond model derived from allostery explains force-activated bacterial adhesion. *Biophysical Journal, 90*(3), 753–764.
36. Crabbe, A., De Boever, P., Van Houdt, R., Moors, H., Mergeay, M., & Cornelis, P. (2008). Use of the rotating wall vessel technology to study the effect of shear stress on growth behaviour of *Pseudomonas aeruginosa* PA01. *Environmental Microbiology, 10*(8), 2098–2110.
37. Crabbe, A., Pycke, B., Van Houdt, R., Monsieurs, P., Nickerson, C., Leys, N., et al. (2010). Response of *Pseudomonas aeruginosa* PAO1 to low shear modelled microgravity involves AlgU regulation. *Environmental Microbiology, 12*(6), 1545–1564.
38. Crabbe, A., Schurr, M. J., Monsieurs, P., Morici, L., Schurr, J., Wilson, J. W., et al. (2011). Transcriptional and proteomic response of *Pseudomonas aeruginosa* PAO1 to spaceflight conditions involves Hfq regulation and reveals a role for oxygen. *Applied and Environmental Microbiology, 77*(4), 1221–1230.
39. Kim, W., Tengra, F. K., Young, Z., Shong, J., Marchand, N., Chan, H. K., et al. (2013). Spaceflight promotes biofilm formation by *Pseudomonas aeruginosa. PLoS One, 8*(4), e62437.
40. Kim, W., Tengra, F. K., Shong, J., Marchand, N., Chan, H. K., Young, Z., et al. (2013). Effect of spaceflight on *Pseudomonas aeruginosa* final cell density is modulated by nutrient and oxygen availability. *BMC Microbiology, 13*, 241.
41. Sheehan, K. B., McInnerney, K., Purevdorj-Gage, B., Altenburg, S. D., & Hyman, L. E. (2007). Yeast genomic expression patterns in response to low-shear modeled microgravity. *BMC Genomics, 8*, 3.
42. Altenburg, S. D., Nielsen-Preiss, S. M., & Hyman, L. E. (2008). Increased filamentous growth of Candida albicans in simulated microgravity. *Genomics, Proteomics & Bioinformatics, 6*(1), 42–50.
43. Crabbe, A., Nielsen-Preiss, S. M., Woolley, C. M., Barrila, J., Buchanan, K., McCracken, J., et al. (2013). Spaceflight enhances cell aggregation and random budding in Candida albicans. *PLoS One, 8*(12), e80677.
44. Yamazaki, T., Yoshimoto, M., Nishiyama, Y., Okubo, Y., & Makimura, K. (2012). Phenotypic characterization of *Aspergillus niger* and Candida albicans grown under simulated microgravity using a three-dimensional clinostat. *Microbiology and Immunology, 56*, 441–446.
45. Allen, C. A., Niesel, D. W., & Torres, A. G. (2008). The effects of low-shear stress on adherent-invasive *Escherichia coli. Environmental Microbiology, 10*(6), 1512.
46. Allen, C. A., Galindo, C. L., Pandya, U., Watson, D. A., Chopra, A. K., & Niesel, D. W. (2007). Transcription profiles of *Streptococcus pneumoniae* grown under different conditions of normal gravitation. *Acta Astronautica, 60*, 433–444.
47. Castro, S. L., Nelman-Gonzalez, M., Nickerson, C. A., & Ott, C. M. (2011). Induction of attachment-independent biofilm formation and repression of Hfq expression by low-fluid-shear culture of *Staphylococcus aureus. Applied and Environmental Microbiology, 77*(18), 6368–6378.

48. Rosado, H., Stapleton, P. D., & Taylor, P. W. (2006). Effect of simulated microgravity on the virulence properties of the opportunistic bacterial pathogen Staphylococcus aureus. In *Proceedings of the 57th International Astronautical Congress.*
49. Rosado, H., Doyle, M., Hinds, J., & Taylor, P. W. (2010). Low-shear modelled microgravity alters expression of virulence determinants of *Staphylococcus aureus. Acta Astronautica, 66*, 408–416.
50. Rosado, H., O'Neill, A. J., Blake, K. L., Walther, M., Long, P. F., Hinds, J., et al. (2012). Rotating wall vessel exposure alters protein secretion and global gene expression in *Staphylococcus aureus. International Journal of Astrobiology, 11*(2), 71–81.
51. Pierson, D. L., Mehta, S. K., Magee, B. B., & Mishra, S. K. (1995). Person-to-person transfer of Candida albicans in the spacecraft environment. *Journal of Medical and Veterinary Mycology, 33*, 145–150.
52. Taylor, G. R. (1974). Recovery of medically important microorganisms from Apollo astronauts. *Aerospace Medicine, 45*(8), 824–828.

Chapter 11
Using Spaceflight and Spaceflight Analogue Culture for Novel Mechanistic Insight into *Salmonella* Pathogenesis

Jennifer Barrila, James W. Wilson, Anjali Soni, Jiseon Yang, C. Mark Ott, and Cheryl A. Nickerson

Historical Landmarks

1967—*Salmonella enterica* serovar Typhimurium (*S.* Typhimurium) BS-5 (P-22)/P-22 launched aboard U.S. Biosatellite II. Alterations in bacterial growth and prophage induction were observed as a function of spaceflight and gamma irradiation [1, 2].

2000—Nickerson et al., discover that Low Shear Modeled Microgravity (LSMMG) is a novel environmental regulatory factor of virulence, pathogenesis-related stress responses, and protein synthesis in *S.* Typhimurium [3].

2002—Wilson et al., use global microarray analysis to identify *S.* Typhimurium genes belonging to the LSMMG regulon [4].

2002—Wilson et al., demonstrate that the *rpoS* gene is not required for LSMMG-induced phenotypes in *S.* Typhimurium during late exponential phase culture [5].

(continued)

J. Barrila • J. Yang
The Biodesign Institute, Center for Infectious Diseases and Vaccinology, Arizona State University, Tempe, AZ, USA

J.W. Wilson • A. Soni
Department of Biology, Villanova University, Villanova, PA, USA

C. Mark Ott
Biomedical Research and Environmental Sciences Division, NASA Johnson Space Center, 2101 NASA Parkway, Mail Code: SK4, Houston, TX 77058, USA

C.A. Nickerson (✉)
School of Life Sciences, The Biodesign Institute, Center for Infectious Diseases and Vaccinology, Arizona State University, Tempe, AZ, USA
email: cheryl.nickerson@asu.edu

C.A. Nickerson et al. (eds.), *Effect of Spaceflight and Spaceflight Analogue Culture on Human and Microbial Cells*, DOI 10.1007/978-1-4939-3277-1_11

(continued)

2007—Wilson et al., report results from the MICROBE spaceflight experiment (Shuttle mission STS-115), which was the first investigation of the impact of true spaceflight culture on the virulence and global gene expression of any microbial pathogen. Spaceflight increased the virulence of *S.* Typhimurium and caused global changes in gene and protein expression in unique ways not observed using traditional approaches on Earth. The evolutionarily conserved RNA chaperone protein Hfq was identified as a likely global regulator involved in the response to this environment [6].

2007—Nauman et al., use the Rotating Wall Vessel (RWV) bioreactor to show that incremental increases in fluid shear lead to incremental alterations in pathogenesis-related stress responses and gene expression in *S.* Typhimurium [7].

2008—MDRV, a follow-up to the MICROBE experiment is performed on Shuttle mission STS-123. Wilson et al., independently validated the *Salmonella* virulence findings from the STS-115 experiment and further discovered that increases in the inorganic ion concentration in the growth medium during spaceflight culture prevent the spaceflight-associated increase in *Salmonella* virulence [8].

2010—First on-orbit infection of a human cell line is conducted during the STL-IMMUNE spaceflight experiment aboard Shuttle mission STS-131 [9]. Human intestinal epithelial cells (HT-29) were infected with *S.* Typhimurium and evaluated for global changes in gene and protein expression, and morphological changes.

2012—Pacello et al., show that catalases KatG and KatN are important for increased oxidative stress resistance of several *Salmonella* strains in response to LSMMG culture, but were independent of the *rpoS, rpoE, oxyR*, and *hfq* genes [10].

2015—The Micro-5/PHOENIX spaceflight experiment launches aboard SpaceX CRS-5 to the International Space Station (ISS) [11]. The study is a follow-up to the MICROBE and MDRV spaceflight experiments and is the first on-orbit infection ever to be monitored in real time. The model host organism *Caenorhabditis elegans* was infected on-orbit with *S.* Typhimurium cultured under different media ion concentrations.

11.1 Introduction

The genus *Salmonella* comprises a diverse group of Gram-negative enteric bacteria that are a major cause of morbidity and mortality worldwide [12–15]. It has been estimated that *Salmonella* species cause 93.8 million cases of gastroenteritis and 21 million cases of enteric fever annually, resulting in 350,000–750,000 deaths [12–14]. *Salmonella* serotypes are widely disseminated in the environment and are routinely found in mammals, birds, reptiles, and amphibians. Clinical manifestation of *Salmonella* infections can vary depending on the specific serovar and host, and can include gastroenteritis, fever, bacteremia, and chronic asymptomatic carriage [15]. The continual and rapid emergence of multidrug resistant strains continues to present challenges for the treatment of salmonellosis [16–19].

Transmission of *Salmonella* through contaminated food and water presents a serious health threat for astronauts during spaceflight missions, and this risk of exposure is expected to increase with longer mission duration and increased use of regenerative life support systems. Previous studies have demonstrated that the microbial populations aboard Mir and the International Space Station (ISS) are largely dominated by common environmental and commensal flora; however, clear routes for the transmission of pathogens have been identified, including in-flight exchange of microbial flora by the crew [20–22]. Similarly, while pre-flight evaluation of spaceflight food indicates that most foods do not contain medically significant organisms, pathogens (including *Salmonella* spp.) have been detected and have disqualified production lots from flight [20–22]. Even with strict microbial monitoring, *Salmonella* species were still found to be present in the crew refuse aboard Space Shuttle mission STS-108 [23]. This discovery, in combination with the decreased immune function observed in astronauts during spaceflight [24–38], further emphasizes the importance of ongoing and future investigations into the impact of the microgravity environment on the virulence of *Salmonella* and into possible countermeasures for the prevention and treatment of salmonellosis to mitigate infectious disease risk for the crew during spaceflight missions.

11.2 Low Shear Modeled Microgravity Alters the Virulence, Pathogenesis-Related Stress Responses, and Global Gene Expression of *S.* Typhimurium

***S.* Typhimurium was the first bacterial pathogen to be profiled for changes in virulence in response to either spaceflight or spaceflight analogue culture** [3, 6]. Prior to these studies, evidence had mounted showing that spaceflight and/or spaceflight analogue culture led to alterations in specific bacterial phenotypes including changes in growth, antibiotic resistance, and secondary metabolite expression and secretion [39–47]. However, none of these studies evaluated the ability for either spaceflight or spaceflight analogue culture to induce alterations in microbial virulence. Since *S.* Typhimurium is one of the best-characterized bacterial pathogens and routinely serves as a model system for studying fundamental mechanisms of microbial pathogenesis, it was a logical first candidate for evaluating the potential of the microgravity environment to induce alterations in microbial virulence.

The RWV bioreactor (Fig. 11.1) serves as a powerful ground-based tool for culturing cells under conditions that mimic aspects of the quiescent, low fluid shear culture environment found in microgravity (for a more detailed discussion of the operational principles of the RWV—see Chap. 2). This rotary culture system, when oriented in the low shear modeled microgravity (LSMMG) orientation (Fig. 11.1),

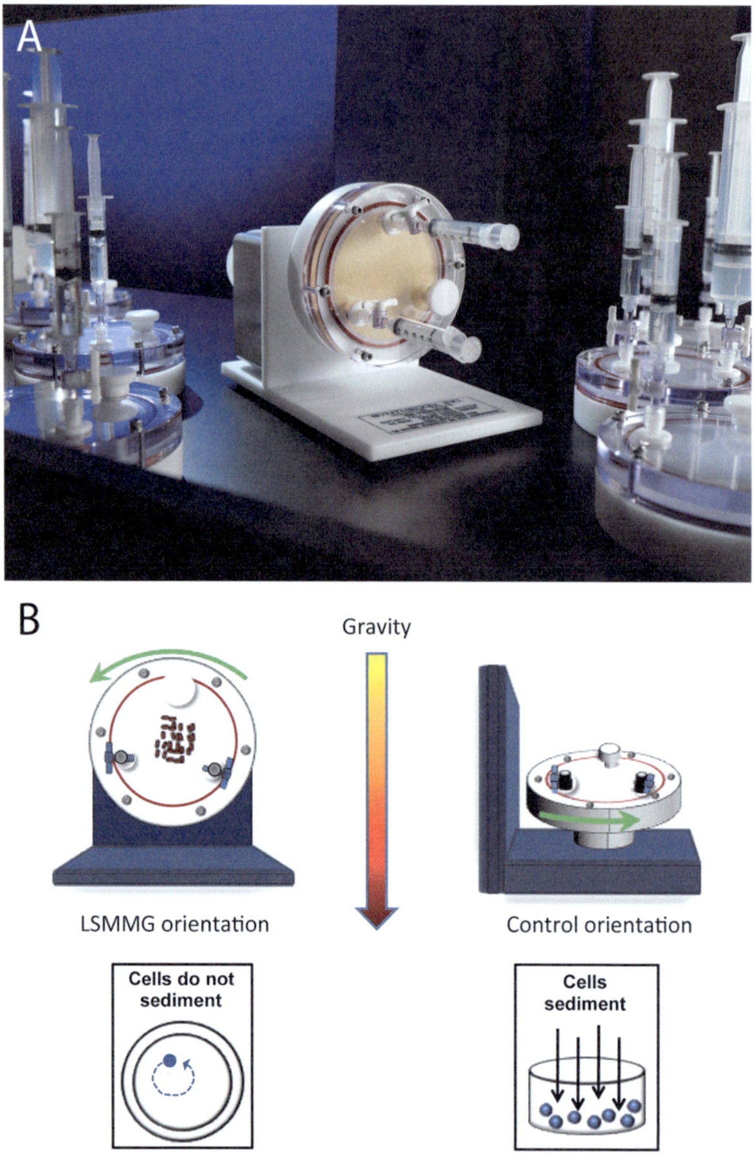

Fig. 11.1 Spaceflight analogue culture in the RWV bioreactor. (**a**) The RWV bioreactor provides a ground-based tool used for culturing cells under conditions that mimic aspects of the low fluid shear culture environment found in microgravity. The RWV is filled completely with cells and medium with zero headspace allowed to minimize fluid shear forces on the cells. (**b**) RWV bioreactor oriented in the LSMMG and control orientations, which rotate along axes perpendicular and parallel to the gravitational force vector, respectively. During LSMMG culture, there is a solid-body rotation of the medium, and thus cells do not sediment

provides an optimized form of suspension culture in which sedimentation is offset by the rotation of the medium, allowing the cells to grow under low fluid shear conditions similar to those encountered in the microgravity environment, as well as in certain areas of the human body [40, 48]. In the initial microgravity analogue studies performed with *S.* Typhimurium by Nickerson et al., [3], it was demonstrated that LSMMG culture led to decreased time-to-death, decreased LD_{50}, and increased tissue colonization relative to control cultures in the mouse model of infection. LSMMG cultures of *S.* Typhimurium also displayed increased survival in macrophages and increased resistance to acid stress, two critical stressors that *Salmonella* naturally encounters in vivo. Two-dimensional (2-D) protein gel electrophoresis indicated that 38 proteins were downregulated threefold or greater in LSMMG relative to the control cultures. Collectively, these observations established the paradigm that the LSMMG environment could alter the virulence, pathogenesis-related stress responses, and global protein expression profiles of microbial pathogens like *Salmonella* [3]. Consequently, these findings had important implications for the potential of true spaceflight culture to induce similar phenotypic responses, which if confirmed, would reveal a potentially dangerous gap in the known risks for the immunocompromised astronauts during spaceflight missions.

The next step was to begin to understand what genes and groups of genes were altered in response to LSMMG culture in order to identify the underlying molecular mechanisms involved in these phenotypic changes. To identify the *S.* Typhimurium genes that were altered in expression in response to LSMMG culture, whole genome microarray analysis was performed in a subsequent study using RNA harvested from *S.* Typhimurium cultures grown under LSMMG or control conditions [4]. The results demonstrated that 163 genes distributed globally across the *S.* Typhimurium genome were differentially regulated, and represented a variety of functional groups including lipopolysaccharide (LPS) synthesis, protein secretion systems, starvation/stress response, ribosomal subunits, virulence factors, iron utilization enzymes, transcriptional regulation, and several of unknown function. Secondary assays including RT-PCR analysis of differentially regulated genes and LPS gels confirmed the hits obtained from the microarray analysis. The majority of genes were downregulated in expression (97 genes vs. 68 genes) during LSMMG culture. Surprisingly, despite the LSMMG-induced increases in virulence, none of the up-regulated genes corresponded to known virulence factors. This finding suggests that LSMMG may alter the virulence of *Salmonella* by a previously uncharacterized mechanism(s). Alternatively, the increase in virulence may be the cumulative result of contributions from multiple genes of different functions that are differentially regulated as part of the global reprogramming of *Salmonella* under LSMMG conditions.

Since ferric uptake regulator (Fur) protein binding sites were found to be associated with many of the genes differentially regulated in the microarray analysis, an isogenic *S.* Typhimurium *fur* mutant strain was constructed and evaluated for LSMMG-induced alterations in acid stress resistance, with the results compared to those obtained with the wild type strain. Unlike the wild type, the *fur* mutant displayed no difference in acid stress resistance between the LSMMG and control cultures, indicating that the *fur* gene may play a role in the response of *S.* Typhimurium to LSMMG.

Given the global alterations in molecular genetic and phenotypic responses of *S.* Typhimurium to LSMMG culture, with strong correlations between changes in select environmental stress responses and virulence, it was hypothesized that the *rpoS* gene was a likely candidate for playing a role in LSMMG signal transmission, as it is a master global regulator of virulence and for a wide variety of stress responses in *Salmonella* [49, 50]. To test this hypothesis, an isogenic *rpoS* mutant derivative was extensively and systematically analyzed for LSMMG-induced alterations in several pathogenesis-related stress responses, and the results compared to the wild type parent strain [5]. This study provided several key pieces of information: (1) the *rpoS* gene is not required for *S.* Typhimurium to display LSMMG-induced phenotypes during the late exponential/early stationary phase of culture; (2) LSMMG alters the resistance of *S.* Typhimurium to a variety of other stresses that had not been previously tested, including osmotic, thermal, and oxidative stress; and (3) cells cultured in LSMMG in minimal media display a shorter lag phase and doubling time compared to control cultures.

As it became increasingly clear that LSMMG culture profoundly impacted the overall physiology and gene expression of *Salmonella* in ways that affected its ability to cause disease in the host, the next big question for researchers to address was to identify the potential physical stimulus (or stimuli) that is responsible for inducing these changes. One hypothesis, was that the low fluid shear conditions present during LSMMG culture may mimic the low fluid shear environment found in select regions of the human body that are normally encountered by *Salmonella* (e.g., in between brush border microvilli of intestinal epithelial cells [40, 51]) thus triggering a response or set of responses that signal increases in virulence and stress resistance. Based on this hypothesis, one would predict that disruption of the low shear environment in the RWV with higher levels of fluid shear would result in reversal of LSMMG-induced phenotypes. As such, it should follow that the resulting stress or molecular genetic responses would be altered in a manner that is directly proportional to the amount of fluid shear stress introduced into the system. To address this hypothesis, *S.* Typhimurium was exposed to a series of incrementally increasing fluid shear levels in the RWV by the incorporation of different sized inert beads and subsequently profiled for targeted changes in stress resistance and gene expression. As predicted, *S.* Typhimurium cultures were found to display the predicted progressive changes in acid and thermal stress responses and targeted gene expression profiles, including *rtsA*, a regulatory protein implicated in *Salmonella* intestinal invasion (Fig. 11.2). This was the first study to provide evidence that incremental changes in fluid shear can cause corresponding changes in biological responses in *S.* Typhimurium during the infection process. Ongoing studies are focused on further dissecting this relationship.

Fig. 11.2 (continued) its activity measured during growth of the fusion strain in the RWV with and without the 1/8-in. bead. As a control, a Km^R gene promoter predicted to be unaffected by the changes in shear was fused to *lacZ*. The data from three independent trials each performed with triplicate samples are shown for each strain. Standard errors of the means are given. Copyright © American Society for Microbiology, Applied and Environmental Microbiology, 73 (3), 2007, 699–705, doi:10.1128/AEM.02428-06

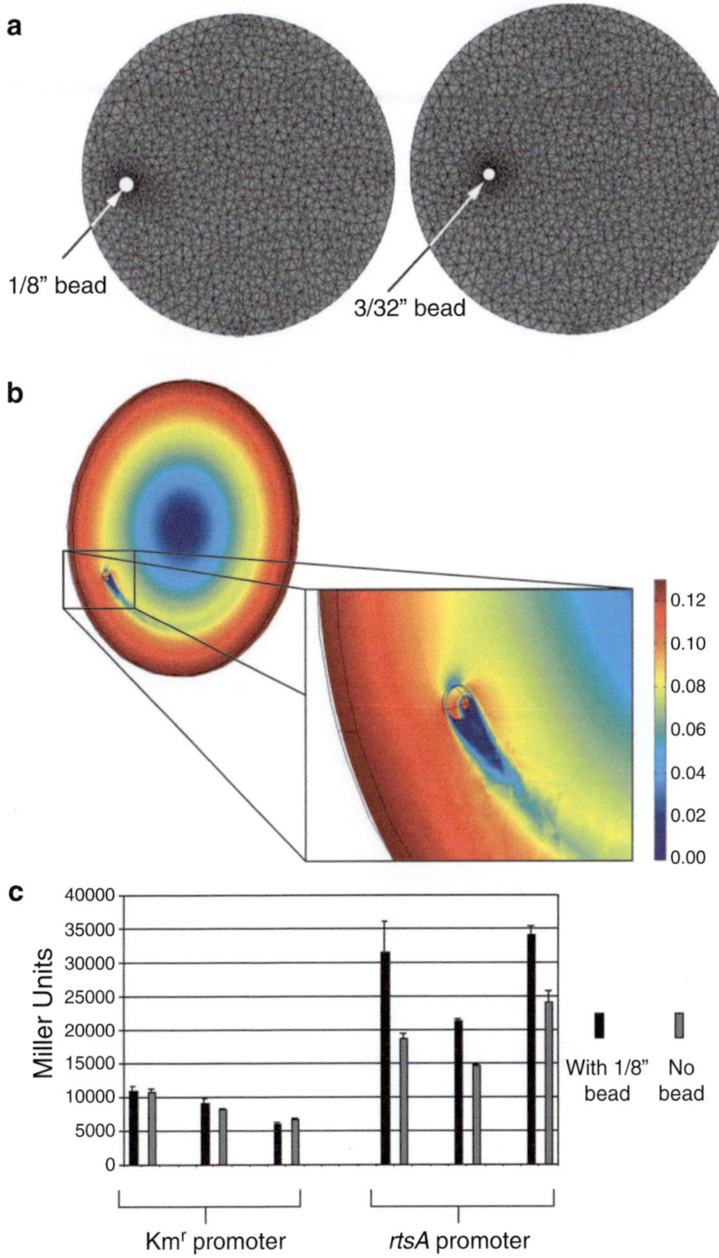

Fig. 11.2 Computational modeling of fluid shear stress in the RWV bioreactor (**a**) To investigate the range of fluid shear forces present during LSMMG culture, beads of different sizes (3/32-in. and 1/8-in.) were incorporated into the RWV and their equilibrium positions monitored during culture and used to build the model geometry. The model was divided into a grid or smaller elements (10,230 for the 1/8-in. bead and 11,218 elements for the 3/32-in. bead). (**b**) Fluid velocity distribution within the RWV bioreactor. Velocity (m/s) increases with radius from the center of the bioreactor, except in the region near the spherical bead (*inset*). (**c**) Gene expression changes in *S.* Typhimurium with and without the addition of a 1/8-in. bead. The promoter for the *rtsA* gene (located in a cluster of shear-regulated genes in *S.* Typhimurium) was fused to a *lacZ* reporter, and

11.3 True Spaceflight Culture Alters the Virulence and Induces Global Changes in Gene Expression in *S.* Typhimurium

The exciting discovery that spaceflight analogue culture could reprogram *S.* Typhimurium in a unique manner that resulted in profound alterations in molecular genetic and phenotypic responses of the bacterium laid the foundation for these experiments to be conducted in the true microgravity environment. In September 2006, the Space Shuttle Atlantis (STS-115) launched into low earth orbit carrying a science payload designated as MICROBE. MICROBE was the first experiment to investigate the effects of the microgravity environment on the virulence and global gene expression profiles of any microbial pathogen [6]. For this study, *S.* Typhimurium was loaded into spaceflight hardware called Group Activation Packs (GAPs; Fig. 11.3). Each GAP contained eight test tubes called Fluid Processing Apparatuses (FPAs) that were partitioned into three chambers

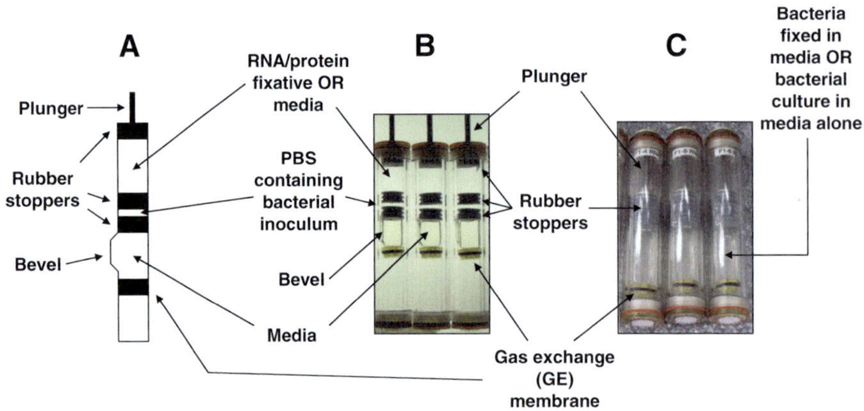

Fig. 11.3 Fluid Processing Apparatus (FPA). (**a**) Schematic diagram of an FPA. An FPA consists of a glass barrel that contains a short bevel on one side and stoppers inside that separate individual chambers containing fluids used in the experiment. The glass barrel loaded with stoppers and fluids is housed inside a Lexan sheath containing a plunger that pushes on the top stopper to facilitate mixing of fluids at the bevel. The bottom stopper in the glass barrel (and also the bottom of the Lexan sheath) is designed to contain a gas-permeable membrane that allows air exchange during bacterial growth. In the STS-115, STS-123, and STS-131 spaceflight experiments, the bottom chamber contained media, the middle chamber contained the bacterial inoculum suspended in PBS, and the top chamber contained either RNA/protein fixative or additional media. Upon activation, the plunger was pushed down so that only the middle chamber fluid was mixed with the bottom chamber to allow media inoculation and bacterial growth. At this step, the plunger was pushed until the bottom of the middle rubber stopper was at the top part of the bevel. After the 25-hour growth period, the plunger was pushed until the bottom of the top rubber stopper was at the top part of the bevel such that the top chamber fluid was added. (**b**) Photograph of FPAs in pre-flight configuration. (**c**) Photograph of FPAs in post-flight configuration showing that all stoppers have been pushed together and the entire fluid sample is in the bottom chamber. Figure reproduced from [8] in accordance with the Creative Commons Attribution (CC BY) license

(Fig. 11.3). The bottom chamber contained bacterial media; the middle chamber contained *S*. Typhimurium in a stasis buffer that allowed the bacteria to stay viable but not actively growing; and the upper chamber contained either bacterial media or fixative. Identical sets of hardware were also prepared for synchronous ground controls that were housed in the Orbital Environmental Simulation (OES) room at the Kennedy Space Center (which maintained identical temperature and humidity conditions to those found aboard the Space Shuttle in real time). Once on orbit, NASA astronaut Heidemarie Stefanyshyn-Piper (Fig. 11.4) activated the growth of the bacteria by turning a hand crank on the GAP, thereby combining the bottom and middle chambers of the individual FPAs. Similar operations were performed synchronously with ground-based controls. After 24 h of growth, the crewmember turned the hand crank a second time to combine the top FPA chamber with the rest of the sample, which either added fixative to preserve the samples for post-flight gene expression studies, or added fresh media for samples to be brought back viable and used for infection studies post-landing. Upon shuttle landing, viable (non-fixed) samples were combined and immediately used to infect BALB/c mice with a range of doses in order to obtain LD_{50} and time-to-death profiles. It is important to note that no statistical differences in final cell densities were observed between flight and ground cultures of *S*. Typhimurium.

The results of the virulence study using *S*. Typhimurium cultured under true microgravity conditions closely paralleled the earlier findings obtained with LSMMG cultures of *S*. Typhimurium, demonstrating that microgravity culture led to decreased LD_{50} (2.7-fold), decreased time-to-death, and increased percent mortality across multiple infectious dosages in a murine model of infection, relative to ground control cultures (Fig. 11.5) [3, 6]. It is important to note that these increased virulence findings were also independently validated in a second spaceflight experiment aboard STS-123 (Space Shuttle Endeavour, see below), in which a 6.9-fold decrease in LD_{50} was observed in mice infected with the spaceflight cultures as compared to those infected with the synchronous ground controls [8]. Moreover, as for the MICROBE study, no statistical differences in the final cell densities of *S*. Typhimurium were observed.

Global transcriptomic and proteomic analyses from the MICROBE experiment revealed that spaceflight culture led to unique changes in gene and protein expression in *S*. Typhimurium (including virulence factors) that are not observed using traditional experimental approaches on Earth, suggesting unexpected ways that *Salmonella* may cause disease in the host. Specifically, 167 transcripts and 73 proteins were identified to change expression in response to culture of *S*. Typhimurium in spaceflight (Fig. 11.5 and Table 11.1). As with the ground-based microarray findings in the RWV, there were a larger number of downregulated transcripts as compared to up-regulated (98 versus 69, respectively), and these genes were globally distributed across the *S*. Typhimurium genome and belonged to a variety of functional groups, including biofilm formation, iron utilization, flagellar biosynthesis, outer membrane porins, ribosomal subunits, conjugative transfer, ABC transporters, and small non-coding RNAs. The finding that genes important for biofilm formation were altered in expression was supported by the morphologi-

Fig. 11.4 *Salmonella* spaceflight experiments. Astronauts who performed on-orbit *Salmonella* experiments aboard the Space Shuttle (Stefanyshyn-Piper, Gorie, Wilson, Yamazaki, and Magnus) and the International Space Station (Virts) are shown alongside the Nickerson laboratory science payload mission patches (STS-123 not shown). Photo credit: NASA. Astronauts Stefanyshyn-Piper; Gorie; Wilson; Yamazaki; Magnus; Virts

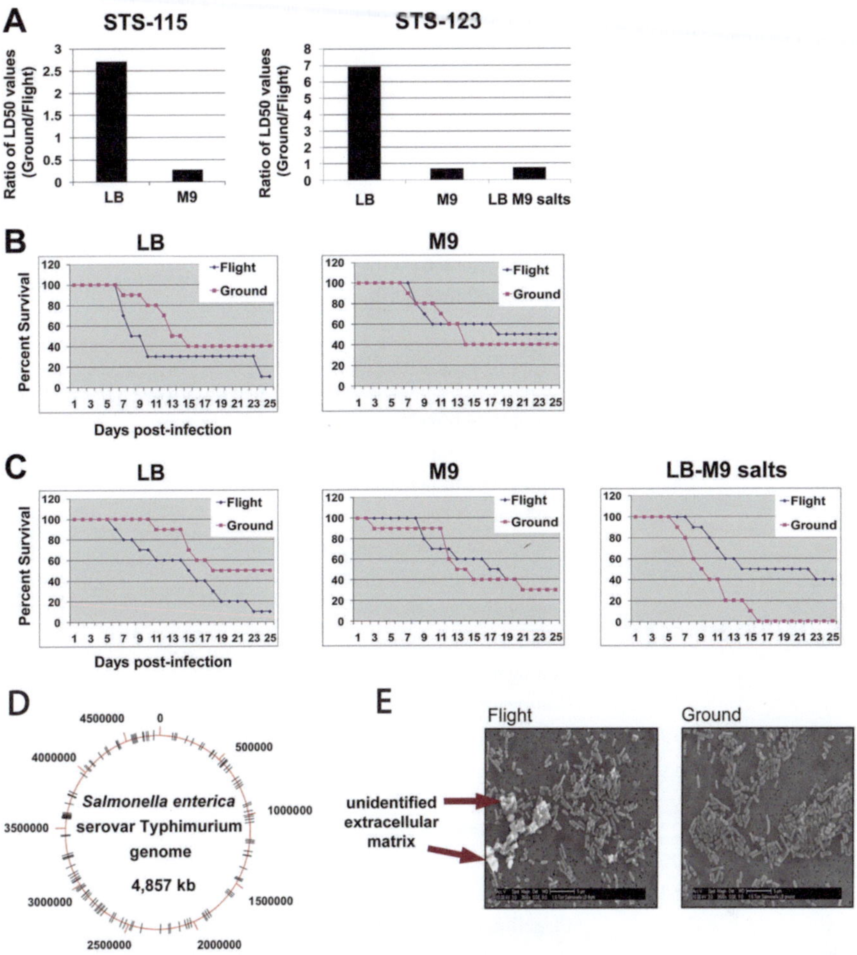

Fig. 11.5 *S.* Typhimurium virulence in LB, M9, and LB–M9 spaceflight cultures. (**a**) Ratio of LD_{50} values of *S.* Typhimurium spaceflight and ground cultures grown in LB (Lennox Broth), M9, or LB–M9 salts media. Female Balb/c mice were perorally infected with a range of bacterial doses from either spaceflight or ground cultures and monitored over a 30-day period for survival. (**b**) Time-to-death curves of mice infected with spaceflight and ground cultures from STS-115 (infectious dosage: 10^7 bacteria for both media). (**c**) Time-to-death curves of mice infected with spaceflight and ground cultures from STS-123 (infectious dosage: 10^6 bacteria for LB and 10^7 bacteria for M9 and LB–M9 salts). Infectious dosages were selected such that the rates in time-to-death facilitated normalized comparisons across the different media. (**d**) Map of the 4.8-Mb circular *S.* Typhimurium genome with the locations of the genes belonging to the space flight transcriptional stimulon indicated as black hash marks. (**e**) SEM of spaceflight and ground *S.* Typhimurium bacteria showing the formation of an extracellular matrix and associated cellular aggregation of space flight cells. (Magnification: ×3500). Panels A–C, from [8]. Panels D–E were reproduced with permission from [6]. Copyright (2007) National Academy of Sciences, U.S.A.

Table 11.1 Spaceflight stimulon genes belonging to Hfq regulon or involved with iron utilization or biofilm formation

Gene	Fold change	Function
Hfq regulon genes		
Up-regulated		
Outer membrane proteins		
ompA	2.05	Outer membrane porin
ompC	2.44	Outer membrane porin
ompD	3.34	Outer membrane porin
Plasmid transfer apparatus		
traB	4.71	Conjugative transfer, assembly
traN	4.24	Conjugative transfer, aggregate formation
trbA	3.14	Conjugative transfer
traK	2.91	Conjugative transfer
traD	2.87	Conjugative transfer, DNA transport
trbC	2.68	Conjugative transfer
traH	2.59	Conjugative transfer, assembly
traX	2.37	Conjugative transfer, fimbrial acetylation
traT	2.34	Conjugative transfer
trbB	2.32	Conjugative transfer
traG	2.21	Conjugative transfer, assembly
traF	2.11	Conjugative transfer
traR	1.79	Conjugative transfer
Various cellular functions		
gapA	7.67	Glyceraldehyde-3-phosphate dehydrogenase A
sipC	6.27	Cell invasion protein
adhE	4.75	Iron-dependent alcohol dehydrogenase of AdhE
glpQ	2.58	Glycerophosphodiester phosphodiesterase, periplasmic
fliC	2.11	Flagellin, filament structural protein
sbmA	1.67	Putative ABC superfamily transporter
Downregulated		
Small RNAs		
alpha RBS	0.305	Small RNA
rnaseP	0.306	Small RNA regulatory
csrB	0.318	Small RNA regulatory
tke1	0.427	Small RNA
oxyS	0.432	Small RNA regulatory
RFN	0.458	Small RNA
rne5	0.499	Small RNA
Ribosomal proteins		
rpsL	0.251	30S ribosomal subunit protein S12
rpsS	0.289	30S ribosomal subunit protein S19
rplD	0.393	50S ribosomal subunit protein L4

(continued)

Table 11.1 (continued)

Gene	Fold change	Function
rpsF	0.401	30S ribosomal subunit protein S6
rplP	0.422	50S ribosomal subunit protein L16
rplA	0.423	50S ribosomal subunit protein L1
rpme	2 0.473	50S ribosomal protein L31 (second copy)
rplY	0.551	50S ribosomal subunit protein L25
Various cellular functions		
ynaF	0.201	putative universal stress protein
ygfE	0.248	Putative cytoplasmic protein
dps	0.273	Stress response DNA-binding protein
hfq	0.298	Host factor for phage replication, RNA chaperone
osmY	0.318	Hyperosmotically inducible periplasmic protein
mysB	0.341	Suppresses protein export mutants
rpoE	0.403	Sigma E (sigma 24) factor of RNA polymerase
cspD	0.421	Similar to CspA but not cold shock induced
nlpb	0.435	Lipoprotein-34
ygaC	0.451	Putative cytoplasmic protein
ygaM	0.453	Putative inner membrane protein
gltI	0.479	ABC superfamily, glutamate/aspartate transporter
ppiB	0.482	Peptidyl-prolyl cis-trans isomerase B (rotamase B)
atpE	0.482	Membrane-bound ATP synthase, F0 sector, subunit c
yfiA	0.482	Ribosome associated factor, stabilizes against dissociation
trxA	0.493	Thioredoxin 1, redox factor
nifU	0.496	NifU homologs involved in Fe-S cluster formation
rbfA	0.506	Ribosome-binding factor, role in processing of 10S rRNA
rseB	0.514	Anti-sigma E factor
yiaG	0.528	Putative transcriptional regulator
ompX	0.547	Outer membrane protease, receptor for phage OX2
rnpA	0.554	RNase P, protein component (protein C5)
hns	0.554	DNA-binding protein; pleiotropic regulator
lamB	0.566	Phage lambda receptor protein; maltose high-affinity receptor
rmf	0.566	Ribosome modulation factor
tpx	0.566	Thiol peroxidase
priB	0.571	Primosomal replication protein N
Iron utilization/storage genes		
adhE	4.76	Iron-dependent alcohol dehydrogenase of AdhE
entE	2.24	2,3-Dihydroxybenzoate-AMP ligase
hydN	2.03	Electron transport protein (FeS senter) from formate to hydrogen
dmsC	0.497	Anaerobic dimethyl sulfoxide reductase, subunit C
nifU	0.495	Nifu homologs involved in Fe-S cluster formation
fnr	0.494	Transcriptional regulator, iron-binding

(continued)

Table 11.1 (continued)

Gene	Fold change	Function
fdnH	0.458	Formate dehydrogenase-N, Fe-S beta subunit, nitrate-inducible
frdC	0.411	Fumarate reductase, anaerobic, membrane anchor polypeptide
bfr	0.404	Bacterioferrin, an iron storage homoprotein
ompW	0.276	Outer membrane protein W; colicin S4 receptor
dps	0.273	Stress response DNA-binding protein and ferritin
Genes implicated in/associated with biofilm formation		
wza	2.30	Putative polysaccharide export protein, outer membrane
wcaI	2.07	Putative glycosyl transferase in colanic acid biosynthesis
ompA	2.06	Outer membrane protein
wcaD	1.82	Putative colanic acid polymerase
wcaH	1.76	GDP-mannose mannosyl hydrolase in colanic acid biosynthesis
manC	1.71	Mannose-1-phosphate guanylyltransferase
wcaG	1.68	Bifunctional GDP fucose synthetase in colanic acid biosynthesis
wcaB	1.64	Putative acyl transferase in colanic acid biosynthesis
fimH	1.61	Fimbrial subunit
fliS	0.339	Flagellar biosynthesis
flgM	0.343	Flagellar biosynthesis
flhD	0.356	Flagellar biosynthesis
fliE	0.438	Flagellar biosynthesis
fliT	0.444	Flagellar biosynthesis
cheY	0.461	Chemotaxic response
cheZ	0.535	Chemotaxic response

cal differences discovered between the flight and ground cultures of *S*. Typhimurium using scanning electron microscopy (SEM), which revealed enhanced bacterial cellular aggregation and clumping associated with an extracellular matrix in the spaceflight samples [6].

Further analysis of the gene expression data revealed that nearly a third of the transcripts identified were genes whose expression levels are post-transcriptionally regulated by the RNA-binding protein Hfq [6]. Moreover, the *hfq* gene was also downregulated, supporting the previous microarray trends for this gene observed for *S*. Typhimurium in response to microgravity analogue culture in the RWV [4, 6]. Hfq is an RNA chaperone that is evolutionarily conserved in a wide range of organisms across all three domains of life [52]. Hfq exerts post-transcriptional control over gene expression by facilitating the pairing between small non-coding RNAs with cognate mRNAs, thereby affecting the translation and turnover rates of a wide array of transcripts [53–59]. As mentioned above, a number of small non-coding RNAs were differentially regulated in *S*. Typhimurium in response to spaceflight. Consequently, Hfq plays an important role in regulating bacterial virulence and physiology in response to stress, and while it does so largely through regulating the expression of several major sigma factors such as σ^{70} (*rpoE*-encoded) and σ^{38} (*rpoS*-encoded), the protein also exerts its global regulatory effects independent of these sigma proteins [53–60]. In order to validate the role of Hfq in the responses of *S.* Typhimurium observed in response to LSMMG and spaceflight culture, an *hfq* mutant was constructed and assessed for LSMMG-induced alterations in resistance to acid stress and macrophage survival [6]. The *hfq* mutant no longer displayed the differences in acid stress resistance or macrophage survival that were observed for the wild type in response to LSMMG culture (Fig. 11.6).

A recent study evaluated a panel of *S. enterica* strains for LSMMG-induced alterations in pathogenesis-related stress responses, and explored the mechanistic role of several candidate genes, including *hfq*, in the resistance of *S.* Typhimurium to oxidative stress following RWV culture. This was the first report to demonstrate that the ability for the *Salmonella* genus to respond to LSMMG culture was conserved across multiple serovars, including Typhimurium (strains 14028, DT104 and LT2), Enteriditis (strain LK5), and Choleraesuis (strain A50) [10]. In addition, this study also determined that the catalases KatG and KatN were important for the

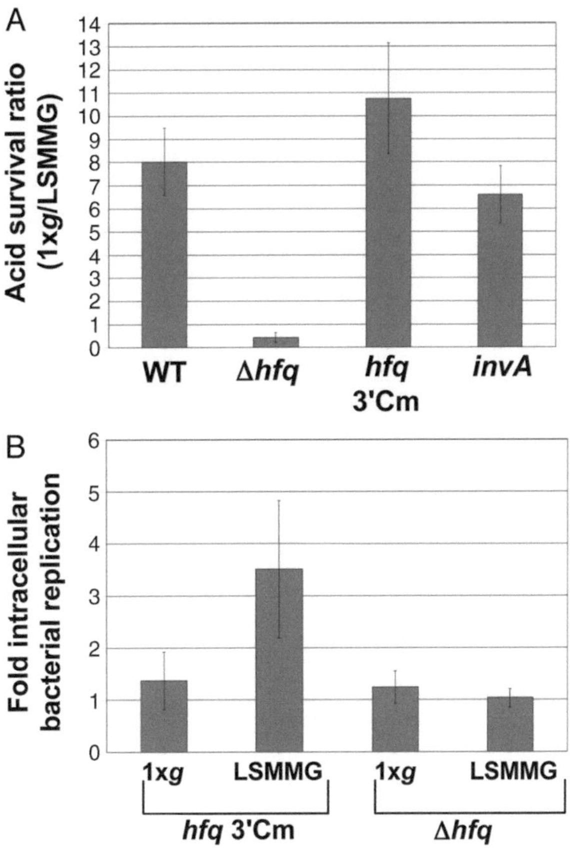

Fig. 11.6 Hfq is required for *S.* Typhimurium LSMMG-induced phenotypes in RWV culture. (**a**) The survival ratio of WT and isogenic *hfq*, *hfq* 3′Cm, and *invA* mutant strains in acid stress after RWV culture in the LSMMG and control positions is plotted ($P<0.05$, ANOVA). (**b**) Fold intracellular replication of *S.* Typhimurium strains *hfq* 3′Cm and Δ*hfq* in J774 macrophages after RWV culture as above. Intracellular bacteria were quantitated at 2 h and 24 h after infection, and the fold increase in bacterial numbers between those two time periods was calculated ($P<0.05$, ANOVA). Figure reproduced with permission from [6]. Copyright (2007) National Academy of Sciences, U.S.A.

alterations in the oxidative stress resistance phenotype observed between the LSMMG and control cultures under the conditions tested. Interestingly, in this study the authors found that mutations in the *rpoS, rpoE, oxyR*, or *hfq* genes did not alter the LSMMG-induced resistance to oxidative stress induced via the addition of hydrogen peroxide. The observation that *rpoS* was not required for LSMMG alteration of stress resistance was consistent with previous studies [5]. However, the finding that *hfq* was not required for the oxidative stress resistance phenotype was intriguing given the findings described above, wherein an *hfq* mutation eliminated the differences normally observed between LSMMG and control cultures for both the acid stress resistance and macrophage resistance phenotypes [6]. The discrepancy between the two studies is likely due to the use of different *S*. Typhimurium strains and bacterial culture methodologies, as well as the assessment of different pathogenesis-related phenotypes (oxidative stress [10] versus acid stress resistance and macrophage survival [6]). Ongoing studies are focused on further exploring the mechanistic role of Hfq in the response to LSMMG and spaceflight culture.

11.4 Mitigating the Spaceflight-Induced Increases in *Salmonella* Virulence: Ions as a Potential Countermeasure

The STS-115 spaceflight study and previous ground-based RWV experiments indicated that culture of *S*. Typhimurium in these low fluid shear environments led to increased virulence and global alterations in gene expression [3, 4, 6, 8]. These results were obtained when *S*. Typhimurium was cultured in Lennox Broth (LB), a nutrient rich medium. During the STS-115 flight experiment, *S*. Typhimurium was also cultured separately in a minimal medium (M9) and assessed for spaceflight-induced changes in virulence and global changes in gene expression [8]. The results with the M9 medium showed striking differences from those observed when *S*. Typhimurium was grown in LB. Specifically, when *S*. Typhimurium was cultured in M9 medium, no differences in virulence were observed between mice infected with either the spaceflight or ground control cultures.

Global microarray analysis of the *Salmonella* cultures grown in M9 during spaceflight and compared to synchronous ground control cultures revealed differential expression of a much smaller set of transcripts relative to what had been observed for the cultures grown in LB (38 vs. 167 genes, respectively) [6]. Genes that were differentially regulated in the M9 cultures included those important for motility, the formation of Hyc hydrogenase, and membrane transporters. Consistent with what was observed for LB medium, 18 % of the transcripts included small non-coding RNAs and mRNAs belonging to the Hfq regulon. The proteomics analysis supported this finding, as 34 % of the proteins that were differentially expressed belonged to the Hfq regulon or were part of a directly related functional group of proteins that are regulated by Hfq.

Since it was previously known that media ion composition could alter the virulence of *S.* Typhimurium independently of the spaceflight environment, a comparative analysis was performed to ascertain whether spaceflight had any additional impact on the relative differences in the LD_{50} values between the LB and M9 cultured *Salmonella*. Notably, the spaceflight-cultured *Salmonella* displayed a 56.8-fold difference in LD_{50} values between the two media types, while the ground control cultures only showed a 5.7-fold difference. These findings demonstrated that although the media composition impacted the LD_{50} values for both spaceflight and ground cultures, the differences between media types were greatly exacerbated by the spaceflight environment.

In order to identify components of the M9 media that could have led to the suppression of virulence in the spaceflight-cultured *S.* Typhimurium, a trace elemental analysis of the LB and M9 media was performed [8]. The analysis revealed five inorganic salts as candidates responsible for this suppressed virulence phenotype, including: (1) sodium phosphate (NaH_2PO_4), (2) potassium phosphate (KH_2PO_4), (3) ammonium chloride (NH_4Cl), (4) sodium chloride (NaCl), and (5) magnesium sulfate ($MgSO_4$). All five of these salts were present at much higher concentrations in the M9 media relative to LB, including phosphate (61-fold), magnesium (18-fold), sulfate (3.6-fold), chloride (3-fold), and potassium (2.4-fold). To validate that the increased concentrations of one or more of these ions were responsible for the virulence phenotype observed with the M9 media, the five salts were supplemented into LB media at the same concentrations they were present in M9, creating a new media that was designated as LB–M9 [8].

On March 11, 2008, Space Shuttle Endeavour launched carrying the experimental science payload known as MDRV, with Space Shuttle Commander Dominic Gorie performing the on-orbit experiments (Fig. 11.4). The MDRV experiment was a follow-up to MICROBE, and had two major experimental objectives with respect to *S.* Typhimurium: (1) validate the virulence discoveries made during the MICROBE study with LB and M9 media; and (2) test the hypothesis that the increased concentration of the five inorganic salts in the hybrid LB–M9 media would be sufficient to reverse the spaceflight-associated increases in virulence previously observed with LB alone. As for the MICROBE study, no statistical differences in the final cell densities of *S.* Typhimurium were observed for any media type. The results of the MDRV experiment confirmed the enhanced virulence phenotype previously observed for *S.* Typhimurium cultured in LB media, with a 6.9-fold decrease in LD_{50} observed in mice infected with spaceflight-grown Salmonella relative to the control cultures (an even greater difference in virulence than what was observed during the STS-115 mission) [6, 8]. Similarly, *S.* Typhimurium grown in M9 medium displayed no differences in LD_{50} or time-to-death between flight and ground cultures. As predicted, the *S.* Typhimurium cultured in the LB–M9 hybrid medium displayed a similar virulence profile to that observed for the M9 cultures—no increase in virulence was observed for spaceflight grown cultures.

In an effort to identify the ion or set of ions responsible for these phenotypes, post-flight studies were conducted using the RWV to investigate ion-induced

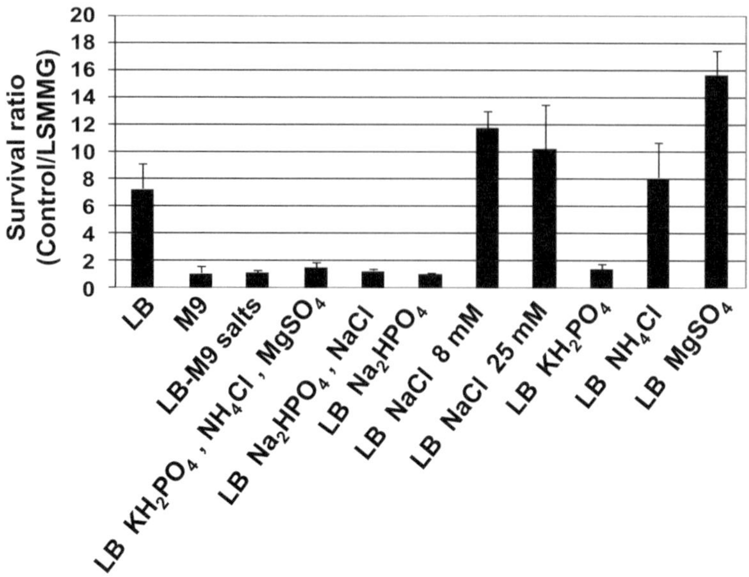

Fig. 11.7 Increased phosphate ion concentration prevents altered *S.* Typhimurium acid tolerance in ground-based spaceflight analogue culture. Cultures of *S.* Typhimurium grown in the indicated medium in the RWV oriented in the LSMMG or control positions were subjected to acid stress (pH 3.5) immediately upon removal from the apparatus. A ratio of percent survival of the bacteria cultured at each orientation in each media is shown. Figure reproduced from [8] in accordance with the Creative Commons Attribution (CC BY) license

alterations in the acid stress response, since this stress had been previously shown to be consistently and reproducibly altered in response to LSMMG culture [3, 5, 6, 8]. When *S.* Typhimurium was cultured in LB, the LSMMG and control cultures displayed significant differences in their acid stress resistance profiles; however, when cultured in M9, this difference was eliminated [8]. Using these findings as a benchmark, a systematic analysis was performed, wherein different combinations of the inorganic salts were incorporated into the LB medium during RWV culture and then the LSMMG cultures and the controls profiled for acid stress resistance (Fig. 11.7). Results from these studies led to the identification of phosphate as the key ion responsible for repressing the differences normally observed between LSMMG and the control cultures, and a likely candidate for the ion in the M9 and LB–M9 media responsible for suppressing the spaceflight-associated increases in virulence. These findings have important implications for the design of potential nutritional countermeasures for the astronauts, and also for the general public on Earth. Since spaceflight and LSMMG culture are able to mimic key aspects of the low fluid shear environment present in certain regions of the human body (e.g., between the apical brush border microvilli along the intestinal tract) [40], these findings also have applications towards a better fundamental understanding of the role that fluid shear stress and nutritional homeostasis in the body play in regulating the virulence of *Salmonella*.

11.5 Using Spaceflight to Forge New Frontiers in the Realm of Host–Pathogen Interactions and Vaccine Development

The collective body of work described above has paved the way for novel, complex microbiological studies to be conducted with *Salmonella* and other microorganisms in the unique environment of spaceflight in order to improve human health, both on Earth and for astronauts during spaceflight missions. Since MICROBE and MDRV, two additional spaceflight studies have explored the impact of spaceflight on *Salmonella* pathogenesis, this time focusing on the host–pathogen interaction. The first experiment, designated as STL-IMMUNE, was launched in April 2010 aboard Space Shuttle Discovery (STS-131). NASA astronaut Stephanie Wilson and JAXA astronaut Naoko Yamazaki helped to monitor this automated experiment on orbit (Fig. 11.4). STL-IMMUNE was the first fundamental space biology experiment to conduct an in-flight infection of human cells. During this study, HT-29 intestinal epithelial cells were cultured in hollow fiber bioreactors during spaceflight or as matched synchronous controls on the ground at the Kennedy Space Center. After several days of growth, the cell cultures were briefly infected with *S.* Typhimurium, and samples were fixed for gene/protein expression analysis or for microscopy. Although the results from these studies have not yet been published, key differences were observed in the infected host response during spaceflight infection with *Salmonella* as compared to synchronous ground-based controls. Specifically, it was determined that spaceflight-culture of uninfected and *Salmonella* infected HT-29 cells uniquely altered cellular morphology, transcriptomic and proteomic profiles as compared to synchronous ground controls (manuscript in preparation).

The Micro-5/PHOENIX (Pathogen HOst ENteric Infection eXperiment) study, which launched in January 2015 aboard SpaceX Dragon (SpaceX-5), advanced the MICROBE, MDRV, and STL-IMMUNE experiments another step further, in that it explored the combined impact of spaceflight culture and ions (with a focus on phosphate) on the *Salmonella*–host interaction using a whole organism as the host: *C. elegans*. *C. elegans* is a well-characterized model host organism with which to study host–pathogen interactions, define virulence mechanisms, and test novel therapeutic strategies to prevent infectious disease caused by a variety of pathogens [62–91]. The organism has flown on several spaceflight missions to understand the effects of microgravity on cellular physiology and development [92–98]. During the Micro-5 study, *C. elegans* and *S.* Typhimurium were transported to the ISS in a stasis buffer that maintained them in a viable, but non-replicating state. In addition, *Escherichia coli* (*E. coli*) OP50 was included in the study as a non-pathogenic control, as it is the normal laboratory food source for *C. elegans*. Once the experiment arrived to the ISS, NASA astronaut Terry Virts (Fig. 11.4) initiated growth of *S.* Typhimurium and *E. coli* separately in LB media, as well as in three other different media types, each containing different combinations of ions. After a period of growth, the bacteria were separately added to *C. elegans* cultures. Cultures were monitored in real time by video to assess time-to-death 50 (TD_{50}), with a subset also fixed 48 h post-infection for gene expression and microscopic analysis. Post-flight analyses are ongoing.

In addition to the two recent studies listed above, the results from the MICROBE and MDRV studies with *Salmonella* also introduced the possibility that the spaceflight platform could potentially be used as a tool to assist in the development and optimization of live Recombinant Attenuated *Salmonella* Vaccines (RASVs). RASVs are genetically engineered *Salmonella* strains that act as a "cargo ship" to deliver protective antigens against different microbial pathogens to the immune system [100–104]. This approach takes advantage of the powerful invasive and colonization properties of *Salmonella* in order to elicit immunogenic responses from all three branches of the immune system (humoral, cellular, and mucosal) against the antigen of interest [100–104]. There are several characteristics of RASVs that have been incorporated during their genetic engineering in order to enhance their safety and efficacy. These include: (1) regulated delayed attenuation, which requires tight control of the expression of key virulence factors to allow RASV strains to display features of wild type *Salmonella* at the time of immunization for efficient invasion and colonization of the host, subsequently followed by a delayed attenuation, in order to prevent the induction of disease symptoms; and (2) regulated delayed lysis, both for attenuation and biological containment [100–104].

Based on the results from the previous spaceflight experiments, which demonstrated that spaceflight could increase the virulence of *Salmonella*, the hypothesis for the RASV spaceflight experiment was that the microgravity environment of spaceflight could be used to accelerate the development of these vaccine strains by (1) enhancing their ability to safely induce a potent and protective immune response, and (2) unveiling novel gene targets to develop new and improved existing vaccine strains. The RASV strains used for this experiment carried antigens against pneumococcal pneumonia—which causes life-threatening diseases (pneumonia, meningitis, bacteremia) that kill over 10 million people annually—particularly children and elderly who are less responsive to current anti-pneumococcal vaccines [102–104]. During the experiment, RASV strains were cultured during spaceflight and as synchronous ground controls. A subset was also fixed on orbit (or on the ground) for post-flight microarray analysis, and the remainder were brought back and immediately used to immunize mice at the Kennedy Space Center upon Shuttle landing. Several weeks post-immunization, the mice were challenged with *Streptococcus pneumoniae* and then monitored for survival. Post-flight analyses are ongoing.

11.6 Summary

The aforementioned studies have launched a novel line of investigations into the importance of studying microbial pathogenesis within the context of physiologically relevant biomechanical forces. Following the initial LSMMG and spaceflight discoveries with *S.* Typhimurium, several other microbial pathogens have been examined and found to alter their global gene expression and/or pathogenesis-related stress response profiles in response to LSMMG and/or spaceflight culture. Two of these opportunistic pathogens, *Pseudomonas aeruginosa* (see Chap. 12) [104, 105] and *Staphylococcus aureus* (see Chap. 14) [106], have also implicated Hfq in their response to the spaceflight and/or LSMMG environments. Moreover, it was recently shown that Hfq plays an important role in the mutualistic symbiotic relationship between the marine bacterium, *Vibrio fischeri*, and the bobtail squid, *Euprymna scolopes*, when both were co-cultured under LSMMG conditions [107]. This study highlights the important role of Hfq in non-pathogenic bacteria in response to LSMMG culture. Ongoing studies are now focused on elucidating the various mechanism(s) and pathways by which both pathogenic and commensal microbes are able to sense and respond to LSMMG and spaceflight culture.

Questions for Future Research

1. Is there a receptor present on the surface of *Salmonella* and/or other bacteria that is involved in the ability to sense and respond to the LSMMG and spaceflight culture environments?
2. What are the detailed regulatory pathways that are involved in the sensing and transmission of the LSMMG and spaceflight environmental signals?
3. In addition to the previously identified bacteria that possess the capability to sense and respond to LSMMG culture, what other bacterial genera/species are able to do so? Do microorganisms localized predominantly on solid surfaces display a difference in this capability relative to those found mainly in liquid environments?
4. Can the spaceflight platform and RWV bioreactor be successfully used for the development and optimization of live, attenuated *Salmonella* vaccine strains?

References

1. Mishra, S. K., & Pierson, D. L. (1992). Space flight, effects on microorganisms. In J. Lederberg (Ed.), *Encyclopedia of microbiology* (pp. 53–60). San Diego: Academic.
2. Dickson, K. J. (1991). Summary of biological spaceflight experiments with cells. *ASGSB Bulletin, 4*, 151–260.
3. Nickerson, C. A., Ott, C. M., Mister, S. J., Morrow, B. J., Burns-Keliher, L., et al. (2000). Microgravity as a novel environmental signal affecting *Salmonella enterica* serovar Typhimurium virulence. *Infection and Immunity, 68*, 3147–3152.
4. Wilson, J. W., Ramamurthy, R., Porwollik, S., McClelland, M., Hammond, T., et al. (2002). Microarray analysis identifies Salmonella genes belonging to the low-shear modeled microgravity regulon. *Proceedings of the National Academy of Sciences of the United States of America, 99*, 13807–13812.
5. Wilson, J. W., Ott, C. M., Ramamurthy, R., Porwollik, S., McClelland, M., et al. (2002). Low-shear modeled microgravity alters the *Salmonella enterica* serovar typhimurium stress

response in an RpoS-independent manner. *Applied and Environmental Microbiology, 68*, 5408–5416.
6. Wilson, J. W., Ott, C. M., Honer zu Bentrup, K., Ramamurthy, R., Quick, L., et al. (2007). Space flight alters bacterial gene expression and virulence and reveals a role for global regulator Hfq. *Proceedings of the National Academy of Sciences of the United States of America, 104*, 16299–16304.
7. Nauman, E. A., Ott, C. M., Sander, E., Tucker, D. L., Pierson, D., et al. (2007). Novel quantitative biosystem for modeling physiological fluid shear stress on cells. *Applied and Environmental Microbiology, 73*, 699–705.
8. Wilson, J. W., Ott, C. M., Quick, L., Davis, R., Honer zu Bentrup, K., et al. (2008). Media ion composition controls regulatory and virulence response of *Salmonella* in spaceflight. *PLoS One, 3*, e3923.
9. NASA, Space Tissue Loss (STS-131) (2010), http://www.nasa.gov/ames/research/space-biosciences/space-tissue-loss-sts-131/%20-%20.VIz2i0tCVKo
10. Pacello, F., Rotilio, G., & Battistoni, A. (2012). Low-shear modeled microgravity enhances *Salmonella enterica* resistance to hydrogen peroxide through a mechanism involving KatG and KatN. *The Open Microbiology Journal, 6*, 53–64.
11. NASA, Investigation of host-pathogen interactions, conserved cellular responses, and countermeasure efficacy during spaceflight using the human surrogate model *Caenorhabditis elegans* (Micro-5) (2012), http://www.nasa.gov/mission_pages/station/research/experiments/Micro5.html
12. Majowicz, S. E., Musto, J., Scallan, E., Angulo, F. J., Kirk, M., et al. (2010). The global burden of nontyphoidal *Salmonella* gastroenteritis. *Clinical Infectious Diseases, 50*, 882–889.
13. Bhutta, Z. A., & Threlfall, J. (2009). Addressing the global disease burden of typhoid fever. *JAMA, 302*, 898–899.
14. Crump, J. A., Luby, S. P., & Mintz, E. D. (2004). The global burden of typhoid fever. *Bulletin of the World Health Organization, 82*, 346–353.
15. Coburn, B., Grassl, G. A., & Finlay, B. B. (2007). *Salmonella*, the host and disease: A brief review. *Immunology and Cell Biology, 85*, 112–118.
16. Kingsley, R. A., Msefula, C. L., Thomson, N. R., Kariuki, S., Holt, K. E., et al. (2009). Epidemic multiple drug resistant *Salmonella* Typhimurium causing invasive disease in sub-Saharan Africa have a distinct genotype. *Genome Research, 19*, 2279–2287.
17. Gordon, M. A., Graham, S. M., Walsh, A. L., Wilson, L., Phiri, A., et al. (2008). Epidemics of invasive *Salmonella enterica* serovar enteritidis and *S. enterica* serovar typhimurium infection associated with multidrug resistance among adults and children in Malawi. *Clinical Infectious Diseases, 46*, 963–969.
18. Feasey, N. A., Dougan, G., Kingsley, R. A., Heyderman, R. S., & Gordon, M. A. (2012). Invasive non-typhoidal salmonella disease: An emerging and neglected tropical disease in Africa. *Lancet, 379*, 2489–2499.
19. Reddy, E. A., Shaw, A. V., & Crump, J. A. (2010). Community-acquired bloodstream infections in Africa: A systematic review and meta-analysis. *The Lancet Infectious Diseases, 10*, 417–432.
20. Castro, V. A., Trasher, A. N., Healy, M., Ott, C. M., & Pierson, D. L. (2004). Microbial characterization during the early habitation of the International Space Station. *Microbial Ecology, 47*, 119–126.
21. Ott, C. M., Bruce, R. J., & Pierson, D. L. (2004). Microbial characterization of free floating condensate aboard the Mir space station. *Microbial Ecology, 47*, 133–136.
22. Ott, C. M. (2004). Human immune function and microbial pathogenesis in human spaceflight. Cancun.
23. Kish, A. L., Hummerick, M., Roberts, M. S., Garland, J. L., Maxwell, S., & Mills, A. L., Biostability and microbiological analysis of Shuttle crew refuse. SAE Technical Paper #2002-01-2356, 2002.
24. Crucian, B., Stowe, R., Mehta, S., Uchakin, P., Quiriarte, H., et al. (2012). Immune system dysregulation occurs during short duration spaceflight on board the space shuttle. *Journal of Clinical Immunology, 33*, 456–465.

25. Crucian, B., Stowe, R., Quiriarte, H., Pierson, D., & Sams, C. (2011). Monocyte phenotype and cytokine production profiles are dysregulated by short-duration spaceflight. *Aviation, Space, and Environmental Medicine, 82*, 857–862.
26. Crucian, B. E., Stowe, R. P., Pierson, D. L., & Sams, C. F. (2008). Immune system dysregulation following short- vs long-duration spaceflight. *Aviation, Space, and Environmental Medicine, 79*, 835–843.
27. Crucian, B., Stowe, R., Mehta, S., Uchakin, P., Quiriarte, H., et al. (2013). Immune system dysregulation occurs during short duration spaceflight on board the space shuttle. *Journal of Clinical Immunology, 33*, 456–465.
28. Kaur, I., Simons, E. R., Castro, V. A., Mark Ott, C., & Pierson, D. L. (2004). Changes in neutrophil functions in astronauts. *Brain, Behavior, and Immunity, 18*, 443–450.
29. Kaur, I., Simons, E. R., Castro, V. A., Ott, C. M., & Pierson, D. L. (2005). Changes in monocyte functions of astronauts. *Brain, Behavior, and Immunity, 19*, 547–554.
30. Kaur, I., Simons, E. R., Kapadia, A. S., Ott, C. M., & Pierson, D. L. (2008). Effect of spaceflight on ability of monocytes to respond to endotoxins of gram-negative bacteria. *Clinical and Vaccine Immunology, 15*, 1523–1528.
31. Sonnenfeld, G. (2005). The immune system in space, including Earth-based benefits of space-based research. *Current Pharmaceutical Biotechnology, 6*, 343–349.
32. Sonnenfeld, G., Butel, J. S., & Shearer, W. T. (2003). Effects of the space flight environment on the immune system. *Reviews on Environmental Health, 18*, 1–17.
33. Sonnenfeld, G., & Shearer, W. T. (2002). Immune function during space flight. *Nutrition, 18*, 899–903.
34. Taylor, G. R. (1993). Immune changes during short-duration missions. *Journal of Leukocyte Biology, 54*, 202–208.
35. Taylor, G. R., & Dardano, J. R. (1983). Human cellular immune responsiveness following space flight. *Aviation, Space, and Environmental Medicine, 54*, S55–S59.
36. Taylor, G. R., Konstantinova, I., Sonnenfeld, G., & Jennings, R. (1997). Changes in the immune system during and after spaceflight. *Advances in Space Biology and Medicine, 6*, 1–32.
37. Taylor, G. R., & Janney, R. P. (1992). In vivo testing confirms a blunting of the human cell-mediated immune mechanism during space flight. *Journal of Leukocyte Biology, 51*, 129–132.
38. Gueguinou, N., Huin-Schohn, C., Bascove, M., Bueb, J. L., Tschirhart, E., et al. (2009). Could spaceflight-associated immune system weakening preclude the expansion of human presence beyond Earth's orbit? *Journal of Leukocyte Biology, 86*, 1027–1038.
39. Nickerson, C. A., Ott, C. M., Wilson, J. W., Ramamurthy, R., LeBlanc, C. L., et al. (2003). Low-shear modeled microgravity: A global environmental regulatory signal affecting bacterial gene expression, physiology, and pathogenesis. *Journal of Microbiological Methods, 54*, 1–11.
40. Nickerson, C. A., Ott, C. M., Wilson, J. W., Ramamurthy, R., & Pierson, D. L. (2004). Microbial responses to microgravity and other low-shear environments. *Microbiology and Molecular Biology Reviews, 68*, 345–361.
41. Demain, A. L., & Fang, A. (2001). Secondary metabolism in simulated microgravity. *The Chemical Record, 1*, 333–346.
42. Fang, A., Pierson, D. L., Koenig, D. W., Mishra, S. K., & Demain, A. L. (1997). Effect of simulated microgravity and shear stress on microcin B17 production by *Escherichia coli* and on its excretion into the medium. *Applied and Environmental Microbiology, 63*, 4090–4092.
43. Fang, A., Pierson, D. L., Mishra, S. K., & Demain, A. L. (2000). Growth of *Streptomyces hygroscopicus* in rotating-wall bioreactor under simulated microgravity inhibits rapamycin production. *Applied Microbiology and Biotechnology, 54*, 33–36.
44. Fang, A., Pierson, D. L., Mishra, S. K., & Demain, A. L. (2000). Relief from glucose interference in microcin B-17 biosynthesis by growth in a rotating-wall bioreactor. *Letters in Applied Microbiology, 31*, 39–41.
45. Fang, A., Pierson, D. L., Mishra, S. K., Koenig, D. W., & Demain, A. L. (1997). Gramicidin S production by Bacillus brevis in simulated microgravity. *Current Microbiology, 34*, 199–204.

46. Fang, A., Pierson, D. L., Mishra, S. K., Koenig, D. W., & Demain, A. L. (1997). Secondary metabolism in simulated microgravity: B-Lactam production by *Streptomyces clavuligerus. Journal of Industrial Microbiology, 18*, 22–25.
47. Gao, Q., Fang, A., Pierson, D. L., Mishra, S. K., & Demain, A. L. (2001). Shear stress enhances microcin B17 production in a rotating wall bioreactor, but ethanol stress does not. *Applied Microbiology and Biotechnology, 56*, 384–387.
48. Barrila, J., Radtke, A. L., Crabbe, A., Sarker, S. F., Herbst-Kralovetz, M. M., et al. (2010). Organotypic 3D cell culture models: Using the rotating wall vessel to study host-pathogen interactions. *Nature Reviews Microbiology, 8*, 791–801.
49. Dong, T., & Schellhorn, H. E. (2010). Role of RpoS in virulence of pathogens. *Infection and Immunity, 78*, 887–897.
50. Hengge-Aronis, R. (2000). The general stress response in *Escherichia coli*. In G. Storz & R. Hengge-Aronis (Eds.), *Bacterial stress responses*. Washington: ASM Press.
51. Guo, P., Weinstein, A. M., & Weinbaum, S. (2000). A hydrodynamic mechanosensory hypothesis for brush border microvilli. *American Journal of Physiology. Renal Physiology, 279*, F698–F712.
52. Vogel, J., & Luisi, B. F. (2011). Hfq and its constellation of RNA. *Nature Reviews Microbiology, 9*, 578–589.
53. Gottesman, S. (2004). The small RNA regulators of *Escherichia coli*: Roles and mechanisms. *Annual Review of Microbiology, 58*, 303–328.
54. Gottesman, S., McCullen, C. A., Guillier, M., Vanderpool, C. K., Majdalani, N., et al. (2006). Small RNA regulators and the bacterial response to stress. *Cold Spring Harbor Symposia on Quantitative Biology, 71*, 1–11.
55. Guisbert, E., Rhodius, V. A., Ahuja, N., Witkin, E., & Gross, C. A. (2007). Hfq modulates the sigmaE-mediated envelope stress response and the sigma32-mediated cytoplasmic stress response in *Escherichia coli. Journal of Bacteriology, 189*, 1963–1973.
56. Majdalani, N., Vanderpool, C. K., & Gottesman, S. (2005). Bacterial small RNA regulators. *Critical Reviews in Biochemistry and Molecular Biology, 40*, 93–113.
57. Pfeiffer, V., Sittka, A., Tomer, R., Tedin, K., Brinkmann, V., et al. (2007). A small non-coding RNA of the invasion gene island (SPI-1) represses outer membrane protein synthesis from the *Salmonella* core genome. *Molecular Microbiology, 66*, 1174–1191.
58. Sittka, A., Lucchini, S., Papenfort, K., Sharma, C. M., Rolle, K., et al. (2008). Deep sequencing analysis of small noncoding RNA and mRNA targets of the global post-transcriptional regulator, Hfq. *PLoS Genetics, 4*, e1000163.
59. Sittka, A., Pfeiffer, V., Tedin, K., & Vogel, J. (2007). The RNA chaperone Hfq is essential for the virulence of *Salmonella* typhimurium. *Molecular Microbiology, 63*, 193–217.
60. Muffler, A., Traulsen, D. D., Fischer, D., Lange, R., & Hengge-Aronis, R. (1997). The RNA-binding protein HF-I plays a global regulatory role which is largely, but not exclusively, due to its role in expression of the sigmaS subunit of RNA polymerase in *Escherichia coli. Journal of Bacteriology, 179*, 297–300.
61. Barrila, J., Sarker, S.F., Hansmeier, N., Briones, N., Park, J., Ott, C.M., et al. (2015). Microgravity uniquely alters the host pathogen interaction between human intestinal cells and Salmonella enterica serovar Typhimurium. In 115th General Meeting of the American Society for Microbiology (Abstr #128), New Orleans, LA.
62. Aballay, A., & Ausubel, F. M. (2001). Programmed cell death mediated by ced-3 and ced-4 protects *Caenorhabditis elegans* from *Salmonella* Typhimurium-mediated killing. *Proceedings of the National Academy of Sciences of the United States of America, 98*, 2735–2739.
63. Aballay, A., Drenkard, E., Hilbun, L. R., & Ausubel, F. M. (2003). *Caenorhabditis elegans* innate immune response triggered by *Salmonella enterica* requires intact LPS and is mediated by a MAPK signaling pathway. *Current Biology, 13*, 47–52.
64. Aballay, A., Yorgey, P., & Ausubel, F. M. (2000). Salmonella typhimurium proliferates and establishes a persistent infection in the intestine of *Caenorhabditis elegans. Current Biology, 10*, 1539–1542.

65. Ikeda, T., Yasui, C., Hoshino, K., Arikawa, K., & Nishikawa, Y. (2007). Influence of lactic acid bacteria on longevity of *Caenorhabditis elegans* and host defense against *Salmonella enterica* serovar enteritidis. *Applied and Environmental Microbiology, 73*, 6404–6409.
66. Labrousse, A., Chauvet, S., Couillault, C., Kurz, C. L., & Ewbank, J. J. (2000). *Caenorhabditis elegans* is a model host for *Salmonella* typhimurium. *Current Biology, 10*, 1543–1545.
67. Sem, X., & Rhen, M. (2012). Pathogenicity of *Salmonella enterica* in *Caenorhabditis elegans* relies on disseminated oxidative stress in the infected host. *PLoS One, 7*, e45417.
68. Tenor, J. L., McCormick, B. A., Ausubel, F. M., & Aballay, A. (2004). *Caenorhabditis elegans*-based screen identifies *Salmonella* virulence factors required for conserved host-pathogen interactions. *Current Biology, 14*, 1018–1024.
69. Wang, C., Wang, J., Gong, J., Yu, H., Pacan, J. C., et al. (2011). Use of *Caenorhabditis elegans* for preselecting *Lactobacillus* isolates to control *Salmonella* Typhimurium. *Journal of Food Protection, 74*, 86–93.
70. Zaborin, A., Romanowski, K., Gerdes, S., Holbrook, C., Lepine, F., et al. (2009). Red death in *Caenorhabditis elegans* caused by *Pseudomonas aeruginosa* PAO1. *Proceedings of the National Academy of Sciences of the United States of America, 106*, 6327–6332.
71. Cooper, V. S., Carlson, W. A., & Lipuma, J. J. (2009). Susceptibility of *Caenorhabditis elegans* to Burkholderia infection depends on prior diet and secreted bacterial attractants. *PLoS One, 4*, e7961.
72. Irazoqui, J. E., Troemel, E. R., Feinbaum, R. L., Luhachack, L. G., Cezairliyan, B. O., et al. (2010). Distinct pathogenesis and host responses during infection of *C. elegans* by *P. aeruginosa* and *S. aureus*. *PLoS Pathogen, 6*, e1000982.
73. Begun, J., Sifri, C. D., Goldman, S., Calderwood, S. B., & Ausubel, F. M. (2005). *Staphylococcus aureus* virulence factors identified by using a high-throughput *Caenorhabditis elegans*-killing model. *Infection and Immunity, 73*, 872–877.
74. Bogaerts, A., Beets, I., Temmerman, L., Schoofs, L., & Verleyen, P. (2010). Proteome changes of *Caenorhabditis elegans* upon a *Staphylococcus aureus* infection. *Biology Direct, 5*, 11.
75. JebaMercy, G., Pandian, S. K., & Balamurugan, K. (2011). Changes in *Caenorhabditis elegans* life span and selective innate immune genes during *Staphylococcus aureus* infection. *Folia Microbiologica (Praha), 56*, 373–380.
76. Sifri, C. D., Begun, J., Ausubel, F. M., & Calderwood, S. B. (2003). *Caenorhabditis elegans* as a model host for *Staphylococcus aureus* pathogenesis. *Infection and Immunity, 71*, 2208–2217.
77. Wu, K., Conly, J., McClure, J. A., Elsayed, S., Louie, T., et al. (2010). *Caenorhabditis elegans* as a host model for community-associated methicillin-resistant *Staphylococcus aureus*. *Clinical Microbiology and Infection, 16*, 245–254.
78. Long, J., Zaborina, O., Holbrook, C., Zaborin, A., & Alverdy, J. (2008). Depletion of intestinal phosphate after operative injury activates the virulence of *P. aeruginosa* causing lethal gut-derived sepsis. *Surgery, 144*, 189–197.
79. Wu, L., Zaborina, O., Zaborin, A., Chang, E. B., Musch, M., et al. (2004). High-molecular-weight polyethylene glycol prevents lethal sepsis due to intestinal *Pseudomonas aeruginosa*. *Gastroenterology, 126*, 488–498.
80. Alegado, R. A., Campbell, M. C., Chen, W. C., Slutz, S. S., & Tan, M.-W. (2003). Characterization of mediators of microbial virulence and innate immunity using the *Caenorhabditis elegans* host-pathogen model. *Cellular Microbiology, 5*, 435–444.
81. Begun, J., Gaiani, J. M., Rohde, H., Mack, D., Calderwood, S. B., et al. (2007). Staphylococcal biofilm exopolysaccharide protects against *Caenorhabditis elegans* immune defenses. *PLoS Pathogen, 3*, e57.
82. Bolz, D. D., Tenor, J. L., & Aballay, A. (2010). A conserved PMK-1/p38 MAPK is required in *Caenorhabditis elegans* tissue-specific immune response to Yersinia pestis infection. *Journal of Biological Chemistry, 285*, 10832–10840.
83. Gallagher, L. A., & Manoil, C. (2001). *Pseudomonas aeruginosa* PAO1 kills *Caenorhabditis elegans* by cyanide poisoning. *Journal of Bacteriology, 183*, 6207–6214.

84. Kim, D. H., Feinbaum, R., Alloing, G., Emerson, F. E., Garsin, D. A., et al. (2002). A conserved p38 MAP kinase pathway in *Caenorhabditis elegans* innate immunity. *Science, 297*, 623–626.
85. Kim, D. H., Liberati, N. T., Mizuno, T., Inoue, H., Hisamoto, N., et al. (2004). Integration of *Caenorhabditis elegans* MAPK pathways mediating immunity and stress resistance by MEK-1 MAPK kinase and VHP-1 MAPK phosphatase. *Proceedings of the National Academy of Sciences of the United States of America, 101*, 10990–10994.
86. Kurz, C. L., & Ewbank, J. J. (2003). *Caenorhabditis elegans*: An emerging genetic model for the study of innate immunity. *Nature Reviews Genetics, 4*, 380–390.
87. Mallo, G. V., Kurz, C. L., Couillault, C., Pujol, N., Granjeaud, S., et al. (2002). Inducible antibacterial defense system in *C. elegans*. *Current Biology, 12*, 1209–1214.
88. Paulander, W., Pennhag, A., Andersson, D. I., & Maisnier-Patin, S. (2007). *Caenorhabditis elegans* as a model to determine fitness of antibiotic-resistant *Salmonella enterica* serovar typhimurium. *Antimicrobial Agents and Chemotherapy, 51*, 766–769.
89. Tenor, J. L., & Aballay, A. (2008). A conserved Toll-like receptor is required for *Caenorhabditis elegans* innate immunity. *EMBO Reports, 9*, 103–109.
90. Nicholas, H. R., & Hodgkin, J. (2004). Responses to infection and possible recognition strategies in the innate immune system of *Caenorhabditis elegans*. *Molecular Immunology, 41*, 479–493.
91. Zaborina, O., Zaborin, A., Romanowski, K., Babrowski, T., & Alverdy, J. (2011). Host stress and virulence expression in intestinal pathogens: Development of therapeutic strategies using mice and *C. elegans*. *Current Pharmaceutical Design, 17*, 1254–1260.
92. Zaborin, A., Gerdes, S., Holbrook, C., Liu, D. C., Zaborina, O. Y., et al. (2012). *Pseudomonas aeruginosa* overrides the virulence inducing effect of opioids when it senses an abundance of phosphate. *PLoS One, 7*, e34883.
93. Adachi, R., Takaya, T., Kuriyama, K., Higashibata, A., Ishioka, N., et al. (2008). Spaceflight results in increase of thick filament but not thin filament proteins in the paramyosin mutant of *Caenorhabditis elegans*. *Advances in Space Research, 41*, 816–823.
94. Adenle, A. A., Johnsen, B., & Szewczyk, N. J. (2009). Review of the results from the International *C. elegans* first experiment (ICE-FIRST). *Advances in Space Research, 44*, 210–216.
95. Higashibata, A., Higashitani, A., Adachi, R., Kagawa, H., Honda, S., et al. (2007). Biochemical and molecular biological analyses of space-flown nematodes in Japan, the First International *Caenorhabditis elegans* Experiment (ICE-First). *Microgravity Science and Technology, 19*, 159–163.
96. Higashibata, A., Szewczyk, N. J., Conley, C. A., Imamizo-Sato, M., Higashitani, A., et al. (2006). Decreased expression of myogenic transcription factors and myosin heavy chains in *Caenorhabditis elegans* muscles developed during spaceflight. *Journal of Experimental Biology, 209*, 3209–3218.
97. Leandro, L. J., Szewczyk, N. J., Benguria, A., Herranz, R., Lavan, D., et al. (2007). Comparative analysis of Drosophila melanogaster and *Caenorhabditis elegans* gene expression experiments in the European Soyuz flights to the International Space Station. *Advances in Space Research, 40*, 506–512.
98. Selch, F., Higashibata, A., Imamizo-Sato, M., Higashitani, A., Ishioka, N., et al. (2008). Genomic response of the nematode *Caenorhabditis elegans* to spaceflight. *Advances in Space Research, 41*, 807–815.
99. Szewczyk, N. J., Mancinelli, R. L., McLamb, W., Reed, D., Blumberg, B. S., et al. (2005). *Caenorhabditis elegans* survives atmospheric breakup of STS-107, space shuttle Columbia. *Astrobiology, 5*, 690–705.
100. Curtiss, R., III, Wanda, S. Y., Gunn, B. M., Zhang, X., Tinge, S. A., et al. (2009). *Salmonella enterica* serovar typhimurium strains with regulated delayed attenuation in vivo. *Infection and Immunity, 77*, 1071–1082.
101. Curtiss, R., 3rd, Xin, W., Li, Y., Kong, W., Wanda, S. Y., et al. (2010). New technologies in using recombinant attenuated *Salmonella* vaccine vectors. *Critical Reviews in Immunology, 30*, 255–270.

102. Li, Y., Wang, S., Scarpellini, G., Gunn, B., Xin, W., et al. (2009). Evaluation of new generation *Salmonella enterica* serovar Typhimurium vaccines with regulated delayed attenuation to induce immune responses against PspA. *Proceedings of the National Academy of Sciences of the United States of America, 106*, 593–598.
103. Shi, H., Santander, J., Brenneman, K. E., Wanda, S. Y., Wang, S., et al. (2010). Live recombinant *Salmonella* Typhi vaccines constructed to investigate the role of *rpoS* in eliciting immunity to a heterologous antigen. *PLoS One, 5*, e11142.
104. Shi, H., Wang, S., Roland, K. L., Gunn, B. M., & Curtiss, R., 3rd. (2010). Immunogenicity of a live recombinant *Salmonella enterica* serovar typhimurium vaccine expressing pspA in neonates and infant mice born from naive and immunized mothers. *Clinical and Vaccine Immunology, 17*, 363–371.
105. Crabbe, A., Pycke, B., Van Houdt, R., Monsieurs, P., Nickerson, C., et al. (2010). Response of *Pseudomonas aeruginosa* PAO1 to low shear modelled microgravity involves AlgU regulation. *Environmental Microbiology, 12*, 1545–1564.
106. Crabbe, A., Schurr, M. J., Monsieurs, P., Morici, L., Schurr, J., et al. (2011). Transcriptional and proteomic response of *Pseudomonas aeruginosa* PAO1 to spaceflight conditions involves Hfq regulation and reveals a role for oxygen. *Applied and Environmental Microbiology, 77*, 1221–1230.
107. Castro, S. L., Nelman-Gonzalez, M., Nickerson, C. A., & Ott, C. M. (2011). Induction of attachment-independent biofilm formation and repression of Hfq expression by low-fluid-shear culture of *Staphylococcus aureus*. *Applied and Environmental Microbiology, 77*, 6368–6378.
108. Grant, K. C., Khodadad, C. L. M., & Foster, J. S. (2014). Role of Hfq in an animal–microbe symbiosis under simulated microgravity conditions. *International Journal of Astrobiology, 13*, 53–61.

Chapter 12
Response of *Pseudomonas aeruginosa* to Spaceflight and Spaceflight Analogue Culture: Implications for Astronaut Health and the Clinic

Aurélie Crabbé, Maria A. Ledesma, C. Mark Ott, and Cheryl A. Nickerson

Historical Landmarks

1970—*Pseudomonas aeruginosa* causes a severe urinary tract infection during the Apollo 13 spaceflight mission [1].

1999—*P. aeruginosa* shows altered resistance to antibiotics when exposed to long-term spaceflight culture conditions [2].

2001—*P. aeruginosa* forms biofilms in spaceflight conditions [3].

2008—*P. aeruginosa* forms self-aggregative biofilms in LSMMG and induces clinically important virulence factors [4].

2010—AlgU and Hfq regulate a significant part of the LSMMG regulon in *P. aeruginosa* [5].

(continued)

A. Crabbé (✉)
School of Life Sciences, The Biodesign Institute, Center for Infectious Diseases and Vaccinology, Arizona State University, Tempe, AZ, USA

Laboratory of Pharmaceutical Microbiology, Faculty of Pharmaceutical Sciences, Ghent University, Ottergemsesteenweg 460, 9000 Ghent, Belgium
e-mail: aurelie.crabbe@ugent.be

M.A. Ledesma
School of Life Sciences, The Biodesign Institute, Center for Infectious Diseases and Vaccinology, Arizona State University, Tempe, AZ, USA

C. Mark Ott
Biomedical Research and Environmental Sciences Division, NASA/Johnson Space Center, 2101 NASA Parkway, Mail: SK4, Houston, TX 77058, USA

C.A. Nickerson
School of Life Sciences, The Biodesign Institute, Center for Infectious Diseases and Vaccinology, Arizona State University, Tempe, AZ, USA

C.A. Nickerson et al. (eds.), *Effect of Spaceflight and Spaceflight Analogue Culture on Human and Microbial Cells*, DOI 10.1007/978-1-4939-3277-1_12

(continued)

2011—*P. aeruginosa* globally alters its transcriptome and proteome in response to spaceflight conditions. Hfq is identified as an important regulator [6].

2013—*P. aeruginosa* forms column-and-canopy biofilm structures when grown in the microgravity environment of spaceflight [7].

12.1 Introduction

12.1.1 A Brief History of Pseudomonas aeruginosa

***Pseudomonas aeruginosa* is a member of the Gamma Proteobacteria class of bacteria, belonging to the Pseudomonadaceae family**. Carle Gessard, a chemist and bacteriologist from Paris, discovered this microorganism in 1882 [8]. This facultative, motile, non-lactose fermenting Gram-negative rod is a highly adaptable bacterium that can colonize various environmental niches such as plants, animals, soil, and marine environments [9–11]. *P. aeruginosa* is an opportunistic pathogen that is responsible for 10–15 % of nosocomial infections worldwide [12], causing infections of the skin, eyes, ears, urethra, respiratory, and intestinal tracts of immunocompromised patients. In particular, this opportunistic pathogen causes severe infections in burn wound and cystic fibrosis (CF) patients. *P. aeruginosa* infections are difficult to treat because of their natural and acquired resistance to many groups of antimicrobial agents [13, 14]. Table 12.1 presents an overview of virulence factors produced by *P. aeruginosa* that were investigated in the context of microgravity and/or microgravity-analogue studies.

Table 12.1 Brief description of *P. aeruginosa* virulence factors studied in microgravity and/or microgravity-analogue conditions

Virulence factor	Main characteristics	References
Quorum Sensing (QS)	• Interbacterial signaling mechanism to regulate gene expression and production of many virulence factors in a population density-dependent manner. • Bacteria produce and secrete small molecules called autoinducers. When these molecules reach a concentration threshold, they diffuse into the cell and coordinate a response that will allow bacteria to promote their survival as a group. • *P. aeruginosa* uses QS to regulate the production of multiple virulence factors, including motility, biofilm formation, efflux pump expression, elastase, catalase, pyocyanin, rhamnolipid production, lectins, iron chelators, and exotoxin A. • *P. aeruginosa* possesses three quorum-sensing systems: Las, Rhl, and PQS with their corresponding autoinducers *N*-(3-oxo-dodecanoyl)-L-homoserine lactone (3-oxo-C_{12}-HSL), *N*-butanoyl-L-homoserine lactone (C_4-HSL), and 2-heptyl-3-hydroxy-4-quinolone, respectively.	[15–32]

(continued)

Table 12.1 (continued)

Virulence factor	Main characteristics	References
	• The Las system comprises the LasI synthase protein and the LasR transcriptional regulator. LasI produces the AHL signal molecule 3-oxo-C_{12}-HSL. In order for LasR to become an active transcription factor, it binds to 3-oxo-C_{12}-HSL, which allows the resulting multimer complex to bind to DNA and to regulate gene transcription. • For the Rhl system, the RhlI synthase produces C_4-HSL, RhlR is the regulator, and the expression of genes only occurs when RhlR complexes with C_4-HSL. • For the PQS system, the PQS signal is recognized by the cognate receptor PqsR and regulates the production of PQS along with other virulence factors such as pyocyanin, rhamnolipids, and elastase.	
Biofilm formation and exopoly-saccharides	• Multicellular bacterial communities encapsulated in an extracellular polymeric substance (EPS) comprised of nucleic acids, proteins, and polysaccharides. • Biofilm formation can form on abiotic (e.g., plastic) and biotic (e.g., wounds, mucus in CF lung) surfaces and confers resistance against antimicrobial agents and immune mechanisms of defense. • Biofilm formation is initiated by surface attachment of planktonic bacteria and their subsequent growth and aggregation lead to microcolony formation and differentiation into structured, mature, antibiotic-resistant communities (Fig. 12.1b). • Virulence factors involved in *P. aeruginosa* biofilm formation include QS, type IV pili, flagella, extracellular DNA, alginate, Pel and Psl polysaccharides, rhamnolipids. • The *Psl*-encoded exopolysaccharide is rich in mannose and galactose, and is involved in attachment, biofilm maintenance, and self-aggregation. • The Pel-encoded exopolysaccharide is rich in glucose, and is involved in the formation of pellicles, i.e. microbial aggregations at the interface between air and liquid. • Alginate is a polymer of β-D-mannuronic acid and α-L-guluronic acid that protects *P. aeruginosa* from phagocytosis and antibody recognition. *P. aeruginosa* can convert from a non-mucoid to an alginate overproducing mucoid phenotype, including in the lungs of CF patients. • The key player in the regulatory pathway leading to alginate production is the alternative sigma factor AlgU (also known as AlgT). • AlgU is regulated by the algUmucABCD operon. In the absence of a specific stimulus or in non-mucoidal strains, the anti-sigma factor MucA binds AlgU, causing a decrease in the expression of promoters targeted by AlgU.	[7, 31–66]

(continued)

Table 12.1 (continued)

Virulence factor	Main characteristics	References
Exoproteases	• *P. aeruginosa* produces exoproteases including elastases, alkaline protease, and phospholipase C (Plc). • The elastolytic zinc metalloprotease elastase B (LasB), also known as pseudolysin is a major virulence factor in *P. aeruginosa*. LasB is believed to cause tissue damage in the host by hydrolysis of extracellular matrix constituents and by attacking intercellular tight junctions. It has also been shown to degrade elements of the immune system such as TNF-α, IFN-γ, IL-2, IL-8, and surfactant protein A (SP-A) and antibacterial peptides. • In addition to LasB, the LasA protease (also called staphylolysin) is a 20-kDa endopeptidase secreted by *P. aeruginosa* that has the ability to lyse *Staphylococcus aureus*. • LasA has a low level of elastolytic activity, but it is a key enzyme that boosts the elastolytic activity of LasB.	[32, 67–79]
Rhamnolipids	• Bacterial glycolipidic biosurfactants that act as heat-stable extracellular hemolysins, and lyse polymorphonuclear leukocytes (PMNs) and macrophages, causing necrotic cell death. In addition, rhamnolipids have antimicrobial activities against Gram-positive, and negative bacteria, and against fungi. • Rhamnolipid synthesis is carried out by two glycosyl transfer reactions, catalyzed by a specific rhamnosyltransferase. The first rhamnosyltransferase is encoded by the rhlAB operon, and is in charge of the formation of mono-rhamnolipids; the second rhamnosyltransferase, encoded by the rlhC gene, converts mono-rhamnolipids into di-rhamnolipids.	[31, 32, 80–88]
Phenazines	• Secondary metabolites that are known as redox-active and pigmented antibiotics. In *P. aeruginosa*, these compounds include pyocyanin, phenazine-1-carboxylic acid (PCA), 1-hydroxyphenazine (1-OH-PHZ), and phenazine-1-carboxamide (PCN), which are regulated by QS. • The pyocyanin concentration in sputum from CF patients correlated with the decline in lung function and with the degree of severity of the disease.	[31, 89–91]
Lectins	• In *P. aeruginosa*, the adhesion to host tissues is mediated by numerous adhesins such as lectins. • LecA and LecB, are two soluble lectins from *P. aeruginosa* that bind to galactose and fucose, respectively. • LecA has been shown to: (a) be cytotoxic for respiratory epithelial cells through the decrease in their growth rate; b) induce permeability defects in the intestinal epithelium, causing an increase in the absorption of exotoxin A, a significant extracellular virulence factor. • LecB was shown to be involved in protease IV activity and pilus biogenesis.	[24, 92–97]

(continued)

Table 12.1 (continued)

Virulence factor	Main characteristics	References
Hfq	• Highly conserved RNA chaperone that serves to control stability of small regulatory RNAs and mRNAs as well as positive and negative translational regulation of target mRNAs by sRNAs. • It has been suggested to be a global regulator of PAO1 virulence in a murine model of infection and of stress response. • Hfq mutants of *P. aeruginosa* demonstrated a 10- to 50-fold decreased virulence in a murine model of infection. • Hfq is involved in the production of alginate, catalase, pyocyanin, elastase, and QS.	[98–105]

12.2 Importance for Studying the Influence of Microgravity and Microgravity-Analogue Conditions on *P. aeruginosa*

12.2.1 Prevalence of Pseudomonas Species in Spaceflight and Astronaut Safety

As the opportunistic pathogen *P. aeruginosa* is commonly isolated from both normal human flora and the environment [106]**, close monitoring of this microorganism in spacecraft is of importance**. *Pseudomonas* species have been isolated from the potable water system [107], air and surfaces of the International Space Station (ISS) [108, 109], surface, air, and condensate of the Russian Mir station [110], on surfaces of other spacecraft (such as flight ready circuit board of future Europa mission) [109], and from astronaut body samples [1]. Both non-pathogenic and potentially pathogenic species of *Pseudomonas* have been recovered during spacecraft and habitat contamination studies, including *P. aeruginosa*, *P. putida*, *P. stutzeri*, and *P. fluorescens* [1, 107, 108, 110]. *Pseudomonas* species were among the three most abundant genera in the humidity condensate recovery system from Mir (20.8 % of samples tested positive) [110], but were less frequently isolated in air and surface samples of Mir and ISS (1.4–4.3 %) [108, 110]. Overall, *Pseudomonas* species were less prevalent as compared to *Staphylococcus* sp., *Bacillus* sp., and *Ralstonia* sp., which were among the most isolated organisms in air and surface samples of Mir and ISS [107, 108, 110].

P. aeruginosa caused a severe urinary tract infection in a crewmember of Apollo 13 [1]. While the Apollo 13 astronaut tested negative for *P. aeruginosa* pre-flight, this microorganism was recovered in high numbers in the astronaut's urine post-flight and was thus suspected to be at the origin of the infection. It is important to mention that the Apollo 13 mission was characterized by significant stress and unusual environmental conditions, such as low temperature, high moisture, and shortage of potable water due to an oxygen tank explosion during the mission. The astronaut presumably became more prone to infection due to a compromised immune system given these extreme circumstances. While in-flight and post-flight treatment with broad-range antibiotics proved unsuccessful in decreasing the *P. aeruginosa* load in the urinary tract of the astronaut, post-flight antibiotic sensitivity testing was at the origin of an effective treatment. In addition, *P. aeruginosa* was isolated from Apollo 15 and 17 astronauts, and in-flight cross contamination was reported [1]. In the era of the space shuttle program, infectious diseases were reported 26 times in space shuttle missions from April 1981 to January 1998, representing 1.4 % of in-flight medical events, with respiratory infections being the most common [111]. No significant infections with *P. aeruginosa* were reported. It is also relevant to note that biofilm growth of *Pseudomonas* species, such as *P. putida*, can lead to biofouling and corrosion, which could affect the integrity of the spaceflight habitat and the efficacy of the water purification system, as previously reported for Mir [112–114]. In addition, plant pathogens, such as *P. syringae*, could have effects on the functionality of space-based advanced life support systems [115].

12.2.2 Compromised Immune System and Susceptibility to P. aeruginosa Infection

Astronauts experience significant impairments of the immune system during spaceflight [116, 117]. Leucopenia is one of the spaceflight-associated physiological changes observed in astronauts and animals, and has been supported using in vitro cell culture models [117–119]. Experimental data supporting spaceflight-induced decreases in leukocyte cell numbers are (1) induction of apoptosis, (2) inhibited leukocyte blastogenesis, (3) decreased differentiation of stem cells to leukocytes, and (4) altered subset distribution and compartmentalization of leukocytes. As an opportunistic pathogen, *P. aeruginosa* predominantly infects patients with primary and acquired immunodeficiencies, accounting for 11–14 % of nosocomial infections [120]. Of particular importance for spaceflight is the fact that 14–21 % of patients with acute leukemia, leading to leucopenia, suffer from *P. aeruginosa* bacteremia [121, 122]. Whether spaceflight-related leucopenia would render immunocompromised astronauts more prone to *P. aeruginosa* infections, especially in the frame of long-term missions, is unclear. In this context, certain aspects of the altered immune response in space, including leucopenia, are mimicked through hind limb unloading of rodents and bed-rest studies of humans [123]. Interestingly, Aviles and colleagues showed that mice exposed to hind limb unloading were more susceptible to *P. aeruginosa* infection [124], indicating that the decreased immune response during spaceflight missions could potentially lead to a higher risk for infectious disease by this organism.

12.2.3 Clinical Applications

As described in Chaps. 2 **and** 10, **the absence of buoyancy-induced convection and sedimentation in the microgravity environment of spaceflight results in low fluid-shear growth conditions in aqueous habitats**. Fluid-shear has been found to profoundly affect the physiology of *Pseudomonas* species, with important applications for the patient. During their natural life cycles and infection process, bacteria such as *P. aeruginosa* encounter a wide range of fluid-shear levels ranging from low fluid-shear near the surface of epithelial cells, moderate fluid-shear in mucosal secretions (e.g., 0.8 dynes/cm^2 for saliva) to high fluid-shear in the blood stream and catheters (up to 10 dynes/cm^2) [125–127]. In addition, a correlation exists between fluid-shear forces in certain regions of the human body and specific disease states. For example, low fluid-shear zones are believed to be present in the lung mucus of patients with cystic fibrosis (CF) due to the absence of mucociliary clearance [4, 128], which represents the main shear-causing factor in normal lung mucus [129]. Shear forces affect the adhesion of *Pseudomonas* sp. to abiotic surfaces and influence biofilm architecture [130–133]. Indeed, *P. aeruginosa* shows enhanced adhesion to abiotic surfaces, and increased surface coverage and biomass as a function of fluid-shear levels [133, 134]. *P. fluorescens* biofilms grown under high fluid-shear regimes are denser, contain more EPS, but are less thick compared to biofilms formed under laminar flow conditions [135]. Furthermore, the competitive behavior of *P. aeruginosa* is affected by the level of fluid-shear [134]. It is worth noting that most studies on the fluid-shear effects of *Pseudomonas* sp. were conducted by generating surface-associated biofilms using flow-through systems, such as flow cells, which are relevant for specific regions of the human body such as catheters and the blood stream, but are less representative of environments in which biofilms are not attached to a surface. Specifically, *P. aeruginosa* has been shown to form self-aggregative biofilms in the lung mucus of patients with CF that are not attached to a surface [33, 136]. In this regard, *P. aeruginosa* uses mucins and other components of the viscous mucus layer that covers the lung epithelium to develop highly antibiotic-resistant biofilms, so-called microcolonies [33, 34]. Therefore, it is relevant to complement flow-through biofilm research with systems that allow one to mimic the self-aggregative biofilm phenotype of *P. aeruginosa* observed in vivo, using for example microgravity-analogue culture systems (see Chaps. 2 and 10).

12.3 Microgravity and Microgravity-Analogue Responses of *Pseudomonas aeruginosa*

12.3.1 Spaceflight-Triggered Responses of P. aeruginosa

Growth

While the experimental set-up and current limitations of spaceflight hardware precluded the determination of growth profiles for *P. aeruginosa* (or any microorganism) during culture in microgravity conditions, final cell densities were

recently reported after 72 h of growth in modified artificial urine medium under varying carbon sources, phosphate concentrations, and oxygen levels [137]. Kim and colleagues found that the combined effect of low phosphate levels and no aeration increased final cell density of *P. aeruginosa* PA14 in spaceflight compared to ground control cultures [137]. When bacteria were grown with aeration and/or high phosphate levels, no differences in growth profiles were observed between spaceflight and control conditions, which correspond to results obtained for *S.* Typhimurium [138, 139]. Since flow cytometry analysis was used to determine final cell counts in the study of Kim et al., the role of potential differences in cell viability between spaceflight and control conditions remains to be defined.

Biofilm Formation

McClean et al. published the first report on biofilm formation by *P. aeruginosa* in the microgravity environment of spaceflight [3]. In this study, *P. aeruginosa* PAO1 was grown on polycarbonate membranes in microgravity conditions and formed biofilms that were highly resistant to mechanical disruption. This finding is intriguing, since low fluid-shear conditions typically result in the formation of loosely attached biofilms that are easily disrupted by mechanical forces [4, 128, 134]. Recently, microgravity was found to increase biofilm formation by *P. aeruginosa* PA14 and to affect the three-dimensional architecture [7]. Specifically, spaceflight-grown biofilms of *P. aeruginosa* on cellulose ester membrane discs demonstrated an increased number of viable cells, biomass, and mean thickness. A novel spaceflight-induced biofilm architecture was proposed, comprised of column-shaped structures overlaid by canopies (Fig. 12.1). In contrast, biofilms formed under ground control conditions were flat, which reflects the previously reported *P. aeruginosa* biofilm structure in static conditions. The observed microgravity-associated column-and-canopy biofilms show resemblance with *P. aeruginosa* biofilms grown under hydrodynamic conditions (such as in flow cell systems), which are mushroom-shaped. In a similar fashion as these mushroom-shaped biofilms, the column-and-canopy biofilms of spaceflight-cultured *P. aeruginosa* were found to be dependent on flagellar motility but not on type IV pili dependent motility (Fig. 12.1). Oxygen limitation in microgravity growth conditions was proposed to play a role in the spaceflight-associated biofilm architecture since culture aeration (through insertion of a gas permeable membrane) outweighed the differences between biofilms grown under normal and microgravity conditions.

Gene Expression Related to Virulence Factor Production

Microarray analysis of *P. aeruginosa* PAO1 grown in the microgravity environment of spaceflight revealed differential regulation of 167 genes, among which 52 genes were upregulated and 115 genes downregulated [6]. The gene encoding the RNA-binding protein Hfq and a significant part of the Hfq regulon were

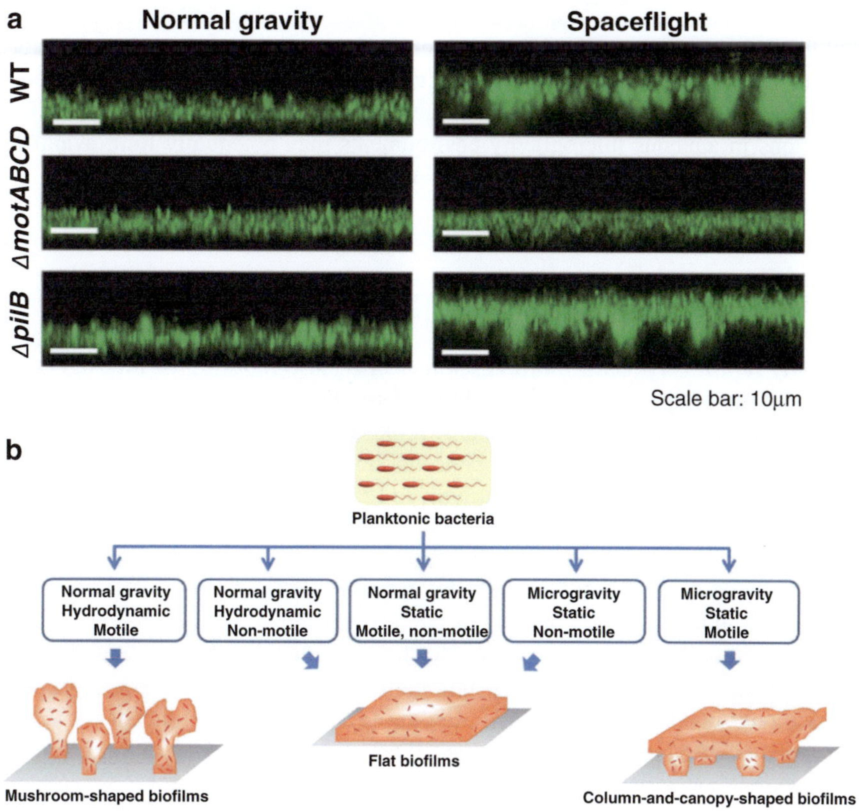

Fig. 12.1 Spaceflight-cultured biofilms of *P. aeruginosa*. (**a**) Confocal laser scanning micrographs of 3-day-old *P. aeruginosa* biofilms formed by wild type, Δ*motABCD* (deficient in flagellar motility), and Δ*pilB* (deficient in pili-mediated motility) in spaceflight and ground control cultures. (**b**) Illustration of the differences between *P. aeruginosa* biofilms grown under spaceflight conditions and under hydrodynamic and static conditions in normal gravity [7]

differentially expressed in spaceflight cultures of *P. aeruginosa*. Since Hfq was also identified as a key regulator in the spaceflight response of *S.* Typhimurium and in the microgravity-analogue responses of *S. Typhimurium*, *P. aeruginosa*, and *S. aureus* [5, 138–141], these results could entail common spaceflight-induced regulatory mechanisms across different bacterial species. In response to spaceflight culturing, *P. aeruginosa* induced the expression of genes encoding lectins (*lecA* and *lecB*), chitinase (*chiC*), and rhamnolipids (*rhlA*); all are involved in the virulence of this microorganism. On the other hand, downregulation of genes encoding heat shock proteins and the *N*-butanoyl-L-homoserine lactone synthase *rhlI* was observed in response to spaceflight culture. A major portion of the spaceflight regulon in *P. aeruginosa* (60 %) was found to be involved in anaerobic metabolism, among which genes involved in denitrification showed the highest fold induction. While in this study, biofilm formation was not assessed, the spaceflight-induced increases in biofilm formation and column-and-canopy architecture in the paper from Kim and colleagues [7] were also suggested to be due to growth in oxygen-limiting conditions. Taken together, these data suggest that the extremely low fluid-shear and mixing conditions in aqueous microgravity environments may potentially limit available oxygen, resulting in major transcriptional and phenotypic changes in *P. aeruginosa*.

Antibiotic Resistance

The only report on the antibiotic resistance of *P. aeruginosa* in the microgravity environment of spaceflight was published by Juergensmeyer and colleagues [2]. In this study, *P. aeruginosa* was grown on solid media for 4 months in microgravity and ground control conditions. Antibiotic disc tests were performed following sample recovery on Earth. Susceptibility to 12 antibiotics was tested, representing five different modes of action, i.e. cell wall synthesis inhibition (ampicillin, penicillin G, cephalothin), protein synthesis inhibition through binding to the 30S ribosomal subunit (gentamycin, kanamycin, streptomycin, and tetracycline), protein synthesis inhibition through binding to the 50S ribosomal subunit (chloramphenicol, erythromycin), plasma membrane disruption (colistin and polymyxin B), and DNA-dependent RNA polymerase inhibition (rifampicin). Following culture in microgravity conditions, *P. aeruginosa* showed statistically significant increased resistance to colistin; and demonstrated increased susceptibility to cephalothin, polymyxin B, and rifampicin. Since the antibiotic resistance profile of *P. aeruginosa* in microgravity conditions was tested when grown on solid media, possible indirect effects of microgravity were not considered in this study. This could be of importance to understand the antibiotic resistance of *P. aeruginosa* in aqueous habitats of the spacecraft, such as drinking water, rinse less shampoo, toothpaste, mouthwash, and water for laundry. In addition, other spaceflight-associated environmental factors, such as irradiation, could have contributed to the altered antibiotic resistance profile of *P. aeruginosa*.

12.3.2 LSMMG Triggered Responses of P. aeruginosa

Growth

Initial studies with *P. aeruginosa* in the RWV bioreactor revealed that the growth profile of this bacterium was not affected by LSMMG culture conditions [142, 143]. In agreement with these studies, no significant differences in final cell densities were found when *P. aeruginosa* was grown for 24 h in LSMMG compared to control conditions both at 28° and 37° [4, 5].

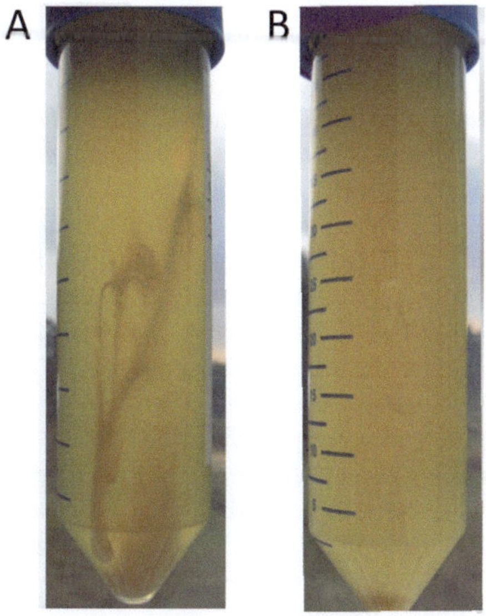

Fig. 12.2 *P. aeruginosa* biofilm formation in LSMMG. (**a**) Photograph of *P. aeruginosa* self-aggregative biofilms formed under LSMMG growth conditions. (**b**) Higher fluid-shear control that contained no self-aggregative biofilms [4]. With permission from Wiley

Biofilm Formation and Resistance to Environmental Stressors

Differences in the biofilm phenotype of *P. aeruginosa* in LSMMG as compared to higher fluid-shear growth conditions when grown at 37° were reported [4] (Fig. 12.2). More specifically, LSMMG induced a self-aggregative biofilm phenotype in *P. aeruginosa*, and induced the expression of genes involved in mannose-rich polysaccharide biosynthesis (*psl* genes). Specifically, bacteria adhered to each other in LSMMG, and aggregates were loosely associated with the gas permeable membrane of the RWV bioreactor. In contrast, when *P. aeruginosa* was cultured under higher fluid-shear conditions (by addition of a ceramic bead to the LSMMG condition), biofilms were generated that were strongly attached to the membrane of the RWV bioreactor [4]. The observed self-aggregative biofilm phenotype is relevant to that of *P. aeruginosa* during chronic lung infections in patients with CF (see above) [144]. This study indicated that low fluid-shear conditions in the lungs of CF patients could be at least partially involved in the development of biofilms that are highly resistant to environmental stressors and antibiotics. In a follow-up study, *P. aeruginosa* showed increased resistance to oxidative and thermal stressors when grown in LSMMG at 28° compared to control conditions [5]. The increased stress resistance could be due to the observed higher production of the exopolysaccharide alginate in LSMMG. Microarray profiling of *P. aeruginosa* cultures grown at 28° in LSMMG versus control conditions confirmed the induction of genes involved in stress resistance and alginate production of this microorganism [5]. The alternative sigma factor *AlgU*, involved in alginate production, was identified as a potential regulator in the LSMMG response of *P. aeruginosa*, since *algU* transcription and a significant portion of its regulon were upregulated.

Gene Expression and Virulence Factor Production

P. aeruginosa **cultured in LSMMG versus higher fluid-shear conditions induced the production of rhamnolipids (*rhlA*), the *N*-butanoyl-l-homoserine lactone quorum-sensing system (*rhlI*), and elastase at the gene expression and/or phenotypic level** [4]. The expression of these virulence genes/factors is of clinical relevance in the context of chronic lung infections in patients with CF [144]. In a later study, the complete transcriptional profile of *P. aeruginosa* in response to LSMMG versus control conditions was determined, and revealed differential expression of 134 genes, including genes involved in the production of virulence factors [5]. In addition to the identification of AlgU as a potential regulator of the LSMMG response of *P. aeruginosa* (see above), Hfq and 50 genes of the Hfq regulon were differentially expressed in LSMMG-cultured *P. aeruginosa*, indicating a possible role for this regulator as well. A hierarchical ranking of the main regulatory pathways induced in LSMMG was proposed, and suggested that AlgU could regulate Hfq and other LSMMG-induced transcriptional regulators (such as Anr, RpoH, RsmA, AlgR) with downstream effects on stress resistance, motility, and microaerophilic metabolism (Fig. 12.3). A notable number of genes upregulated in LSMMG are involved in the growth of *P. aeruginosa* in oxygen-limiting conditions. *P. aeruginosa* predominantly induced the expression of genes involved in microaerophilic growth, such as terminal oxidases with high affinity for oxygen, suggesting limited oxygen availability in the LSMMG culture environment. In accordance with the gene expression data, modeling of the oxygen transfer rate in LSMMG suggested that it was significantly lower as compared to control cultures. The increased alginate concentration in LSMMG cultures was ruled out as the causative factor for the modeled lower oxygen transfer rate, and a cell-associated compound was proposed to be at the origin of a decreased transfer of oxygen in this test condition [5].

12.4 Conclusions

Microgravity and microgravity-analogue culture conditions profoundly affect the gene expression and phenotype of *P. aeruginosa*. Whether the increased expression of virulence factors, different biofilm mode of growth, enhanced resistance to specific antibiotics lead to a higher risk for infectious disease caused by this organism in spaceflight compared to on Earth remains to be determined. While thus far *P. aeruginosa* caused one infection in-flight, the inability to treat it during the mission questions the effectiveness of antibiotics available onboard spacecraft. This incident illustrates the necessity for in-flight clinical microbiology analysis as well as for the assessment of antimicrobial treatment efficiency, especially in the frame of long-term missions.

Of particular importance is the immunocompromised nature of astronauts, which has strong resemblance with the patient population that is susceptible to

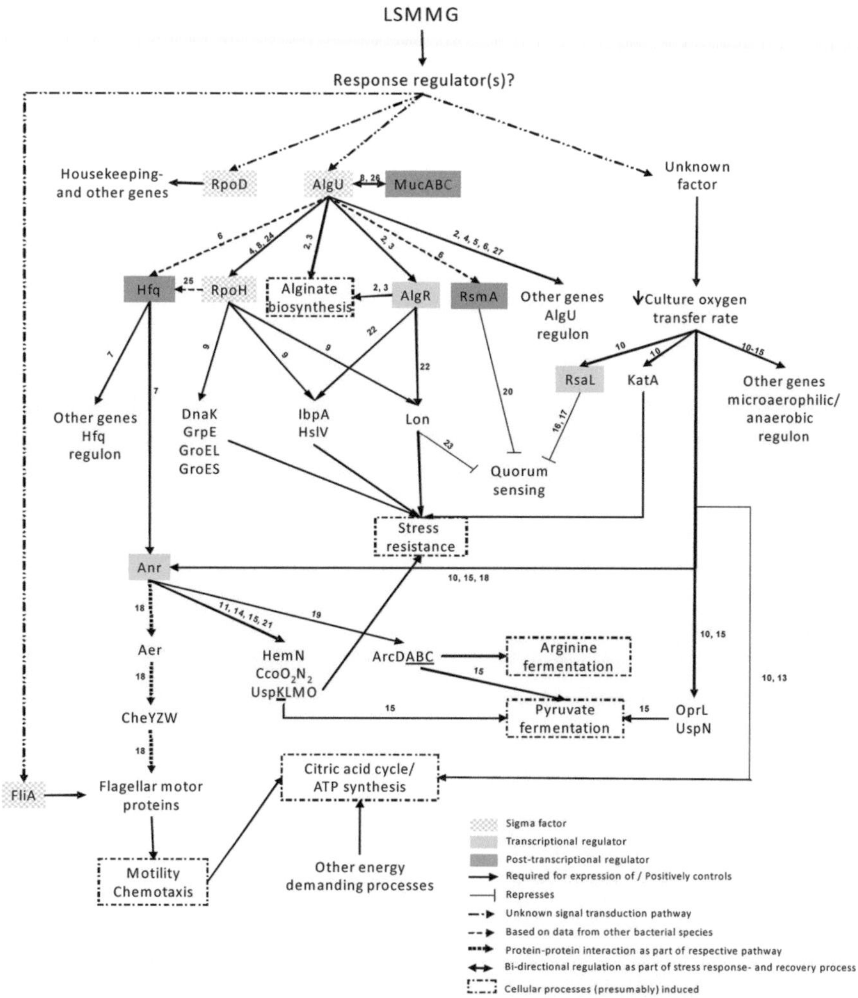

Fig. 12.3 Hierarchical ranking of LSMMG regulatory pathways of *P. aeruginosa* [5]. With permission from Wiley

P. aeruginosa infection on Earth. Therefore, in addition to close monitoring of *P. aeruginosa* on the astronaut body and in the space habitat, and decontamination when possible, strategies to counterbalance the astronaut's affected immune system will presumably be most effective at mitigating the risk for infectious disease with this organism.

In addition to understanding the risk for infectious disease during spaceflight missions, studying *P. aeruginosa* in microgravity and microgravity-analogue conditions has advanced our knowledge on the physical factors that affect the phenotype and virulence of this opportunistic pathogen. Expanding our knowledge on the environmental factors in the host that trigger expression of key virulence factors in *P. aeruginosa*, could help design novel strategies to fight infectious diseases caused by this microorganism, both during spaceflight missions and in the clinic.

Questions for Future Research

In vitro and/or in vivo model systems can help address the following questions to further enhance our understanding of the infectious disease risk by *P. aeruginosa* during current and future manned spaceflight missions:

(a) Does the compromised immune system of astronauts enhance the risk for infection by opportunistic pathogens, such as *P. aeruginosa*?
(b) What is the effectiveness of antibiotics during in-flight infections?
(c) Does microgravity have *direct* or *indirect* effects on virulence and antibiotic resistance of *P. aeruginosa*?
(d) In the frame of future missions to Mars and other celestial bodies, what are the long-term effects of microgravity conditions on the virulence of *P. aeruginosa*?
(e) How can microgravity-induced changes in *P. aeruginosa* phenotype and virulence characteristics be altered to minimize risk for infectious diseases?

References

1. Taylor, G. R. (1974). Recovery of medically important microorganisms from Apollo astronauts. *Aerospace Medicine, 45*(8), 824–828.
2. Juergensmeyer, M. A., Juergensmeyer, E. A., & Guikema, J. A. (1999). Long-term exposure to spaceflight conditions affects bacterial response to antibiotics. *Microgravity Science and Technology, 12*(1), 41–47.
3. McLean, R. J., Cassanto, J. M., et al. (2001). Bacterial biofilm formation under microgravity conditions. *FEMS Microbiology Letters, 195*(2), 115–119.
4. Crabbé, A., De Boever, P., et al. (2008). Use of the rotating wall vessel technology to study the effect of shear stress on growth behaviour of *Pseudomonas aeruginosa* PA01. *Environmental Microbiology, 10*(8), 2098–2110.
5. Crabbé, A., Pycke, B., et al. (2010). Response of *Pseudomonas aeruginosa* PAO1 to low shear modelled microgravity involves AlgU regulation. *Environmental Microbiology, 12*(6), 1545–1564.
6. Crabbé, A., Schurr, M. J., et al. (2011). Transcriptional and proteomic responses of *Pseudomonas aeruginosa* PAO1 to spaceflight conditions involve Hfq regulation and reveal a role for oxygen. *Applied and Environmental Microbiology, 77*(4), 1221–1230.
7. Kim, W., Tengra, F. K., et al. (2013). Spaceflight promotes biofilm formation by *Pseudomonas aeruginosa. PLoS One, 8*(4), e62437.
8. Gessard, C. (1882). Sur les colorations bleue et verte des lignes a pansements. *Comptes Rendus de l'Académie des Sciences – Series D, 94*, 536–538.
9. Khan, N. H., Ahsan, M., et al. (2010). Culturability and survival of marine, freshwater and clinical *Pseudomonas aeruginosa. Microbes and Environments, 25*(4), 266–274.

10. Kimata, N., Nishino, T., et al. (2004). *Pseudomonas aeruginosa* isolated from marine environments in Tokyo Bay. *Microbial Ecology, 47*(1), 41–47.
11. Stover, C. K., Pham, X. Q., et al. (2000). Complete genome sequence of *Pseudomonas aeruginosa* PAO1, an opportunistic pathogen. *Nature, 406*(6799), 959–964.
12. Blanc, D. S., Petignat, C., et al. (1998). Frequency and molecular diversity of *Pseudomonas aeruginosa* upon admission and during hospitalization: A prospective epidemiologic study. *Clinical Microbiology and Infection, 4*(5), 242–247.
13. Pechere, J. C., & Kohler, T. (1999). Patterns and modes of beta-lactam resistance in *Pseudomonas aeruginosa. Clinical Microbiology and Infection, 5*(Suppl 1), S15–S18.
14. McGowan, J. E., Jr. (2006). Resistance in nonfermenting gram-negative bacteria: Multidrug resistance to the maximum. *American Journal of Infection Control, 34*(5 Suppl 1), S29–S37. discussion S64–73.
15. Stevens, A. M., Schuster, M., & Rumbaugh, K. P. (2012). Working together for the common good: Cell-cell communication in bacteria. *Journal of Bacteriology, 194*(9), 2131–2141.
16. Williams, P., & Camara, M. (2009). Quorum sensing and environmental adaptation in *Pseudomonas aeruginosa*: A tale of regulatory networks and multifunctional signal molecules. *Current Opinion in Microbiology, 12*(2), 182–191.
17. Gambello, M. J., Kaye, S., & Iglewski, B. H. (1993). LasR of *Pseudomonas aeruginosa* is a transcriptional activator of the alkaline protease gene (apr) and an enhancer of exotoxin A expression. *Infection and Immunity, 61*(4), 1180–1184.
18. Toder, D. S., Ferrell, S. J., et al. (1994). lasA and lasB genes of *Pseudomonas aeruginosa*: Analysis of transcription and gene product activity. *Infection and Immunity, 62*(4), 1320–1327.
19. Ochsner, U. A., & Reiser, J. (1995). Autoinducer-mediated regulation of rhamnolipid biosurfactant synthesis in *Pseudomonas aeruginosa. Proceedings of the National Academy of Sciences of the United States of America, 92*(14), 6424–6428.
20. Hassett, D. J., Ma, J. F., et al. (1999). Quorum sensing in *Pseudomonas aeruginosa* controls expression of catalase and superoxide dismutase genes and mediates biofilm susceptibility to hydrogen peroxide. *Molecular Microbiology, 34*(5), 1082–1093.
21. Brint, J. M., & Ohman, D. E. (1995). Synthesis of multiple exoproducts in *Pseudomonas aeruginosa* is under the control of RhlR-RhlI, another set of regulators in strain PAO1 with homology to the autoinducer-responsive LuxR-LuxI family. *Journal of Bacteriology, 177*(24), 7155–7163.
22. Latifi, A., Winson, M. K., et al. (1995). Multiple homologues of LuxR and LuxI control expression of virulence determinants and secondary metabolites through quorum sensing in *Pseudomonas aeruginosa* PAO1. *Molecular Microbiology, 17*(2), 333–343.
23. Pearson, J. P., Pesci, E. C., & Iglewski, B. H. (1997). Roles of *Pseudomonas aeruginosa* las and rhl quorum-sensing systems in control of elastase and rhamnolipid biosynthesis genes. *Journal of Bacteriology, 179*(18), 5756–5767.
24. Winzer, K., Falconer, C., et al. (2000). The *Pseudomonas aeruginosa* lectins PA-IL and PA-IIL are controlled by quorum sensing and by RpoS. *Journal of Bacteriology, 182*(22), 6401–6411.
25. Gambello, M. J., & Iglewski, B. H. (1991). Cloning and characterization of the *Pseudomonas aeruginosa* lasR gene, a transcriptional activator of elastase expression. *Journal of Bacteriology, 173*(9), 3000–3009.
26. Pearson, J. P., Gray, K. M., et al. (1994). Structure of the autoinducer required for expression of *Pseudomonas aeruginosa* virulence genes. *Proceedings of the National Academy of Sciences of the United States of America, 91*(1), 197–201.
27. Kiratisin, P., Tucker, K. D., & Passador, L. (2002). LasR, a transcriptional activator of *Pseudomonas aeruginosa* virulence genes, functions as a multimer. *Journal of Bacteriology, 184*(17), 4912–4919.
28. Ochsner, U. A., Koch, A. K., et al. (1994). Isolation and characterization of a regulatory gene affecting rhamnolipid biosurfactant synthesis in *Pseudomonas aeruginosa. Journal of Bacteriology, 176*(7), 2044–2054.

29. Pearson, J. P., Passador, L., et al. (1995). A second N-acylhomoserine lactone signal produced by *Pseudomonas aeruginosa. Proceedings of the National Academy of Sciences of the United States of America, 92*(5), 1490–1494.
30. Gallagher, L. A., McKnight, S. L., et al. (2002). Functions required for extracellular quinolone signaling by *Pseudomonas aeruginosa. Journal of Bacteriology, 184*(23), 6472–6480.
31. Tashiro, Y., et al. (2013). Interspecies interaction between *Pseudomonas aeruginosa* and other microorganisms. *Microbes and Environments, 28*(1), 13–24.
32. Balasubramanian, D., et al. (2013). A dynamic and intricate regulatory network determines *Pseudomonas aeruginosa* virulence. *Nucleic Acids Research, 41*(1), 1–20.
33. Singh, P. K., Schaefer, A. L., et al. (2000). Quorum-sensing signals indicate that cystic fibrosis lungs are infected with bacterial biofilms. *Nature, 407*(6805), 762–764.
34. Lam, J., Chan, R., et al. (1980). Production of mucoid microcolonies by *Pseudomonas aeruginosa* within infected lungs in cystic fibrosis. *Infection and Immunity, 28*(2), 546–556.
35. Flemming, H. C., & Wingender, J. (2010). The biofilm matrix. *Nature Reviews. Microbiology, 8*(9), 623–633.
36. Donlan, R. M., & Costerton, J. W. (2002). Biofilms: Survival mechanisms of clinically relevant microorganisms. *Clinical Microbiology Reviews, 15*(2), 167–193.
37. Harmsen, M., Yang, L., et al. (2010). An update on *Pseudomonas aeruginosa* biofilm formation, tolerance, and dispersal. *FEMS Immunology and Medical Microbiology, 59*(3), 253–268.
38. O'Toole, G. A., & Kolter, R. (1998). Flagellar and twitching motility are necessary for *Pseudomonas aeruginosa* biofilm development. *Molecular Microbiology, 30*(2), 295–304.
39. Costerton, J. W., Stewart, P. S., & Greenberg, E. P. (1999). Bacterial biofilms: A common cause of persistent infections. *Science, 284*(5418), 1318–1322.
40. Stapper, A. P., Narasimhan, G., et al. (2004). Alginate production affects *Pseudomonas aeruginosa* biofilm development and architecture, but is not essential for biofilm formation. *Journal of Medical Microbiology, 53*(Pt 7), 679–690.
41. Lopez, D., Vlamakis, H., & Kolter, R. (2010). Biofilms. *Cold Spring Harbor Perspectives in Biology, 2*(7), a000398.
42. Alhede, M., Kragh, K. N., et al. (2011). Phenotypes of non-attached *Pseudomonas aeruginosa* aggregates resemble surface attached biofilm. *PLoS One, 6*(11), e27943.
43. Kim, W., Tengra, F. K., et al. (2003). Biofilm formation by *Pseudomonas aeruginosa* wild type, flagella and type IV pili mutants. *Molecular Microbiology, 48*(6), 1511–1524.
44. Boles, B. R., Thoendel, M., & Singh, P. K. (2005). Rhamnolipids mediate detachment of *Pseudomonas aeruginosa* from biofilms. *Molecular Microbiology, 57*(5), 1210–1223.
45. Friedman, L., & Kolter, R. (2004). Two genetic loci produce distinct carbohydrate-rich structural components of the *Pseudomonas aeruginosa* biofilm matrix. *Journal of Bacteriology, 186*(14), 4457–4465.
46. Jackson, K. D., Starkey, M., et al. (2004). Identification of psl, a locus encoding a potential exopolysaccharide that is essential for *Pseudomonas aeruginosa* PAO1 biofilm formation. *Journal of Bacteriology, 186*(14), 4466–4475.
47. Matsukawa, M., & Greenberg, E. P. (2004). Putative exopolysaccharide synthesis genes influence *Pseudomonas aeruginosa* biofilm development. *Journal of Bacteriology, 186*(14), 4449–4456.
48. Haussler, S., Ziegler, I., et al. (2003). Highly adherent small-colony variants of *Pseudomonas aeruginosa* in cystic fibrosis lung infection. *Journal of Medical Microbiology, 52*(Pt 4), 295–301.
49. Sarkisova, S., Patrauchan, M. A., et al. (2005). Calcium-induced virulence factors associated with the extracellular matrix of mucoid *Pseudomonas aeruginosa* biofilms. *Journal of Bacteriology, 187*(13), 4327–4337.
50. Ma, L., Jackson, K. D., et al. (2006). Analysis of *Pseudomonas aeruginosa* conditional psl variants reveals roles for the psl polysaccharide in adhesion and maintaining biofilm structure postattachment. *Journal of Bacteriology, 188*(23), 8213–8221.

51. Smith, E. E., Buckley, D. G., et al. (2006). Genetic adaptation by *Pseudomonas aeruginosa* to the airways of cystic fibrosis patients. *Proceedings of the National Academy of Sciences of the United States of America, 103*(22), 8487–8492.
52. Yamamoto, K., Haruta, S., et al. (2010). Determinative factors of competitive advantage between aerobic bacteria for niches at the air-liquid interface. *Microbes and Environments, 25*(4), 317–320.
53. Yamamoto, K., Arai, H., et al. (2011). Trade-off between oxygen and iron acquisition in bacterial cells at the air-liquid interface. *FEMS Microbiology Ecology, 77*(1), 83–94.
54. Yamamoto, K., Arai, H., et al. (2012). Involvement of flagella-driven motility and pili in *Pseudomonas aeruginosa* colonization at the air-liquid interface. *Microbes and Environments, 27*(3), 320–323.
55. Pressler, T., Bohmova, C., et al. (2011). Chronic *Pseudomonas aeruginosa* infection definition: EuroCareCF Working Group report. *Journal of Cystic Fibrosis, 10*(Suppl 2), S75–S78.
56. Evans, L. R., & Linker, A. (1973). Production and characterization of the slime polysaccharide of *Pseudomonas aeruginosa. Journal of Bacteriology, 116*(2), 915–924.
57. Mathee, K., Ciofu, O., et al. (1999). Mucoid conversion of *Pseudomonas aeruginosa* by hydrogen peroxide: A mechanism for virulence activation in the cystic fibrosis lung. *Microbiology, 145*(Pt 6), 1349–1357.
58. Govan, J. R., & Fyfe, J. A. (1978). Mucoid *Pseudomonas aeruginosa* and cystic fibrosis: Resistance of the mucoid from to carbenicillin, flucloxacillin and tobramycin and the isolation of mucoid variants in vitro. *Journal of Antimicrobial Chemotherapy, 4*(3), 233–240.
59. Speert, D. P., Farmer, S. W., et al. (1990). Conversion of *Pseudomonas aeruginosa* to the phenotype characteristic of strains from patients with cystic fibrosis. *Journal of Clinical Microbiology, 28*(2), 188–194.
60. Terry, J. M., Pina, S. E., & Mattingly, S. J. (1991). Environmental conditions which influence mucoid conversion *Pseudomonas aeruginosa* PAO1. *Infection and Immunity, 59*(2), 471–477.
61. Martin, D. W., Holloway, B. W., & Deretic, V. (1993). Characterization of a locus determining the mucoid status of *Pseudomonas aeruginosa*: AlgU shows sequence similarities with a Bacillus sigma factor. *Journal of Bacteriology, 175*(4), 1153–1164.
62. DeVries, C. A., & Ohman, D. E. (1994). Mucoid-to-nonmucoid conversion in alginate-producing *Pseudomonas aeruginosa* often results from spontaneous mutations in algT, encoding a putative alternate sigma factor, and shows evidence for autoregulation. *Journal of Bacteriology, 176*(21), 6677–6687.
63. Balasubramanian, D., Kong, K. F., et al. (2011). Co-regulation of {beta}-lactam resistance, alginate production and quorum sensing in *Pseudomonas aeruginosa. Journal of Medical Microbiology, 60*(Pt 2), 147–156.
64. Balloy, V., et al. (2007). The role of flagellin versus motility in acute lung disease caused by *Pseudomonas aeruginosa. Journal of Infectious Diseases, 196*(2), 289–296.
65. Arora, S. K., et al. (2001). A genomic island in *Pseudomonas aeruginosa* carries the determinants of flagellin glycosylation. *Proceedings of the National Academy of Sciences of the United States of America, 98*(16), 9342–9347.
66. Engel, J. N. (2003). Molecular pathogenesis of acute *Pseudomonas aeruginosa* infections. In A. Hauser & J. Rello (Eds.), *Severe infections caused by Pseudomonas Aeruginosa* (pp. 201–230). New York: Kluwer Academic/Plenum Press.
67. Morihara, K. (1964). Production of elastase and proteinase by *Pseudomonas aeruginosa. Journal of Bacteriology, 88*, 745–757.
68. McIver, K. S., Kessler, E., & Ohman, D. E. (2004). Identification of residues in the *Pseudomonas aeruginosa* elastase propeptide required for chaperone and secretion activities. *Microbiology, 150*(Pt 12), 3969–3977.
69. Azghani, A. O. (1996). *Pseudomonas aeruginosa* and epithelial permeability: Role of virulence factors elastase and exotoxin A. *American Journal of Respiratory Cell and Molecular Biology, 15*(1), 132–140.

70. de Bentzmann, S., Polette, M., et al. (2000). *Pseudomonas aeruginosa* virulence factors delay airway epithelial wound repair by altering the actin cytoskeleton and inducing overactivation of epithelial matrix metalloproteinase-2. *Laboratory Investigation, 80*(2), 209–219.
71. Horvat, R. T., Clabaugh, M., et al. (1989). Inactivation of human gamma interferon by *Pseudomonas aeruginosa* proteases: Elastase augments the effects of alkaline protease despite the presence of alpha 2-macroglobulin. *Infection and Immunity, 57*(6), 1668–1674.
72. Leidal, K. G., Munson, K. L., et al. (2003). Metalloproteases from *Pseudomonas aeruginosa* degrade human RANTES, MCP-1, and ENA-78. *Journal of Interferon and Cytokine Research, 23*(6), 307–318.
73. Parmely, M., Gale, A., et al. (1990). Proteolytic inactivation of cytokines by *Pseudomonas aeruginosa. Infection and Immunity, 58*(9), 3009–3014.
74. Theander, T. G., Kharazmi, A., et al. (1988). Inhibition of human lymphocyte proliferation and cleavage of interleukin-2 by *Pseudomonas aeruginosa* proteases. *Infection and Immunity, 56*(7), 1673–1677.
75. Schad, P. A., Bever, R. A., et al. (1987). Cloning and characterization of elastase genes from *Pseudomonas aeruginosa. Journal of Bacteriology, 169*(6), 2691–2696.
76. Kessler, E., Safrin, M., et al. (1993). Secreted LasA of *Pseudomonas aeruginosa* is a staphylolytic protease. *Journal of Biological Chemistry, 268*(10), 7503–7508.
77. Kessler, E. (1995). beta-lytic endopeptidases. *Methods in Enzymology, 248*, 740–756.
78. Kessler, E., & Ohman, D. E. (1998). Staphylolysin. In A. J. Barrett, N. D. Rawlings, & J. F. Woessner (Eds.), *Handbook of proteolytic enzymes* (1st ed., pp. 1476–1478). London: Academic.
79. Kessler, E., Safrin, M., et al. (1997). Inhibitors and specificity of *Pseudomonas aeruginosa* LasA. *Journal of Biological Chemistry, 272*(15), 9884–9889.
80. Abdel-Mawgoud, A. M., Lepine, F., & Deziel, E. (2010). Rhamnolipids: Diversity of structures, microbial origins and roles. *Applied Microbiology and Biotechnology, 86*(5), 1323–1336.
81. Johnson, M. K., & Boese-Marrazzo, D. (1980). Production and properties of heat-stable extracellular hemolysin from *Pseudomonas aeruginosa. Infection and Immunity, 29*(3), 1028–1033.
82. Jensen, P. O., Bjarnsholt, T., et al. (2007). Rapid necrotic killing of polymorphonuclear leukocytes is caused by quorum-sensing-controlled production of rhamnolipid by *Pseudomonas aeruginosa. Microbiology, 153*(Pt 5), 1329–1338.
83. McClure, C. D., & Schiller, N. L. (1992). Effects of *Pseudomonas aeruginosa* rhamnolipids on human monocyte-derived macrophages. *Journal of Leukocyte Biology, 51*(2), 97–102.
84. Read, R. C., Roberts, P., et al. (1992). Effect of *Pseudomonas aeruginosa* rhamnolipids on mucociliary transport and ciliary beating. *Journal of Applied Physiology, 72*(6), 2271–2277.
85. Burger, M. M., Glaser, L., & Burton, R. M. (1963). The enzymatic synthesis of a rhamnose-containing glycolipid by extracts of *Pseudomonas aeruginosa. Journal of Biological Chemistry, 238*, 2595–2602.
86. Deziel, E., Lepine, F., et al. (2003). rhlA is required for the production of a novel biosurfactant promoting swarming motility in *Pseudomonas aeruginosa*: 3-(3-hydroxyalkanoyloxy) alkanoic acids (HAAs), the precursors of rhamnolipids. *Microbiology, 149*(Pt 8), 2005–2013.
87. Haba, E., Pinazo, A., et al. (2003). Physicochemical characterization and antimicrobial properties of rhamnolipids produced by *Pseudomonas aeruginosa* 47T2 NCBIM 40044. *Biotechnology and Bioengineering, 81*(3), 316–322.
88. Caiazza, N. C., Shanks, R. M., & O'Toole, G. A. (2005). Rhamnolipids modulate swarming motility patterns of *Pseudomonas aeruginosa. Journal of Bacteriology, 187*(21), 7351–7361.
89. Dietrich, L. E., Teal, T. K., et al. (2008). Redox-active antibiotics control gene expression and community behavior in divergent bacteria. *Science, 321*(5893), 1203–1206.
90. Wilson, R., Sykes, D. A., et al. (1988). Measurement of *Pseudomonas aeruginosa* phenazine pigments in sputum and assessment of their contribution to sputum sol toxicity for respiratory epithelium. *Infection and Immunity, 56*(9), 2515–2517.

91. Paroutis, P., Touret, N., & Grinstein, S. (2004). The pH of the secretory pathway: Measurement, determinants, and regulation. *Physiology (Bethesda), 19*, 207–215.
92. Karlsson, K. A. (2001). Pathogen-host protein-carbohydrate interactions as the basis of important infections. *Advances in Experimental Medicine and Biology, 491*, 431–443.
93. Glick, J., & Garber, N. (1983). The intracellular localization of *Pseudomonas aeruginosa* lectins. *Journal of General Microbiology, 129*(10), 3085–3090.
94. Tielker, D., Hacker, S., et al. (2005). *Pseudomonas aeruginosa* lectin LecB is located in the outer membrane and is involved in biofilm formation. *Microbiology, 151*(Pt 5), 1313–1323.
95. Bajolet-Laudinat, O., Girod-de Bentzmann, S., et al. (1994). Cytotoxicity of *Pseudomonas aeruginosa* internal lectin PA-I to respiratory epithelial cells in primary culture. *Infection and Immunity, 62*(10), 4481–4487.
96. Laughlin, R. S., Musch, M. W., et al. (2000). The key role of *Pseudomonas aeruginosa* PA-I lectin on experimental gut-derived sepsis. *Annals of Surgery, 232*(1), 133–142.
97. Sonawane, A., Jyot, J., & Ramphal, R. (2006). *Pseudomonas aeruginosa* LecB is involved in pilus biogenesis and protease IV activity but not in adhesion to respiratory mucins. *Infection and Immunity, 74*(12), 7035–7039.
98. Valentin-Hansen, P., Eriksen, M., & Udesen, C. (2004). The bacterial Sm-like protein Hfq: A key player in RNA transactions. *Molecular Microbiology, 51*(6), 1525–1533.
99. Majdalani, N., Vanderpool, C. K., & Gottesman, S. (2005). Bacterial small RNA regulators. *Critical Reviews in Biochemistry and Molecular Biology, 40*(2), 93–113.
100. Sonnleitner, E., Hagens, S., et al. (2003). Reduced virulence of a hfq mutant of *Pseudomonas aeruginosa* O1. *Microbial Pathogenesis, 35*(5), 217–228.
101. Passador, L., Cook, J. M., et al. (1993). Expression of *Pseudomonas aeruginosa* virulence genes requires cell-to-cell communication. *Science, 260*(5111), 1127–1130.
102. Suh, S. J., Silo-Suh, L., et al. (1999). Effect of rpoS mutation on the stress response and expression of virulence factors in *Pseudomonas aeruginosa. Journal of Bacteriology, 181*(13), 3890–3897.
103. Chugani, S. A., Whiteley, M., et al. (2001). QscR, a modulator of quorum-sensing signal synthesis and virulence in *Pseudomonas aeruginosa. Proceedings of the National Academy of Sciences of the United States of America, 98*(5), 2752–2757.
104. Gallagher, L. A., & Manoil, C. (2001). *Pseudomonas aeruginosa* PAO1 kills Caenorhabditis elegans by cyanide poisoning. *Journal of Bacteriology, 183*(21), 6207–6214.
105. Usher, L. R., Lawson, R. A., et al. (2002). Induction of neutrophil apoptosis by the *Pseudomonas aeruginosa* exotoxin pyocyanin: A potential mechanism of persistent infection. *Journal of Immunology, 168*(4), 1861–1868.
106. Erb-Downward, J. R., Thompson, D. L., et al. (2011). Analysis of the lung microbiome in the "healthy" smoker and in COPD. *PLoS One, 6*(2), e16384.
107. Bruce, R. J., Ott, C. M., et al. (2005). Microbial surveillance of potable water sources of the International Space Station. *SAE Transactions, 114*(1), 283–292.
108. Novikova, N., De Boever, P., et al. (2006). Survey of environmental biocontamination on board the International Space Station. *Research in Microbiology, 157*(1), 5–12.
109. La Duc, M. T., Kern, R., & Venkateswaran, K. (2004). Microbial monitoring of spacecraft and associated environments. *Microbial Ecology, 47*(2), 150–158.
110. Novikova, N. D. (2004). Review of the knowledge of microbial contamination of the Russian manned spacecraft. *Microbial Ecology, 47*(2), 127–132.
111. Ball, J. R., & Evans, C. H., Jr. (Eds.). (2001). *Safe passage: Astronaut care for exploration missions* (p. 105). Washington: National Academy Press.
112. Gu, J. D., Roman, M., et al. (1998). The role of microbial biofilms in deterioration of space station candidate materials. *International Biodeterioration and Biodegradation, 41*(1), 25–33.
113. Zaloguyev, S. N., & Viktorov, A. N. (1985). Results of microbiological studies performed during the use of Salyut-6 Space Station. *Kosmicheskaya Biologiya i Aviakosmicheskaya Meditsina, 3*, 64–66.

114. Viktorov, A. N. (1992). Microflora of cabins of manned space objects and the problem of biological damage to the structural materials used in them. *Biologiya Aviakosmicheskaia i Ekologicheskaia Meditsina, 26*(3), 41–48.
115. Schuerger, A. C. (1998). Microbial contamination of advanced life support (ALS) systems poses a moderate threat to the long-term stability of space-based bioregenerative systems. *Life Support & Biosphere Science, 5*(3), 325–337.
116. Crucian, B., Stowe, R., et al. (2013). Immune system dysregulation occurs during short duration spaceflight on board the space shuttle. *Journal of Clinical Immunology, 33*(2), 456–465.
117. Gueguinou, N., Huin-Schohn, C., et al. (2009). Could spaceflight-associated immune system weakening preclude the expansion of human presence beyond Earth's orbit? *Journal of Leukocyte Biology, 86*(5), 1027–1038.
118. Battista, N., Meloni, M. A., et al. (2012). 5-Lipoxygenase-dependent apoptosis of human lymphocytes in the International Space Station: Data from the ROALD experiment. *FASEB Journal, 26*(5), 1791–1798.
119. Sonnenfeld, G. (2005). The immune system in space, including Earth-based benefits of space-based research. *Current Pharmaceutical Biotechnology, 6*(4), 343–349.
120. Driscoll, J. A., Brody, S. L., & Kollef, M. H. (2007). The epidemiology, pathogenesis and treatment of *Pseudomonas aeruginosa* infections. *Drugs, 67*(3), 351–368.
121. Funada, H., & Matsuda, T. (1998). Changes in the incidence and etiological patterns of bacteremia associated with acute leukemia over a 25-year period. *Internal Medicine, 37*(12), 1014–1018.
122. Chatzinikolaou, I., Abi-Said, D., et al. (2000). Recent experience with *Pseudomonas aeruginosa* bacteremia in patients with cancer: Retrospective analysis of 245 episodes. *Archives of Internal Medicine, 160*(4), 501–509.
123. Wei, L. X., Zhou, J. N., et al. (2003). Lymphocyte reduction induced by hindlimb unloading: Distinct mechanisms in the spleen and thymus. *Cell Research, 13*(6), 465–471.
124. Aviles, H., Belay, T., et al. (2003). Increased susceptibility to *Pseudomonas aeruginosa* infection under hindlimb-unloading conditions. *Journal of Applied Physiology, 95*(1), 73–80.
125. Beeson, J. G., Rogerson, S. J., et al. (2000). Adhesion of Plasmodium falciparum-infected erythrocytes to hyaluronic acid in placental malaria. *Nature Medicine, 6*(1), 86–90.
126. Friedman, M. H., & Giddens, D. P. (2005). Blood flow in major blood vessels-modeling and experiments. *Annals of Biomedical Engineering, 33*(12), 1710–1713.
127. Guo, P., Weinstein, A. M., & Weinbaum, S. (2000). A hydrodynamic mechanosensory hypothesis for brush border microvilli. *American Journal of Physiology. Renal Physiology, 279*(4), F698–F712.
128. Goeres, D. M., Hamilton, M. A., et al. (2009). A method for growing a biofilm under low shear at the air-liquid interface using the drip flow biofilm reactor. *Nature Protocols, 4*(5), 783–788.
129. Blake, J. (1973). A note on mucus shear rates. *Respiration Physiology, 17*(3), 394–399.
130. Fonseca, A. P., & Sousa, J. C. (2007). Effect of shear stress on growth, adhesion and biofilm formation of *Pseudomonas aeruginosa* with antibiotic-induced morphological changes. *International Journal of Antimicrobial Agents, 30*(3), 236–241.
131. Purevdorj, B., Costerton, J. W., & Stoodley, P. (2002). Influence of hydrodynamics and cell signaling on the structure and behavior of *Pseudomonas aeruginosa* biofilms. *Applied and Environmental Microbiology, 68*(9), 4457–4464.
132. Stoodley, P., Lewandowski, Z., et al. (1999). Structural deformation of bacterial biofilms caused by short-term fluctuations in fluid shear: An in situ investigation of biofilm rheology. *Biotechnology and Bioengineering, 65*(1), 83–92.
133. Lecuyer, S., Rusconi, R., et al. (2011). Shear stress increases the residence time of adhesion of *Pseudomonas aeruginosa. Biophysical Journal, 100*(2), 341–350.
134. Zhang, W., Sileika, T., & Packman, A. I. (2013). Effects of fluid flow conditions on interactions between species in biofilms. *FEMS Microbiology Ecology, 84*(2), 344–354.

135. Pereira, M. O., Kuehn, M., et al. (2002). Effect of flow regime on the architecture of a Pseudomonas fluorescens biofilm. *Biotechnology and Bioengineering, 78*(2), 164–171.
136. Sriramulu, D. D., Lunsdorf, H., et al. (2005). Microcolony formation: A novel biofilm model of *Pseudomonas aeruginosa* for the cystic fibrosis lung. *Journal of Medical Microbiology, 54*(Pt 7), 667–676.
137. Kim, W., Tengra, F. K., et al. (2013). Effect of spaceflight on *Pseudomonas aeruginosa* final cell density is modulated by nutrient and oxygen availability. *BMC Microbiology, 13*(1), 241.
138. Wilson, J. W., Ott, C. M., et al. (2007). Space flight alters bacterial gene expression and virulence and reveals a role for global regulator Hfq. *Proceedings of the National Academy of Sciences of the United States of America, 104*(41), 16299–16304.
139. Wilson, J. W., Ott, C. M., et al. (2008). Media ion composition controls regulatory and virulence response of Salmonella in spaceflight. *PLoS One, 3*(12), e3923.
140. Castro, S. L., Nelman-Gonzalez, M., et al. (2011). Low fluid shear culture of *Staphylococcus aureus* induces attachment-independent biofilm formation and represses hfq expression. *Applied and Environmental Microbiology, 77*(18), 6368–6378.
141. Wilson, J. W., Ramamurthy, R., et al. (2002). Microarray analysis identifies Salmonella genes belonging to the low-shear modeled microgravity regulon. *Proceedings of the National Academy of Sciences of the United States of America, 99*(21), 13807–13812.
142. Guadarrama, S., Pulcini, E., et al. (2005). *Pseudomonas aeruginosa* growth and production of Exotoxin A in static and modeled microgravity environments. *Gravitational and Space Biology Bulletin, 18*(2), 85–86.
143. England, L. S., Gorzelak, M., & Trevors, J. T. (2003). Growth and membrane polarization in *Pseudomonas aeruginosa* UG2 grown in randomized microgravity in a high aspect ratio vessel. *Biochimica et Biophysica Acta, 1624*(1–3), 76–80.
144. Ciofu, O., Mandsberg, L. F., et al. (2012). Phenotypes selected during chronic lung infection in cystic fibrosis patients: Implications for the treatment of *Pseudomonas aeruginosa* biofilm infections. *FEMS Immunology and Medical Microbiology, 65*(2), 215–225.

Chapter 13
Cellular Response of *Escherichia coli* to Microgravity and Microgravity Analogue Culture

Rachna Singh and A.C. Matin

Historical Landmarks

1985—Tixador et al. demonstrate an increased bacterial resistance to antibiotics under microgravity conditions [1].

1986—Ciferri et al. report that genetic recombination, especially conjugation, is enhanced in *Escherichia coli* in microgravity [2].

1996—Thevenet et al. demonstrate that the growth kinetics of non-motile, but not motile, *E. coli* strains are altered by exposure to microgravity [3].

1997—Klaus et al. propose a fluid dynamics-based model to explain the observed changes in *E. coli* growth kinetics under microgravity [4].

1997—Fang et al. describe an altered secondary metabolism in *E. coli* following exposure to low shear modeled microgravity (LSMMG) [5].

2004—Lynch et al. report the role and regulation of stationary-phase sigma factor, sigma S, in general resistance conferred by LSMMG in *E. coli* [6].

2006—Lynch et al. devise a method for cultivating bacterial biofilms under LSMMG and show that *E. coli* forms more copious biofilms that are more resistant to disinfectants and antibiotics in LSSMG as compared to control orientation [7].

2007—Tucker et al. report the first transcriptomic profiling of *E. coli* in response to LSMMG culture in rich and minimal media [8].

2008—Allen et al. demonstrate a potential role for indole as a signaling molecule that regulates adherence under LSMMG conditions in a clinical isolate of adherent and invasive *E. coli* [9].

R. Singh • A.C. Matin, PhD (✉)
Department of Microbiology and Immunology, Stanford University School of Medicine,
Sherman Fairchild Science Building D317, 299 Campus Drive,
Stanford 94305 5124, CA, USA
e-mail: a.matin@stanford.edu

C.A. Nickerson et al. (eds.), *Effect of Spaceflight and Spaceflight Analogue Culture on Human and Microbial Cells*, DOI 10.1007/978-1-4939-3277-1_13

13.1 Introduction

Life on Earth arose in the presence of gravity and it is intriguing to consider how living things react to a condition, such as microgravity, which is entirely foreign to them. This is one reason for interest in investigating the effect of microgravity on microbes. There are more practical reasons as well. For instance, it is important to more systematically investigate the early indications that bacteria become more resistant and virulent under LSMMG [6, 7, 10]. The importance of this aspect is further underscored by the increasing evidence that the human immune response is compromised under spaceflight conditions [11, 12]. LSMMG culture conditions reflect the environment of certain routes of microbial infections here on Earth (as described in Chap. 11), so an understanding of pathogenicity under these conditions is relevant to health issues for the general public. Likewise, changes in microbial responses induced by true microgravity culture in spaceflight are important to consider for their potential impact on bioregenerative life support systems. *E. coli* is important in both respects, as (1) uropathogenic *E. coli* (UPEC) is one of the several causative agents of urinary tract infections, which previously have and may affect future spaceflight missions [13] and (2) *E. coli* plays critical roles in ecosystems for renewal of resources, such as oxygen and water, as well as waste recycling. Such ecosystems need to be established in spacecraft and eventually on other planets for effective space exploration and colonization.

The environment beyond the Earth's magnetic field is extremely harsh, involving in addition to microgravity, highly intense solar radiation, extreme temperatures, and high vacuum. Only two missions, Apollo 16 and 17, have carried microbial experiments beyond low Earth orbit that would have been exposed to these conditions, but none of the studies concerned *E. coli* and are therefore not considered here. Most of the studies on *E. coli* and other bacteria have been performed in Earth-orbiting spacecraft, e.g. the Russian Foton satellites [14], space shuttles, and space stations, such as MIR and the International Space Station (ISS) [15–17]. These vehicles are subjected to a gravity-induced free fall around the Earth resulting in a state resembling weightlessness akin to microgravity [14]. Pressurized modules and life support systems have been used in these shuttles and stations that shielded the microbes from many of the harsh stresses of space, exposing them primarily to microgravity and radiation. In most of the experiments conducted during these missions that involved *E. coli*, the effect of both influences was cumulatively studied, but in some experiments, a parallel 1*g* control was included, suggesting that the effect was due to microgravity alone. In other studies, the effect of radiation alone has been examined. Theoretical considerations suggested that bacterial cells were too small to be affected by microgravity [18]. However, it is well documented that culture of bacteria in the microgravity environment of spaceflight alters gene expression and physiology, including stress resistance and virulence properties, as discussed here for *E. coli*. However, whether these responses are due to an indirect and/or direct effect of microgravity remains to be determined.

13.2 Microgravity

13.2.1 Simulated Systems for Generating Microgravity ("Modeled Microgravity")

Given the constraints on equipment and astronaut time, a thorough examination of the effect of microgravity on cellular characteristics is a demanding task during spaceflight. Therefore, cell culture systems have been devised that simulate aspects of the microgravity environment on Earth. A detailed description of these systems and their operational principles can be found in Chap. 2.

The search for spaceflight-analogue cell culture systems using *E. coli* has resulted in several novel approaches. For example, Benoit and Klaus [19] generated a ground-based analogue of microgravity in *E. coli* using genetically engineered gas vesicles, obviating the need for rotation. Gas vesicles are formed by certain bacteria; their protein coat is permeable to ambient gases, permitting such bacteria to stay afloat in nutritionally favorable regions of natural waters. The engineered *E. coli* remained suspended in the medium and exhibited changes very similar to the clinostat-grown isogenic strain of the bacterium not generating the vesicles.

In a different approach, Lynch et al. [7] modified the high-aspect-ratio vessel (HARV)-based rotating wall vessels (RWVs) for the cultivation of bacterial biofilms under LSMMG conditions. The detailed description and operation of HARVs under LSMMG and control orientation is described in Chap. 10. Bacterial biofilms are mainly surface-attached communities encased in a matrix, which is generally composed of extracellular polysaccharides, DNA, and proteins. Biofilm formation is part of the normal growth cycle of most bacteria [20]. In the biofilm phase, bacteria exhibit markedly greater virulence and enhanced stress resistance; consequently, diseases in which biofilms play a major role (e.g., endocarditis, cystitis, and cystic fibrosis) are usually chronic and hard to treat [21]. Bacterial biofilm formation in space is well documented. Heavy colonization by biofilms occurred on the MIR space station, causing extensive corrosion and blockage of the onboard water purification system [22]; and, as mentioned above, the occurrence of urinary tract infections in astronauts has been reported, in which biofilms may have played a role [13]. It is therefore important to investigate the effect of the spaceflight environment on microbial biofilm formation and resistance.

While biofilm formation during true microgravity culture in spaceflight has been characterized multiple times for the bacterium *Pseudomonas aeruginosa* [23, 24] (Chap. 12), studies of *E. coli* biofilm formation have only been investigated in LSMMG culture [7, 24]. In these LSMMG studies, small, low density glass microcarrier beads were used in HARV RWVs to provide a surface for the development of biofilms, and extensive mathematical analysis involving equations governing particle translation, rotation, and fluid dynamics was conducted to ascertain that bacteria growing on these beads did indeed experience LSMMG conditions [7]. Since the bead density was small (10^{-6}, lower than the density of culture medium) and given their dilute distribution, the absence of any significant inter-particle interaction was assumed, so that analysis of the behavior of a single bead was taken to represent the system as a whole. Particle trajectory was described using direct numerical simulation techniques, which allowed the modeling of fluid and bead motion without the need for making assumptions as to the nature of the equations governing fluid flow and the particle trajectory. The calculated path of a representative bead and its final equilibrium point in LSMMG conditions (Fig. 13.1a) predicted

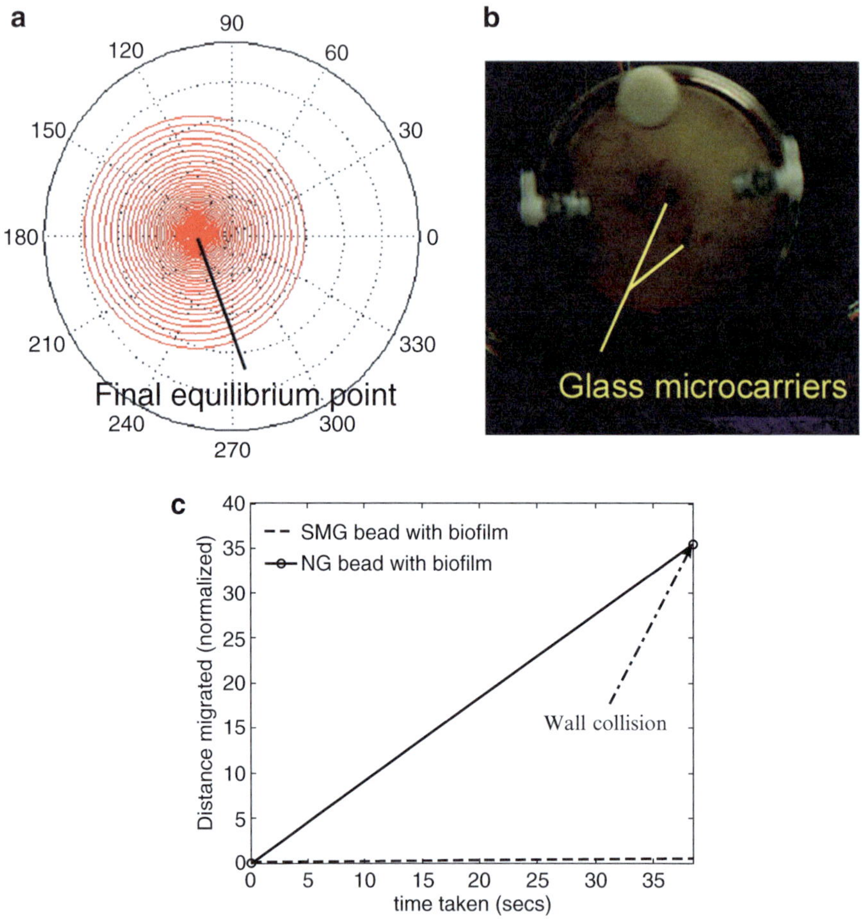

Fig. 13.1 (**a**) Numerically calculated trajectory (*red trace*) of a single microcarrier with a density lighter than the culture medium, as observed in a fluid-filled high-aspect-ratio vessel (HARV) rotated about a horizontal axis. The particle spirals inward toward a fixed equilibrium point. (**b**) The observed behavior of microcarrier beads in a fluid-filled HARV under experimental conditions. Beads rotate in the center of the vessel. (**c**) Numerically calculated migration distance of a heavier bead (due to biofilm formation) under (LSMMG ≈ SMG in the graph) and control orientation (≈normal gravity; NG). The bead in the HARV under control orientation quickly migrates downward along the axis of the HARV, whereas the bead in the LSMMG vessel migrates slowly radially outward toward the cylinder wall. Reproduced with permission from Lynch et al. [7]

that the equilibrium state is attained at the center of the bead's spiral trajectory, ensuring its suspension for long periods. This was confirmed experimentally: beads in the HARV cultured under LSMMG conditions remained suspended for the entire experimental duration of 24 h (Fig. 13.1b).

Once biofilms develop, the beads acquire increased density and become heavier than the culture medium. Therefore, numerical simulations of the motion of a heavier bead under control orientation and LSMMG conditions were also undertaken. The calculated normalized migration distance, scaled by particle diameter under the two conditions, is shown in Fig. 13.1c. It is clear that the calculated radial migration velocity of the LSMMG bead is much lower than the axial migration of the bead under control orientation. For the time interval that is required for the bead to migrate from the top of the bioreactor to its base under control orientation, the LSMMG bead travels a much shorter distance outward, equal approximately to half its diameter. These dissimilar migrations result from the differential effects of centrifugal and gravitational buoyancy [25] under the two conditions, and would result in greater suspension duration for the LSMMG bead than that under control orientation. These predictions were also borne out by the experimental results: the LSMMG beads remained suspended close to the center of the HARV for the 24 h period of the experiment, while in the HARV under control orientation, settling of beads to the bottom of the vessel was consistently seen. Fluid shear stress under LSMMG was calculated to be ca. 0.02 dynes/cm^2.

13.2.2 Growth

How microgravity might affect microbial growth is of fundamental importance and has received considerable attention and is discussed here for *E. coli* [3, 4, 26–34]**.** In a series of experiments carried out on suspension cultures of *E. coli* aboard seven US space shuttle missions (STS-37, -43, -50, -54 -57, -60, and 62), Klaus et al. reported similar growth rates for spaceflight cultured *E. coli* in glucose minimal medium as compared to matched ground controls [4]. However, spaceflight cultures showed a shorter lag phase and a longer exponential phase compared to ground controls. The authors ascribed this effect to altered fluid dynamics in microgravity. Specifically, they proposed that the decreased convection in microgravity results in reduced bulk fluid volume in the cell's immediate environment, permitting a more rapid conditioning of the growth medium that is required for initiating growth, accounting for the shorter lag; the same phenomenon could conceivably also account for the extended exponential phase. The slower removal of the metabolic waste from this microenvironment might form a "pseudo-membrane" in the form of an osmotic solute gradient interfering with nutrient flux to the cell [4], which would consequently be prolonged. Although there is no clear evidence that the bacterial-mediated medium conditioning is always necessary for growth to begin, this view would appear to be supported by the report that *E. coli* growth kinetics showed no difference in microgravity compared to ground controls when cultivation involved semi-solid, agar-based media [34].

A greater growth yield of *E. coli* was seen under microgravity: 7.8×10^8 vs. 4.4×10^8 colony forming units (CFU)/ml in microgravity vs. the ground controls, respectively [4]. Brown et al. [31] also found a 25 % higher cell yield for this bacterium on space shuttle Discovery (STS 95) compared to the Earth controls; and a similar pattern was seen in clinostat-generated culture, although to a lesser extent—9 % higher compared to the control. Other studies conducted aboard MIR space station, space shuttle Discovery, and space shuttle Endeavour also found that microgravity culture decreases the duration of the lag phase and increases cell yield in *E. coli* [26, 27, 29]. Growth in these studies was measured as CFU or by optical density changes. In cases where nutrient utilization was measured [31], the increased cell yield in microgravity reflected increased substrate utilization efficiency generating more cells per unit substrate consumed.

But contradictory findings have also been described in spaceflight cultured *E. coli*, including an unaffected lag phase [28], an unchanged exponential-phase duration [27], similar final population yields [28, 32, 33], and similar substrate utilization efficiency [32]. Based on A_{660} measurements, Lynch et al. [6] found that in glucose-M9 medium, *E. coli* grew more slowly in HARV-based RWVs (3 h generation time vs. 1 h in conventional flask cultures). But there was no difference in this respect between the LSMMG HARV and the control HARV, and in both, growth ceased upon the exhaustion of glucose attaining the same final yield.

Differences between *E. coli* strains and culture-media composition used in these experiments are likely to have influenced the results [3, 28, 30]. For example, based on studies on Columbia mission STS-6, it was suggested that growth of motile and non-motile *E. coli* strains differed in their response to microgravity culture. While the motile strain showed no significant difference in growth between ground and flight samples, the non-motile strain showed a considerably shorter growth lag phase in flight [3]. Unlike most spaceflight experiments conducted with *E. coli*, this study included a 1*g* inflight control; setup using an onboard centrifuge, to provide insight as to whether any observed differences were due to microgravity culture. Vukanti et al. [28] compared *E. coli* growth kinetics in Luria broth (LB) and glucose-M9 medium under LSMMG culture in RWV bioreactors. While cultivation in glucose-M9 medium resulted in increased cell population under LSMMG compared to the control orientation, this effect did not occur in LB medium, which resulted in similar cell yields under both conditions. This finding is in agreement with that previously reported for *S. enterica* serovar Typhimurium, wherein LSMMG culture decreased the generation times of a wild type and isogenic *rpoS* mutant strain in M9 minimal medium [35]. The influence of the culture medium on the bacterial growth response is also shown by the work of Baker et al. [30], although their findings are opposite to those of Vukanti et al. [28], as they observed an increased bacterial cell yield in LSMMG in rich medium (0.2 % nutrient broth) but not to any significant degree in glucose-M9 medium. Vukanti et al. also reported that culturing *E. coli* under LSMMG as compared to control orientation resulted in differences in pH and dissolved oxygen levels depending upon the medium employed [28]. Gene transcription has also been found to be influenced by differences in culture-media composition, with LSMMG cultures grown in minimal media (MOPS-glucose) exhibiting alterations in cell-envelope related genes, and those grown in rich media (LB) showing changes in genes related to translation [8]. In addition, RWV rotation speed used to culture *E. coli* may also be a factor in altering growth phase kinetics and/or final cell counts [3, 27, 28, 30]. The increased yield that was seen [28, 30] under LSMMG in a rich medium occurred when the rotation speed was 40 rpm; but at a rotation speed of 2–20 rpm, no increase was seen. LSMMG is usually generated using rotation speeds in the range of 20–25 rpm, and how well the higher and lower rotation speeds used in these experiments succeeded in generating LSMMG was not addressed. Additionally, variation in storage periods, and limited number of replicates that could be tested during spaceflights, as well as the lack of appropriately modeled fluid and inertial conditions for NG cultures may have also influenced the results [3, 27].

Overall, evidence supports the conclusion that the growth kinetics of *E. coli* are influenced by MG and LSMMG culture, although the underlying mechanisms responsible for these differences remain uncertain. An intriguing possibility suggested by several studies is increased cell yield, possibly due to increased substrate utilization efficiency, i.e., generation of more cells per unit substrate consumed. However, biomass in most of these studies was not measured directly and its increase was assumed based on the CFU count or optical density (O.D.) measurements. It has been proposed that *E. coli* "sees" microgravity and LSMMG as a form of stress (see below), and a common response of this bacterium to stress is fragmentation, a phenomenon in which one cell generates several (smaller) cells [36, 37]; indeed, reduced cell size has been reported under LSMMG culture [30]. Thus, the CFU count as well as O.D. can, in theory, increase without increased biomass formation, and it remains a possibility that increased CFU counts reported in the above experiments may have involved this phenomenon. But if substrate utilization efficiency does increase under microgravity (and LSMMG), it would be important to elucidate its physiological basis. The finding by Huitema et al. [38] that membrane fluidity increases in *E. coli* under LSMMG culture makes it tempting to consider that increased membrane fluidity might spare some of the energy which otherwise would be required for the uptake of nutrients. However, contradictory findings complicate the picture here as well: England et al. [39] found no difference in membrane fluidity in *P. aeruginosa*. Any specific LSMMG-responsive gene also remains to be identified in *E. coli* [8]. Clearly, further work is needed to more clearly establish how microgravity culture affects bacterial growth kinetics.

13.2.3 DNA Repair and Genetic Transfer

Efficient growth in space requires proficiency in repair mechanisms, since DNA synthesis is subject to errors, which may be exacerbated by space microgravity and radiation. The possibility that microgravity may interfere with DNA repair was indicated by findings with *Saccharomyces cerevisiae* on Spacelab IML-1 (STS-42) [40], although the results were contradicted by experiments on another flight, Spacelab mission SMM-03 (STS-76) [41]. Similar contradictions have been reported with *E. coli*. Studies on NASA MIR found that microgravity culture inhibited DNA repair mechanisms in this bacterium [42]. In experiments using the Biorack capability on Spacelab IML-2 (STS-65), the results were different. The aim of these studies was to determine the effect of microgravity culture on repair of DNA strand breaks and the induction of the SOS response on several unicellular organisms including *E. coli*. The SOS response is a tell-tale sign of DNA damage. No difference was found in these respects between the cells in microgravity and ground control [43, 44]. Experiments conducted on space shuttle Discovery (mission STS-91) gave similar results as the SOS response and the frequency of forward and reverse mutations, induced after pre-flight exposure to X rays, UV, and alkylating agents, were unaltered by subsequent culture in microgravity [45].

Bacterial growth and survival also necessitate the ability to gain favorable traits by acquiring beneficial genetic material. There are three types of genetic recombination in bacteria, the exchange of chromosomal DNA between bacterial cells via sex pili (conjugation), the transfer of short DNA segments by bacteriophages (transduction), and the uptake of naked, extracellular DNA fragments by bacterial cells (transformation) [46]. The efficiency of these three mechanisms was tested by Ciferri et al. [2] aboard Spacelab D-1. During conjugation, significantly more DNA was transferred in the spaceflight experiments, resulting in a three-to-four fold higher number of recombinants especially for the late markers; this may have been due to decreased mating interruptions. However, no appreciable changes in transduction were observed in this same study. The increased conjugation observed in *E. coli* during spaceflight is also in agreement with reports showing an increased conjugal transfer rate in *S.* Typhimurium during spaceflight [47] as well as increased expression of numerous genes encoding the plasmid transfer apparatus in this same organism [48]. In addition, alterations in phage induction in *E. coli* have also been reported during spaceflight culture [49].

13.2.4 Applied Aspects

Given that growth characteristics are likely affected by microgravity culture, there has been interest in exploring how bacterial responses to this environment can be practically applied for beneficial purposes. Products of bacterial secondary metabolism, particularly drugs, have received special attention. The effect of microgravity or LSMMG culture on secondary metabolism appears to vary with the organism. For example, LSMMG culture suppresses the production of β-lactam antibiotics by *Streptomyces clavuligerus* and rapamycin by *Streptomyces hygroscopicus*, whereas the production of gramicidin by *Bacillus brevis* remains unaffected [50]. In *E. coli,* this aspect has been examined for microcin B17 (Mcc B17) production [5, 50]. Microcins are low-molecular-weight antibacterial peptides excreted by *Enterobactericeae* which, at nanomolar concentrations, can injure bacteria by causing disruptions ranging from cell membrane damage to DNA gyrase inhibition [51]. The production of Mcc B17 under LSMMG was only 30 % of that in the control orientation. Interestingly, virtually all of the cellular MccB17 was secreted into the medium during LSMMG culture, whereas in conventional flask cultures, this compound accumulates within the cells. As the former location would facilitate microcin recovery, the finding is of considerable applied interest. The extracellular secretion was observed in RWVs under both LSMMG and control orientation, indicating that the effect was due to low shear rather than LSMMG. Indeed, creation of shear by the addition of a single glass or Teflon bead to the LSMMG RWV bioreactor resulted in mostly cellular localization of Mcc B17 [5, 50]. However, under the control orientation, addition of beads to RWVs shifted the site of Mcc B17 production only partially. This is perhaps because the beads remained in the periphery of RWVs during control orientation, whereas they experienced a uniform distribution in the LSMMG culture [5, 50].

Whether recombinant protein production is affected by LSMMG was more recently examined by Xiang et al. [52] for β-D-glucuronidase (β-GUS) in *E. coli*. β-GUS catalyzes the conversion of glycyrrhizin (GL) to glycyrrhetinic acid monoglucuronide (GAMG), which is used in the treatment of inflammatory diseases [53]. The gene from *Penicillium purpurogenum* encoding this enzyme was cloned in *E. coli*. When compared to the control orientation, culture in LSMMG conditions stimulated the recombinant GAMG production in a temperature-dependent fashion, with 18–33 % greater levels in the temperature range of 19–37 °C. A similar effect was reported in the production of recombinant semicarbazide-sensitive amine oxidase (SSAO), a family of glycoproteins that catalyzes the deamination of various exogenous and endogenous monoamines. SSAO is widely distributed in vascular and smooth muscle cells, as well as adipocytes. In this study, SSAO production in cytoplasmic aggregates (inclusion bodies) and soluble fractions increased by 83 and 116 %, respectively, at 18 °C but decreased by 38 and 49 %, respectively, at 37 °C under LSMMG culture as compared with control [54].

13.2.5 Resistance

Microgravity and MMG considerably decrease the susceptibility of *E. coli* broth cultures to various stresses including ethanol, salt, acidity, and antimicrobial agents. In experiments conducted onboard Salyut 7 in July 1982, the minimum inhibitory concentrations of the antibiotics colistin and kanamycin against *E. coli* were found to increase over fourfold, as compared to Earth controls [1]. In this study, post-flight examination of spaceflight cultured *E. coli* did not duplicate the on orbit results, as the bacteria showed normal antibiotic sensitivity when grown on Earth, indicating that the enhanced resistance requires the presence of microgravity [1]. Sensitivity to dihydrostreptomycin and penta-iodide and tri-iodide resins was similarly affected in spaceflight culture [55, 56]. Although the precise mechanism governing the increased drug resistance observed in response to culture in the microgravity environment remains unclear, the authors proposed that factors such as the stimulating effect of microgravity on bacterial multiplication, and/or changes in cell-envelope composition, which could alter antibiotic binding and penetration, may have been involved [1, 55]. An additional contributory factor for the greater sensitivity to the particulate resins in ground controls might have been that normal gravity co-sedimentation of bacteria and resins increases contact with the disinfectant. Under microgravity conditions, the bacteria and resin beads are in free suspension, and thus occupy a larger volume resulting in fewer effective collisions for iodine transfer and bacterial killing [56]. Greater bacterial resistance to various antibiotics in space has been reported by other groups as well [57]. Species-specific differences have also been observed in bacterial response to antibiotics under microgravity. In a flight experiment to determine permanent changes in antibiotic resistance, four species of bacteria were exposed to microgravity on Space Station MIR for 4 months and evaluated after return to Earth to determine sensitivity changes to 12 different antibiotics. In these ground-based studies, most of these bacterial species exhibited greater sensitivity to the antibiotics, while only a few species showed increased resistance [58].

Ground-based flight analogue studies in the RWV bioreactor have also demonstrated enhanced resistance *of E. coli* to antimicrobials under LSMMG as compared to control orientation. Lynch et al. [6] determined the effect of LSMMG on the resistance of exponential- and stationary-phase *E. coli* to high salt (2.5 M NaCl) and acidity (pH 3.5). In this study, growth under LSMMG conditions significantly enhanced the resistance of exponential-phase *E. coli* to each of these stresses. Exposure to salt, for instance, killed nearly all the cells grown under control orientation, but only 40 % of the LSMMG-grown cells (Fig. 13.2a). Stationary-phase bacteria, which intrinsically exhibit enhanced resistance to multiple stresses as compared to exponential-phase cultures [36], acquired even greater resistance to each of the stresses tested under LSMMG as compared to control orientation (Fig. 13.2b). Similar results were found with respect to ethanol, thermal, and oxidative stress sensitivity, wherein LSMMG grown cells exhibited significantly greater resistance compared to the controls [9, 59]. More recently, it was found that LSMMG-culture increased the resistance of *E. coli* stationary-phase cells to gentamicin (16 μg/ml) by 60 % compared to the control orientation. This was σ^s-dependent: in an *rpoS* mutant the LSMMG-cultured cells were virtually completely killed by this antibiotic (R. Singh and A.C. Matin, unpublished).

In contrast, Tucker et al. observed no significant difference in the response of exponential-phase *E. coli*, cultured under LSMMG in both rich and minimal media, to various stresses including antibiotics, acidic and basic conditions, heat shock and oxidative, osmotic, and alcohol stress [8]. Altered stress response has also been reported in other bacteria upon exposure to LSMMG. While *P. aeruginosa* and *S.* Typhimurium have been shown to exhibit increased stress resistance under LSMMG [35, 60], *Staphylococcus aureus* responds in an opposite manner—its

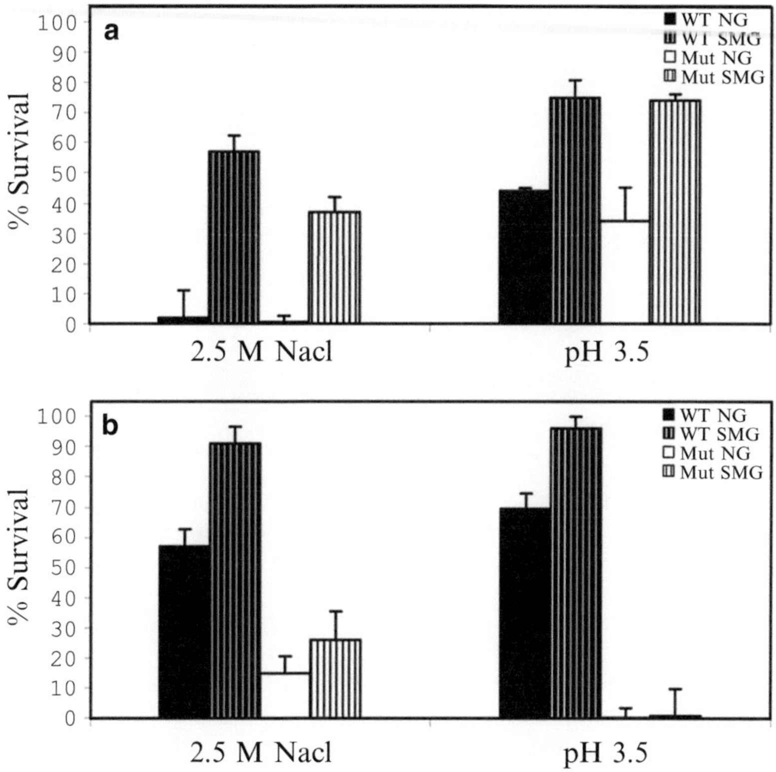

Fig. 13.2 Percentage survival after exposure of exponential (**a**) and stationary (**b**) phase *Escherichia coli* for 1 h to hyperosmotic (2.5 M NaCl) or acid (pH 3.5) stress. *WT*, wild type (AMS6); *Mut*, *rpo*S mutant (AMS150); *NG*, normal gravity ≈ control orientation; SMG (≈ LSMMG), low shear modeled microgravity. Results represent an average of two independent measurements, each analyzed in triplicate; error bars represent standard errors of the mean. Reproduced with permission from Lynch et al. [6]

LSMMG-grown cultures have an increased sensitivity to stresses compared with those grown in control orientation [61].

Stationary phase is the prevalent state for bacteria in most natural environments, including in the infected host [36]. Given that LSMMG cultivation causes some *E. coli* strains to exhibit enhanced resistance in stationary phase above that inherently acquired at this same phase during traditional culture [10], it will be important to characterize *E. coli* stress responses in the true microgravity environment of spaceflight. Recent inflight studies with the bacterial pathogens *S.* Typhimurium and *P. aeruginosa* involving stationary-phase cultures of these organisms are a major step in this direction [48] (see Chaps. 11 and 12, respectively) and need to be extended to other bacteria. As mentioned above, low fluid shear environments relevant to those encountered under microgravity and LSMMG culture conditions are relevant to physiological fluid shear levels encountered by bacteria during their natural

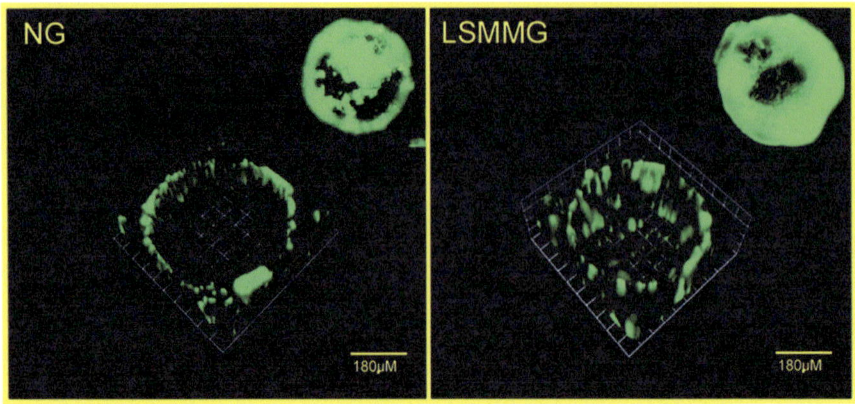

Fig. 13.3 Z-stack assembly of 24 h old *Escherichia coli* biofilms grown under low shear modeled microgravity (LSMMG) and control orientation (≈normal gravity; NG) on microcarrier beads. The cells, AMS6 (pAD123), expressed the green fluorescent protein constitutively. The *upper right* of each frame provides representative x-y images of the respective beads. Reproduced with permission from Lynch et al. [7]

lifecycles on Earth, including in the infected host. This is the case with certain areas of the respiratory, gastrointestinal, and urogenital tracts [62–64]. This not only makes an investigation on the effect of such environments on bacterial pathogenicity urgent for space travelers but also for controlling infectious disease here on Earth.

Stationary-phase bacteria also play a role in the enhanced resistance of bacterial biofilms to eradication by antimicrobial drugs/therapeutics and the immune system. Lynch et al. [7] (Sect. 2.1) found that LSMMG-cultured stationary-phase *E. coli* formed more copious biofilms (Fig. 13.3) that exhibited enhanced resistance to salt and ethanol as compared to their control-grown counterparts (Fig. 13.4). The effect of LSMMG or microgravity culture on *E. coli* virulence has thus far not been much examined, but the closely related bacterium *S.* Typhimurium has demonstrated significantly increased virulence in response to both microgravity and LSMMG culture [10, 48]. Allen et al. reported that culturing a clinical isolate of adherent-invasive *E. coli* under LSMMG as compared to control orientation increased its adherence to Caco-2 cells, a human epithelial colorectal adenocarcinoma cell line. Cell invasion was however unaffected [9]. Indole, a metabolic by-product of tryptophan, was identified as the key signaling molecule regulating adherence under LSMMG conditions [9]. It should also be noted that not all bacteria exhibit enhanced virulence in response to LSMMG. For example, *S. aureus* responds to LSMMG in a contrasting manner, by initiating a biofilm colonization phenotype with reduced virulence [61]. This further emphasizes the need for a thorough examination of the effect of microgravity culture on the virulence and stress resistance of bacteria in various stages of growth and development to determine the potential health hazards of space travel, the results of which will inform the intelligent design of countermeasures.

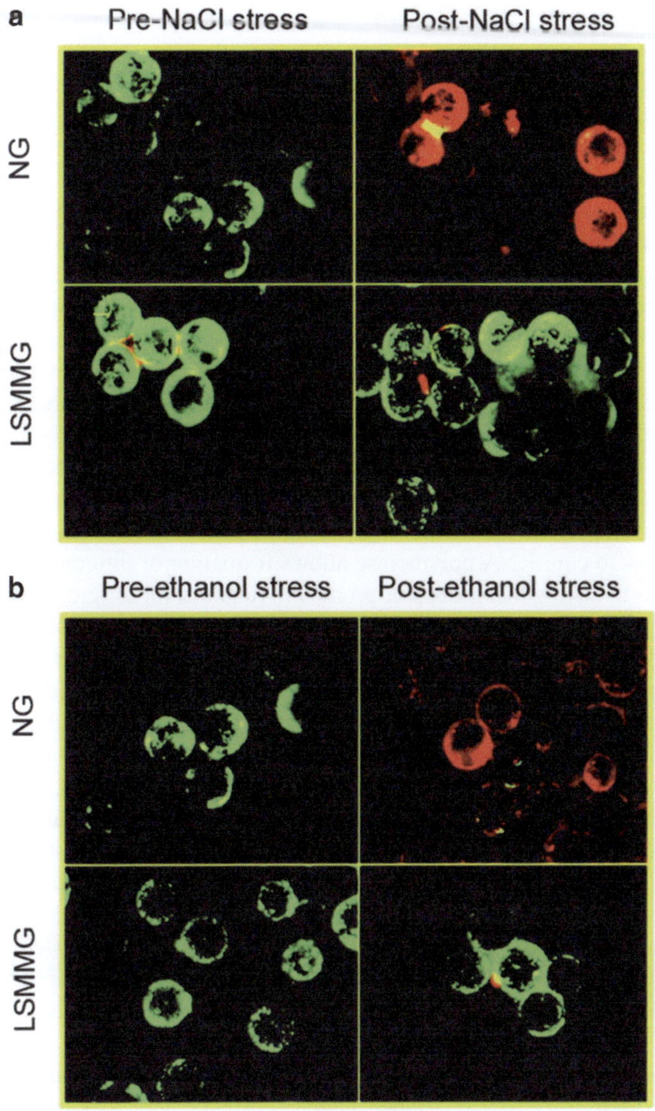

Fig. 13.4 Representative images of *Escherichia coli* AMS6 biofilms grown under low shear modeled microgravity (LSMMG) and control orientation (≈normal gravity; NG) and stained with BacLight viability stain [*green*, viable; *red*, nonviable]) and visualized pre- and post-NaCl (1 M) (**a**) or ethanol (7 %) stress (**b**). Reproduced with permission from Lynch et al. [7]

13.2.6 Molecular Regulation

Role of σ^s

The increased induction of general stress resistance by *E. coli* under LSMMG suggests that this bacterium perceives LSMMG as a stress, and responds to it in a manner similar to that observed for stresses experienced under conventional normal gravity. Exposure to stresses like starvation, heat shock, osmotic, or oxidative stress as well as entry into stationary phase (starvation) activates the general stress response, in which exposure to a given stress confers resistances against multiple stresses [36, 65]. A central regulator of the general stress response in *E. coli* and other Gram negative bacteria is the stationary-phase sigma factor, σ^s or RpoS (product of the *rpoS* gene), which is important for survival of these organisms under extreme conditions and is also a central regulator of virulence [66]. Sigma factors are a family of small proteins that reversibly bind to the RNA polymerase core enzyme (symbol E), resulting in the transcription competent holoenzyme (symbol Eσ^x, where x represents the identity of a sigma factor). This reversible binding of sigma factors to core RNA polymerase allows formation of different holoenzymes that are able to distinguish groups of promoters required for different cellular functions. This results in the coordinated regulation of expression of specialized gene sets in response to environmental stimuli whose products are required for survival in the new environment and is a common method of stress adaptation in bacteria [67].

Lynch et al. [6] examined the role of σ^s in LSMMG-conferred resistance in *E. coli*, using the wild type of this bacterium and its isogenic *rpoS* mutant. It was found that the resistance to high salt (2.5 M NaCl) and acidity (pH 3.5) was independent of σ^s in exponential phase, since both strains exhibited similar responses; however, in the stationary phase of growth the mutant failed to develop stress resistance showing its dependence on σ^s. This was the first time that general stress resistance in any growth phase independent of σ^s was documented in *E. coli*. The results of the spaceflight experiment of Wilson et al. [48] with *S. enterica* serovar Typhimurium, in showing a central role for the Hfq protein on the microgravity effect on this bacterium, would appear to be consistent with the involvement of σ^s in these effects, since the Hfq protein is required for σ^s synthesis.

Consistent with reports that bacterial biofilms share features of stationary-phase cells [68], their increased resistance under LSMMG was also found to be σ^s-dependent [7].

The lack of dependence on σ^s of general resistance in exponential phase *E. coli* under LSMMG culture has additional support. Increased resistance of exponential phase cells under LSMMG culture occurred without an increase in cellular σ^s concentration, unlike stresses experienced by exponential phase *E. coli* under conventional gravity, which generally result in increased cellular levels of this sigma factor. Instead, the LSMMG-grown exponential phase *E. coli* exhibited 30 % lower σ^s levels as compared to their control-grown counterparts (Fig. 13.5). These bacteria also did not show increased expression of σ^s-dependent genes, e.g., *dnaK*, *groEL*,

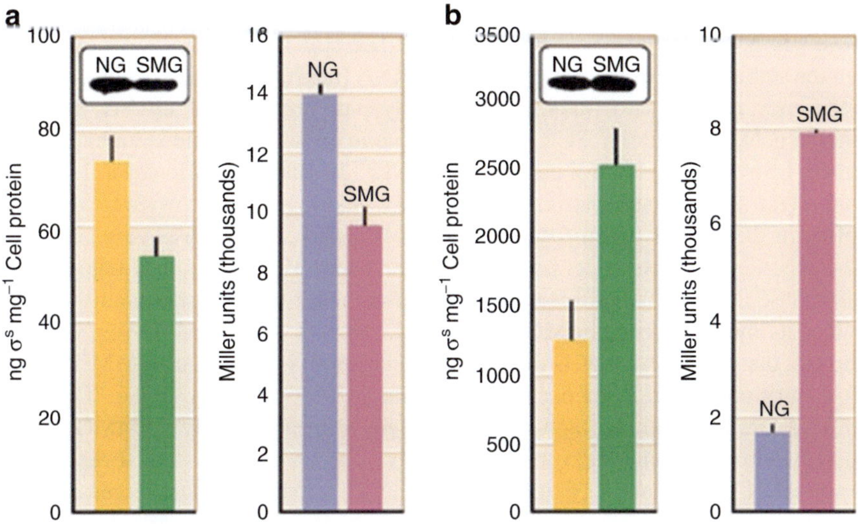

Fig. 13.5 Immunoblots and the corresponding bar charts showing σ^s levels in the exponential (**a**) and stationary (**b**) phases of *Escherichia coli* growth. Bar charts on the *right* in each of the two panels represent the quantification of the sigma factor by another method and confirm the immunoblot results. Reproduced with permission from Matin and Lynch [96]. *SMG*, low shear modeled microgravity (≈LSMMG); *NG*, control orientation (≈normal gravity)

and *pexB* (*dps*), known to be responsible for general resistance under control orientation [69]. In *S*. Typhimurium as well, the increased stress resistance of late exponential phase cells cultured in LSSMG occurred in the absence of induction of genes in the RpoS regulon, including those mentioned above [35].

These findings hint at a new as yet unknown mechanism for increased general resistance in LSMMG-grown exponential phase cells that is independent of σ^s and may not rely on the protective proteins that defend cells against stresses under conventional normal gravity. Increased σ^s levels and induction of the genes that it controls increases *E. coli* resistance, but it also restricts the range of substrates the bacterium can utilize, and some of the protective proteins are mildly toxic [70]. Thus, high levels of this sigma factor represent an impediment to rapid growth [70, 71]. It may be that the σ^s-independent increased resistance that LSMMG-grown exponential phase cells exhibit circumvents this impediment. As mentioned, many microbial niches on Earth are associated with low fluid shear environments, including in the infected host, and it is possible that a combination of increased resistance without restriction of nutritional versatility could be advantageous for the pathogen.

In contrast, σ^s remains of crucial importance in the significantly enhanced resistance of stationary-phase LSMMG-grown cells, wherein this resistance was found to be accompanied with higher levels of this sigma factor than cells grown under control orientation in this phase of growth. In agreement with a role for RpoS in LSMMG-enhanced stress resistance in stationary-phase *E. coli* cultures, induction of σ^s-dependent genes (*csiD*, *katE*, *otsA*, and *tre)* was observed under these conditions [69].

Lynch et al. [6] also investigated the molecular mechanism that determines σ^s levels in *E. coli* under LSMMG conditions. The effect of LSMMG was examined on the stability and transcription of *rpoS* mRNA (mRNA), its translational rate and efficiency, as well as σ^s stability. The steady-state levels of the mRNA, as determined using quantitative PCR, were the same in exponential and stationary phase under LSMMG and control conditions (Tables 13.1 and 13.2). However, LSMMG did affect *rpoS* transcription in the stationary phase, as became evident when stability of the mRNA was measured. The *rpoS* mRNA stability was very similar in the exponential phase under both LSMMG and control orientations. In the stationary phase the stability of this mRNA increased under both growth conditions, but twice as much under LSMMG compared to control conditions. Given, as mentioned, that the mRNA levels remained unchanged, it is clear that LSMMG caused the *rpoS* transcription to decrease two-fold in the stationary phase. The *rpoS* mRNA translational rate and efficiency were also affected by LSMMG. These parameters were calculated from the steady-state σ^s levels, its half-life, and the *rpoS* mRNA copy number (Tables 13.1 and 13.2). LSMMG culture increased these parameters in both growth phases (Table 13.2). The LSMMG growth condition also affected σ^s stability, markedly lowering it in exponential-phase *E. coli* (Fig. 13.6 and Table 13.2). In stationary-phase cultures, σ^s became more stable under both growth conditions, but still remained less stable under LSMMG compared to control orientation. Thus, the growth phase-dependent differential effect of LSMMG on σ^s levels is due mainly to (1) markedly greater instability of the σ^s protein in exponential as compared to stationary phase and (2) increased *rpoS* mRNA translational efficiency in both growth phases. As already noted, LSMMG also decreased the transcriptional rate in stationary phase, which however is compensated by increased mRNA stability. How this regulation is attained is not known. Future study of this question would benefit from the known mechanism of σ^s regulation, which is briefly reviewed below.

Table 13.1 Transcriptional parameters of *rpoS* synthesis in exponential- and stationary-phase *Escherichia coli* cultures grown under LSMMG and control orientation. Reproduced with permission from Lynch et al. [6][a]

	rpoS mRNA[b] [*rpoS*]		mRNA half-life[c] ($\tau_{1/2}{}^{rpoS}$)		Transcriptional rate[d] (K_s^{RNA})	
Growth phase	Control	LSMMG	Control	LSMMG	Control	LSMMG
Exponential	1.3×10^8	1.3×10^8	5.4	5.4	1.6×10^7	1.6×10^7
Stationary	1.4×10^7	1.3×10^7	34.5	72.1	2.8×10^5	1.2×10^5

[a]Each value is an average of at least two independent determinations with analytical triplicates; standard error of the mean was <8 %

[b]Copies of *rpoS* mRNA per microgram of total RNA

[c]Message half-life (minutes)

[d]Copies of *rpoS* synthesized per microgram of total RNA per minute

Table 13.2 Translational parameters of σ^s synthesis of exponential- and stationary-phase *Escherichia coli* cultures grown under LSMMG and control orientation. Reproduced with permission from Lynch et al. [6][a]

Growth phase	σ^s protein[b] $[\sigma^s]$		σ^s half-life[c] ($\tau_{1/2}^{\sigma^s}$)		Translational rate[d] ($K_s^{\sigma^s}$)		Translational efficiency[e] ($K_E^{\sigma^s}$)	
	Control	LSMMG	Control	LSMMG	Control	LSMMG	Control	LSMMG
Exponential	73	5 3	16.4	5.2	3.1	7.0	3.6×10^{11}	8.4×10^{11}
Stationary	1263	2530	28.8	24.6	30.4	71.2	3.3×10^{13}	8.5×10^{13}

[a]Each value is an average of at least two independent determinations with analytical triplicates; standard error of the mean was <8 %
[b]Nanograms σ^s per milligram of protein
[c]Protein half-life (minutes)
[d]Nanograms of σ^s synthesized per milligram of cell protein per minute
[e]Molecules of σ^s synthesized per copy of *rpo*S mRNA per minute

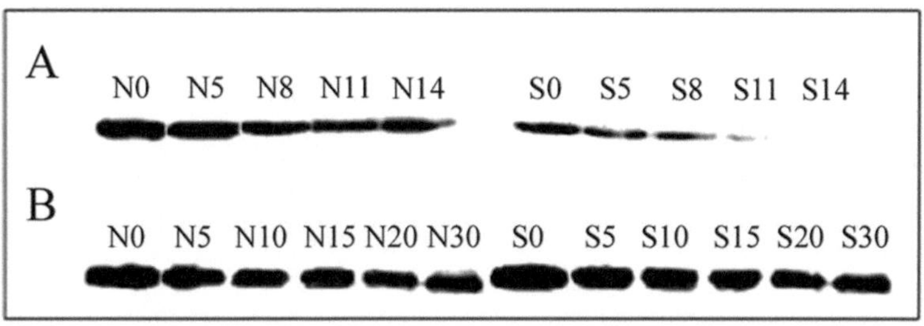

Fig. 13.6 σ^s protein half-life determined by quantitative western blot analysis during the exponential (**a**) or stationary (**b**) phase of *Escherichia coli* growth. N values (N0, N5, etc.) and S values (S0, S5, etc.) indicate sampling time points in minutes under control orientation (NG) and low shear modeled microgravity (LSMMG) conditions, respectively. Reproduced with permission from Lynch et al. [6]

13.2.7 σ^s Regulation

This is schematically summarized in Fig. 13.7 [72]. Under normal gravity, the cellular levels and activity of σ^s are regulated at three levels—transcriptional, translational, and post-translational. Transcriptional regulation has only a minor role in controlling σ^s synthesis, with Fis protein, ArcA-P protein, and cyclic AMP being the negative regulators functioning at this level [73, 74] and BarA-UvrY two-component system being the positive regulator [75]. (Two-component systems are bacterial sensor/regulator systems that sense different environmental cues.) Translational regulation of *rpoS* mRNA is mediated via formation and relaxation of secondary structures and changes in its stability. *rpoS* mRNA possesses a long untranslated region (UTR) at its N-terminal end which can form stem-loop structures with a complementary sequence in the UTR itself or in the coding region [76].

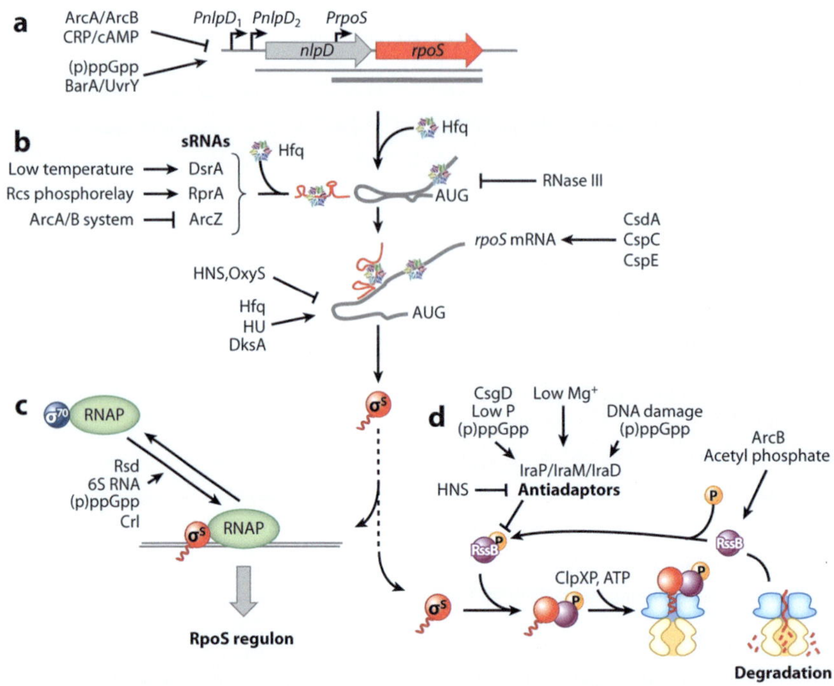

Fig. 13.7 Regulation of RpoS expression, stability and activity. Each *box* indicates different levels of regulation: (**a**) transcriptional regulation, (**b**) translational regulation, (**c**) regulation of RpoS activity, and (**d**) proteolytic regulation. The major transcript for *rpo*S transcription is initiated at *Prpo*S within the *nlp*D open reading frame. This transcript forms a repressive hairpin loop which prevents ribosome binding. Positively regulating small RNAs release this inhibition, with the help of the Hfq chaperone, and promote *rpo*S translation. OxyS small RNA inhibits *rpo*S translation likely by titrating out Hfq. See text for details. Reproduced with permission from Battesti et al. [72]

This blocks the translational initiation region (TIR), preventing its access to the ribosomes [72, 77]. Specific small RNAs (sRNAs), induced in response to different stresses, contain sequences complementary to the UTR stretch of *rpoS* mRNA, and lead to opening of the loop, thereby allowing access of ribosomes to the TIR. The synthesis of Hfq, which has been identified as a regulator of the microgravity and/or LSMMG response in *Salmonella, Pseudomonas*, and *Staph aureus*, and is required for the synthesis of σ^s, may also be regulated by sRNAs; indeed, sRNAs were differentially regulated during spaceflight culture of *S.* Typhimurium [48]. Examples of these positive regulatory sRNAs include DsrA, RprA, ArcZ, GcvB, and the one encoded by the intergenic region of the *pst* operon [72, 78]. CsdA and CspA proteins also aid these sRNAs [79–81]. The sRNA OxyS and the protein LrhA, on the other hand, down-regulate *rpoS* translation [82]. Additionally, CspC and CspE proteins stimulate *rpoS* translation possibly by affecting mRNA structure, whereas RNase III, which degrades *rpoS* mRNA, and the HU protein, which probably blocks the UTR, negatively regulate *rpoS* translation [83].

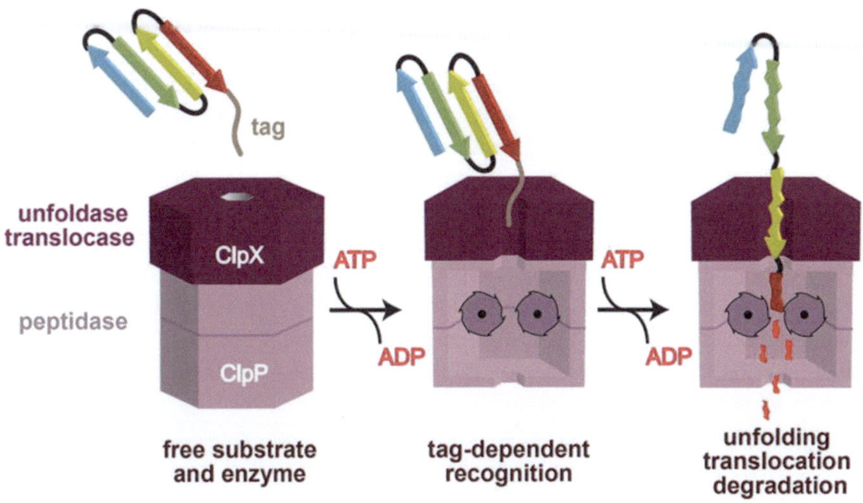

Fig. 13.8 Model describing the substrate recognition and degradation by the ClpXP protease. A peptide tag in a protein substrate binds in the axial pore of the ClpX hexamer, and subsequently, in an ATP-dependent manner, ClpX unfolds the substrate and translocates the unfolded polypeptide into the degradation chamber of ClpP for proteolysis. Reproduced with permission from Baker and Sauer [70]

Post-translational regulation, involving σ^s stability, plays a major role in determining σ^s levels [84]. σ^s is very unstable in unstressed cells, due to its cleavage by the ClpXP protease; however, under stress, it becomes immune to this protease (Fig. 13.8). ClpX is a hexameric AAA+ ATPase (ATPases associated with diverse cellular activities), which binds to one or both ends of the ClpP, a tetradecameric peptidase with 14 active sties. It also binds to the target proteins directly (by recognizing special tags such as ssrA at the target protein's C-terminus) or indirectly (via adapters), and feeds them into the ClpP chamber; the energy to perform this action is derived from the ClpX ATPase activity [70, 85, 86]. RpoS is not directly recognized by ClpX; instead, its association with ClpX is mediated by the adapter protein RssB (or SprE). RssB binding probably alters the conformation of RpoS and exposes the site, situated close to the N-terminus of this protein, which binds to ClpX. A stretch of amino acids spanning K173–K188 residues in RpoS is required for its cleavage by the ClpXP protease [84]. E174 and V177, although not absolutely required, facilitate this degradation [87]. Interestingly, the K173–K188 (corresponding to 519–564 nucleotides of the mRNA) region of RpoS is complementary to the TIR of *rpoS* (the antisense element) and was initially thought to have a role in regulating translation [76]; this still remains a possibility. K173 also appears to be important in transcriptional control. Thus, the K173–K188 stretch may have a role at all three levels of RpoS regulation under different conditions. Why RssB mediates cleavage of RpoS by ClpXP only in unstressed cells and not under stress is a continuing area of investigation. This could probably relate to its de-activation via dephosphorylation in starving cells due to ArcB/ArcA two-component system [88, 89]. Also, three proteins, IraP, IraM, and IraD, referred to as antiadapters, can impair the ability of RssB to promote RpoS degradation. These proteins are synthesized in response to different stresses, but the mechanism of their induction is not known [72]. ppGpp, a global regulator of stringent response, is a positively regulator of the *ira*P promoter [90]; this may further explain the observed up-regulation of σ^s by ppGpp [91]. Additionally, H-NS, a histone-like protein, negatively regulates σ^s by modulating the translational efficiency of *rpo*S mRNA and the stability of newly synthesized σ^s. H-NS is a DNA-binding protein known to regulate genes at transcriptional level. The precise molecular mechanisms by which it controls σ^s at the post-transcriptional level remain to be fully elucidated [65, 92]. Its influence on σ^s stability is likely mediated by the suppression of *ira*M and *ira*D expression [93].

13.2.8 Possibilities for σ^s Regulation in LSMMG

As we have seen, LSMMG profoundly affects the regulation of σ^s synthesis. It decreases the *rpoS* mRNA transcription in stationary phase, increases its rate of translational and efficiency, while decreasing σ^s stability in both growth phases. It is possible that a fundamentally different mechanism compared to the regulation uncovered by the conventional gravity studies accounts for this. Such a mechanism could involve the effect of LSMMG on the folding patterns of macromolecules. This possibility is consistent with the fact that much of the control processes discussed above involve changes in the folding patterns of *rpoS* mRNA, the σ^s protein, and the effectors that act on these molecules. Also, the fact that protein crystallization is significantly facilitated in microgravity [94] could conceivably relate to alteration in the folding pattern. Nevertheless, it would be prudent to address in such studies the known factors that affect σ^s levels based on conventional gravity results. Thus, for instance, it may be useful to determine whether the following have role in the altered mechanism of σ^s levels under LSMMG: Fis and the ArcA-P proteins in transcription; the above-mentioned sRNAs in *rpoS* mRNA translation; and synthesis of antiadapters or phosphorylation state of RssB in σ^s stability. The RNA chaperone protein, Hfq, which plays a crucial role in regulating *rpoS* mRNA translation has been reported to be the key modulator of microgravity and/or LSMMG response in *S.* Typhimurium, *P. aeruginosa*, and *S. aureus* [48, 60, 61]. However, the role of Hfq in LSMMG response of *E. coli* remains to be elucidated. Likewise, the expression of Clp proteases increases in *Staphylococcus aureus* under LSMMG [61, 95]. A similar mechanism might also exist in *E. coli*, and contribute to the reduced stability of σ^s under LSMMG.

13.3 Conclusion

E. coli is an ideal model organism for spaceflight investigations, as it is the best characterized bacterium at the molecular-genetic, biochemical, and physiological levels, and has pathogenic strains that could cause infections during spaceflight. It is also a key component of ecosystems for the renewal of resources critical to supporting life in space. While spaceflight and spaceflight-analogue studies of this bacterium and others still require more research, it is clear that these environments have a major impact on their physiology, stress resistance, and molecular biology. Specifically for *E. coli*, microgravity culture has been shown to alter growth parameters and antibiotic resistance, while studies with LSMMG culture indicate that LSMMG regulates gene expression, protein synthesis, cellular protein composition, biofilm formation, and stress resistance. While spaceflight-analogue research provides useful information, it is important to test these findings under the conditions of true spaceflight. The increasing sophistication of available equipment and instruments onboard space stations and small satellites bode well in this respect for future space research. The danger posed by increased bacterial stress resistance observed in microgravity is exacerbated by the fact that aspects of the human immune response are diminished under microgravity (Chap. 3). Addressing this threat requires understanding at a fundamental level of the effect of microgravity on various life processes of *E. coli* and other bacteria. Thus, for future studies, we need to refine our understanding concerning the effect of microgravity on (1) the growth parameters of bacteria, such as duration of growth phases and growth yield, (2) DNA repair processes and mutation frequency, and (3) potential profound effects on molecular processes, such as protein synthesis. As altered molecular regulation under LSMMG hints at altered macromolecular folding patterns, its in-depth examination is of great importance for space travel as well as for general biological understanding.

Questions for Future Research

- What are the effect(s) of microgravity and/or LSMMG culture on *E. coli*—including growth kinetics and repair mechanisms?
- What is the influence of microgravity and/or LSMMG on the virulence of *E. coli*?
- What are the precise molecular mechanisms governing the altered growth and resistance of planktonic and biofilm forming *E. coli* in microgravity and/or LSMMG?
- What are the underlying mechanisms associated with enhanced antibiotic resistance during spaceflight culture of *E. coli*?
- What are the factors modulating the expression and stability of sigma S under microgravity and/or LSMMG?

References

1. Tixador, R., Richoilley, G., Gasset, G., Planel, H., Moatti, N., Lapchine, L., et al. (1985). *Acta Astronautica, 12*, 131.
2. Ciferri, O., Tiboni, O., Di Pasquale, G., Orlandoni, A. M., & Marchesi, M. L. (1986). *Naturwissenschaften, 73*, 418.
3. Thevenet, D., D'Ari, R., & Bouloc, P. (1996). *Journal of Biotechnology, 47*, 89.
4. Klaus, D., Simske, S., Todd, P., & Stodieck, L. (1997). *Microbiology, 143*(Pt 2), 449.
5. Fang, A., Pierson, D. L., Koenig, D. W., Mishra, S. K., & Demain, A. L. (1997). *Applied and Environmental Microbiology, 63*, 4090.
6. Lynch, S. V., Brodie, E. L., & Matin, A. (2004). *Journal of Bacteriology, 186*, 8207.
7. Lynch, S. V., Mukundakrishnan, K., Benoit, M. R., Ayyaswamy, P. S., & Matin, A. (2006). *Applied and Environmental Microbiology, 72*, 7701.
8. Tucker, D. L., Ott, C. M., Huff, S., Fofanov, Y., Pierson, D. L., Willson, R. C., et al. (2007). *BMC Microbiology, 7*, 15.
9. Allen, C. A., Niesel, D. W., & Torres, A. G. (2008). *Environmental Microbiology, 10*, 1512.
10. Nickerson, C. A., Ott, C. M., Mister, S. J., Morrow, B. J., Burns-Keliher, L., & Pierson, D. L. (2000). *Infection and Immunity, 68*, 3147.
11. Sonnenfeld, G., & Shearer, W. T. (2002). *Nutrition, 18*, 899.
12. Martinelli, L. K., Russomano, T., Dos Santos, M. A., Falcao, F. P., Bauer, M. E., Machado, A., et al. (2009). *IEEE Engineering in Medicine and Biology Magazine, 28*, 85.
13. Barratt, M. R., & Pool, S. L. (Eds.). (2008). *Principles of clinical medicine for space flight* (1st ed.). New York: Springer.
14. Horneck, G., Klaus, D. M., & Mancinelli, R. L. (2010). *Microbiology and Molecular Biology Reviews, 74*, 121.
15. Horneck, G., Bucker, H., Reitz, G., Requardt, H., Dose, K., Martens, K. D., et al. (1984). *Science, 225*, 226.
16. Horneck, G., Bucker, H., Dose, K., Martens, K. D., Bieger, A., Mennigmann, H. D., et al. (1984). *Advances in Space Research, 4*, 19.
17. Rettberg, P., Eschweiler, U., Strauch, K., Reitz, G., Horneck, G., Wanke, H., et al. (2002). *Advances in Space Research, 30*, 1539.
18. Pollard, E. C. (1965). *Journal of Theoretical Biology, 8*, 113.
19. Benoit, M., & Klaus, D. (2005). *Microbiology, 151*, 69.
20. Fux, C. A., Costerton, J. W., Stewart, P. S., & Stoodley, P. (2005). *Trends in Microbiology, 13*, 34.
21. Hall-Stoodley, L., Costerton, J. W., & Stoodley, P. (2004). *Nature Reviews. Microbiology, 2*, 95.
22. Koenig, D. W., & Pierson, D. L. (1997). *Water Science and Technology, 35*, 59.
23. Kim, W., Tengra, F. K., Young, Z., Shong, J., Marchand, N., Chan, H. K., et al. (2013). *PLoS One, 8*, e62437.
24. McLean, R. J., Cassanto, J. M., Barnes, M. B., & Koo, J. H. (2001). *FEMS Microbiology Letters, 195*, 115.
25. Gao, H., Ayyaswamy, P. S., & Ducheyne, P. (1997). *Microgravity Science and Technology, 10*, 154.
26. Kacena, M. A., Smith, E. E., & Todd, P. (1999). *Applied Microbiology and Biotechnology, 52*, 437.
27. Kacena, M. A., Merrell, G. A., Manfredi, B., Smith, E. E., Klaus, D. M., & Todd, P. (1999). *Applied Microbiology and Biotechnology, 51*, 229.
28. Vukanti, R., Model, M. A., & Leff, L. G. (2012). *BMC Microbiology, 12*, 4.
29. Kacena, M. A., Manfredi, B., & Todd, P. (1999). *Microgravity Science and Technology, 12*, 74.
30. Baker, P. W., Meyer, M. L., & Leff, L. G. (2004). *Microgravity Science and Technology, 15*, 39.
31. Brown, R. B., Klaus, D., & Todd, P. (2002). *Microgravity Science and Technology, 13*, 24.
32. Bouloc, P., & D'Ari, R. (1991). *Journal of General Microbiology, 137*, 2839.
33. Gasset, G., Tixador, R., Eche, B., Lapchine, L., Moatti, N., Toorop, P., et al. (1994). *Research in Microbiology, 145*, 111.

34. Kacena, M. A., Leonard, P. E., Todd, P., & Luttges, M. W. (1997). *Aviation, Space, and Environmental Medicine, 68*, 1104.
35. Wilson, J. W., Ott, C. M., Ramamurthy, R., Porwollik, S., McClelland, M., Pierson, D. L., et al. (2002). *Applied and Environmental Microbiology, 68*, 5408.
36. Matin, A. (2009). In M. Schaechter (Ed.), *Encyclopedia of microbiology* (pp. 485–500). Oxford: Academic Press.
37. Adnan, M., Morton, G., & Hadi, S. (2011). *Molecular and Cellular Biochemistry, 357*, 275.
38. Huitema, C., Beaudette, L. A., & Trevors, J. T. (2002). *Rivista di Biologia, 95*, 497.
39. England, L. S., Gorzelak, M., & Trevors, J. T. (2003). *Biochimica et Biophysica Acta, 1624*, 76.
40. Pross, H. D., Kost, M., & Kiefer, J. (1994). *Advances in Space Research, 14*, 125.
41. Pross, H. D., Casares, A., & Kiefer, J. (2000). *Radiation Research, 153*, 521.
42. Harada, K., Sugahara, T., Ohnishi, T., Ozaki, Y., Obiya, Y., Miki, S., et al. (1998). *International Journal of Molecular Medicine, 1*, 817.
43. Horneck, G., Rettberg, P., Kozubek, S., Baumstark-Khan, C., Rink, H., Schafer, M., et al. (1997). *Radiation Research, 147*, 376.
44. Horneck, G., Rettberg, P., Baumstark-Khan, C., Rink, H., Kozubek, S., Schafer, M., et al. (1996). *Journal of Biotechnology, 47*, 99.
45. Takahashi, A., Ohnishi, K., Takahashi, S., Masukawa, M., Sekikawa, K., Amano, T., et al. (2001). *Advances in Space Research, 28*, 555.
46. Cogoli, A., & Gmunder, F. K. (1991). *Advances in Space Biology and Medicine, 1*, 183.
47. Mattoni, R. (1968). *Bioscience, 18*, 602.
48. Wilson, J. W., Ott, C. M., Honer zu Bentrup, K., Ramamurthy, R., Quick, L., Porwollik, S., et al. (2007). *Proceedings of the National Academy of Sciences of the United States of America, 104*, 16299.
49. Alpatov, A. M., I'Lin, E. A., Antipov, V. V., & Tairbekov, M. G. (1989, September–October). *Kosmicheskaia Biologiia i Aviakosmicheskaia Meditsina, 23*, 26.
50. Demain, A. L., & Fang, A. (2001). *Chemical Record, 1*, 333.
51. Duquesne, S., Petit, V., Peduzzi, J., & Rebuffat, S. (2007). *Journal of Molecular Microbiology and Biotechnology, 13*, 200.
52. Xiang, L., Qi, F., Dai, D., Li, C., & Jiang, Y. (2010). *Applied Biochemistry and Biotechnology, 162*, 654.
53. Matsui, S., Matsumoto, H., Sonoda, Y., Ando, K., Aizu-Yokota, E., Sato, T., et al. (2004). *International Immunopharmacology, 4*, 1633.
54. ***Zhang, Y., Lai, C., Duan, J., Guan, N., Ullah, K., & Deng, Y. 55. Tixador, R., Gasset, G., Eche, B., Moatti, N., Lapchine, L., Woldringh, C., et al. (2012). *Applied Microbiology and Biotechnology, 94*, 809.
55. Tixador, R., et al. (1994). *Aviation, Space, and Environmental Medicine, 65*, 551.
56. Marchin, G. L., Silverstein, J., & Brion, G. M. (1997). *Acta Astronautica, 40*, 65.
57. Leys, N. M., Hendrickx, L., De Boever, P., Baatout, S., & Mergeay, M. (2004). *Journal of Biological Regulators and Homeostatic Agents, 18*, 193.
58. Juergensmeyer, M. A., Juergensmeyer, E. A., & Guikema, J. A. (1999). *Microgravity Science and Technology, 12*, 41.
59. Gao, Q., Fang, A., Pierson, D. L., Mishra, S. K., & Demain, A. L. (2001). *Applied Microbiology and Biotechnology, 56*, 384.
60. Crabbe, A., Pycke, B., Van Houdt, R., Monsieurs, P., Nickerson, C., Leys, N. et al. (2010). *Environmental Microbiology, 12*, 1545.
61. Castro, S. L., Nelman-Gonzalez, M., Nickerson, C. A., & Ott, C. M. (2011). *Applied and Environmental Microbiology, 77*, 6368.
62. Guo, P., Weinstein, A. M., & Weinbaum, S. (2000). *American Journal of Physiology. Renal Physiology, 279*, F698.
63. Cai, Z., Xin, J., Pollock, D. M., & Pollock, J. S. (2000). *American Journal of Physiology. Renal Physiology, 279*, F270.

64. Nickerson, C. A., Ott, C. M., Wilson, J. W., Ramamurthy, R., & Pierson, D. L. (2004). *Microbiology and Molecular Biology Reviews, 68*, 345.
65. Barth, M., Marschall, C., Muffler, A., Fischer, D., & Hengge-Aronis, R. (1995). *Journal of Bacteriology, 177*, 3455.
66. Hengge-Aronis, R. (2000). In G. Storz & R. Hengge-Aronis (Eds.), *Bacterial stress responses* (pp. 161–178). Washington, DC: ASM Press.
67. Coleman, C., Baker, C., & Nickerson, C. A. (2006). In C. A. Nickerson & M. J. Schurr (Eds.), *Molecular paradigms of infectious disease: A bacterial perspective* (pp. 438–501). New York: Springer Science & Business Media.
68. Anderl, J. N., Zahller, J., Roe, F., & Stewart, P. S. (2003). *Antimicrobial Agents and Chemotherapy, 47*, 1251.
69. Vukanti, R., & Leff, L. G. (2012). *Microgravity Science and Technology, 24*, 267.
70. Baker, T. A., & Sauer, R. T. (2012). *Biochimica et Biophysica Acta, 1823*, 15.
71. King, T., Ishihama, A., Kori, A., & Ferenci, T. (2004). *Journal of Bacteriology, 186*, 5614.
72. Battesti, A., Majdalani, N., & Gottesman, S. (2011). *Annual Review of Microbiology, 65*, 189.
73. Hirsch, M., & Elliott, T. (2005). *Journal of Bacteriology, 187*, 7204.
74. Hirsch, M., & Elliott, T. (2005). *Journal of Bacteriology, 187*, 1568.
75. Mukhopadhyay, S., Audia, J. P., Roy, R. N., & Schellhorn, H. E. (2000). *Molecular Microbiology, 37*, 371.
76. McCann, M. P., Fraley, C. D., & Matin, A. (1993). *Journal of Bacteriology, 175*, 2143.
77. Cunning, C., Brown, L., & Elliott, T. (1998). *Journal of Bacteriology, 180*, 4564.
78. Majdalani, N., Hernandez, D., & Gottesman, S. (2002). *Molecular Microbiology, 46*, 813.
79. Resch, A., Vecerek, B., Palavra, K., & Blasi, U. (2010). *RNA Biology, 7*, 796.
80. Phadtare, S., & Inouye, M. (2001). *Journal of Bacteriology, 183*, 1205.
81. Cohen-Or, I., Shenhar, Y., Biran, D., & Ron, E. Z. (2010). *Research in Microbiology, 161*, 694.
82. Hussein, R., & Lim, H. N. (2011). *Proceedings of the National Academy of Sciences of the United States of America, 108*, 1110.
83. McCullen, C. A., Benhammou, J. N., Majdalani, N., & Gottesman, S. (2010). *Journal of Bacteriology, 192*, 5559.
84. Schweder, T., Lee, K. H., Lomovskaya, O., & Matin, A. (1996). *Journal of Bacteriology, 178*, 470.
85. Gottesman, S., Roche, E., Zhou, Y., & Sauer, R. T. (1998). *Genes & Development, 12*, 1338.
86. Moore, S. D., & Sauer, R. T. (2007). *Annual Review of Biochemistry, 76*, 101.
87. Becker, G., Klauck, E., & Hengge-Aronis, R. (1999). *Proceedings of the National Academy of Sciences of the United States of America, 96*, 6439.
88. Georgellis, D., Kwon, O., & Lin, E. C. (2001). *Science, 292*, 2314.
89. Malpica, R., Franco, B., Rodriguez, C., Kwon, O., & Georgellis, D. (2004). *Proceedings of the National Academy of Sciences of the United States of America, 101*, 13318.
90. Bougdour, A., & Gottesman, S. (2007). *Proceedings of the National Academy of Sciences of the United States of America, 104*, 12896.
91. Gentry, D. R., Hernandez, V. J., Nguyen, L. H., Jensen, D. B., & Cashel, M. (1993). *Journal of Bacteriology, 175*, 7982.
92. Yamashino, T., Ueguchi, C., & Mizuno, T. (1995). *The EMBO Journal, 14*, 594.
93. Battesti, A., Tsegaye, Y. M., Packer, D. G., Majdalani, N., & Gottesman, S. (2012). *Journal of Bacteriology, 194*, 2470.
94. Lorber, B. (2002). *Biochimica et Biophysica Acta, 1599*, 1.
95. Rosadoa, H., Doylea, M., Hindsb, J., & Taylor, P. W. (2010). *Acta Astronautica, 66*, 408.
96. Matin, A., & Lynch, S. V. (2005). *ASM News, 71*, 235.

Chapter 14
Spaceflight and Spaceflight Analogue Induced Responses in Gram Positive Bacteria

Sarah L. Castro, David W. Niesel, Jennifer Barrila, and C. Mark Ott

Historical Landmarks

1968—Early experiments aboard Gemini and Apollo missions investigate the survival of *Bacillus subtilis* when exposed to the space environment [1].

1982—Experiments aboard the Salyut 7 space station investigating antibiotic sensitivity of *Staphylococcus aureus* indicated the minimum inhibitory concentration when exposed to oxacillin, chloramphenicol, and erythromycin was increased in response to spaceflight culture [2].

1986—*Bacillus subtilis* cultured in the European Space Agency Biorack facility displayed shorter lag growth phase and increases in the rate of cell division and biomass [3].

1999—Studies comparing the response of *B. subtilis* growth on a semi-solid agar and in liquid medium indicate that spaceflight induced alterations in growth profiles and final cell concentrations are the result of fluid dynamics or extracellular transport, rather than a cellular response to gravity [4].

(continued)

S.L. Castro • C. Mark Ott (✉)
Biomedical Research and Environmental Sciences Division, NASA/Johnson Space Center, 2101 NASA Parkway, Mail: SK4, Houston, TX 77058, USA
e-mail: sarah.l.castro@nasa.gov; c.m.ott@nasa.gov

D.W. Niesel
Department of Microbiology and Immunology, University of Texas Medical Branch, 301 University Blvd, Galveston, TX 77555, USA
e-mail: dniesel@utmb.edu

J. Barrila
The Biodesign Institute, Center for Infectious Diseases and Vaccinology, Arizona State University, Tempe, AZ 85287, USA
e-mail: jbarrila@asu.edu

C.A. Nickerson et al. (eds.), *Effect of Spaceflight and Spaceflight Analogue Culture on Human and Microbial Cells*, DOI 10.1007/978-1-4939-3277-1_14

(continued)
2011—*Staphylococcus aureus* cultured in the Rotating Wall Vessel had increased extracellular polymeric substance production, decreased resistance to whole human blood, and decreased carotenoid production compared to control cultures [5].

14.1 Introduction

Over the past 12 years, the interest in microbial responses to spaceflight culture has dramatically increased as a result of (a) the utilization of the NASA designed rotating wall vessel bioreactor as a spaceflight culture analogue for microorganisms [6–8], **(b) advances in and availability of molecular microbiological techniques, and (c) a clear association of spaceflight culture with alterations in virulence and virulence characteristics** [9–11]. Many of these studies have focused on Gram negative pathogens, such as *Salmonella enterica* serovar Typhimurium [9, 10, 12], *Pseudomonas aeruginosa* [11, 13, 14], and *Escherichia coli* [15]. However, recent studies investigating the response of Gram positive organisms have provided intriguing insight into the similarities of the responses among dramatically different species and how evolutionarily conserved these responses may be.

The potential alterations in the disease-associated characteristics of Gram positive organisms during spaceflight missions are of great importance for future human exploration efforts. Gram positive organisms, especially *Staphylococcus* species, are the most prevalent species isolated from the air and surfaces of spacecraft vehicles [16, 17]. The frequent isolation of the opportunistic pathogen, *Staphylococcus aureus* [17, 18], from the environment of the International Space Station (ISS) is not unexpected, as 30–50 % of healthy adults on Earth are colonized with this organism [19]. Passage of these organisms between crewmembers is common as demonstrated by a genetic comparison of *S. aureus* strains isolated from crewmembers aboard the Mir space station [20]. Microbial monitoring also indicated the presence of *Streptococcus* species aboard the Russian space station Mir [21] and from air samples collected during Space Shuttle missions [22]. While *S. pneumoniae* has not been isolated from spacecraft or from a crewmember after flight, this opportunistic bacterium has been isolated after nasopharyngeal sampling from a shuttle crewmember [23] immediately before flight. As potential pathogens, such as *S. aureus* and *S. pneumoniae*, are likely to be carried as part of the normal flora of a crew and may exploit a declining immune system, understanding the mechanisms behind the disease causing potential of Gram positive organisms has tremendous implications for the spaceflight crew.

14.2 *Staphylococcus aureus*

***S. aureus* is a Gram-positive, opportunistic pathogen commonly found on humans and in the environment. This ubiquitous nature is reflected in the common isolation of *S. aureus* from the environment of the space shuttle, Mir space**

station, and ISS [16, 17]. Accordingly, it is important to understand how *S. aureus* responds to the microgravity environment of space and the possible consequences associated with crew health.

In a collaborative effort in 1982, French and Russian crew carried by the Soyuz-T docked with the Salyut 7 space station to carry out a series of biomedical experiments. As part of the Cytos 2 experiment, *S. aureus* was cultured in-flight and assessed for alterations in the minimal inhibitory concentrations (MIC) of oxacillin, chloramphenicol, and erythromycin as compared to ground controls [2]. It was determined that the MIC of all three antibiotics was slightly higher and that the increase in MIC was accompanied by a thickening of the cell wall of spaceflight-cultured *S. aureus* [2]. While this study describes some specific effects of *S. aureus* grown in the microgravity environment of space, it provides only limited characterization of the organism.

The response of several *S. aureus* strains to LSMMG (Low Shear Modeled Microgravity) has been documented by taking advantage of the RWV (Rotaing Wall Vessel) bioreactor [5, 24–27]. The evaluation of *S. aureus* N315 revealed the formation of visible bacterial aggregates in the fluid phase of the LSMMG culture [5]. Analysis of the aggregates by environmental scanning electron microscopy (ESEM) indicated that the LSMMG cultures produced significantly higher amounts of an extracellular polymeric substance (EPS) that enveloped the LSMMG-cultured cells as compared to controls (Fig. 14.1) [5]. When assessed for alterations in antibiotic resistance, the LSMMG-induced aggregates were 1.72-fold more resistant to ciprofloxacin as compared to control-cultured bacteria [5]. The encasement of the cells by an EPS, which confers increased antibiotic resistance, is consistent with characteristics of surface-attached bacterial biofilms [28]. Interestingly, the heavy EPS was in response to LSMMG-culture conditions rather than a more traditional attachment to a solid surface [5].

Fig. 14.1 ESEM images of control- and LSMMG-cultured *S. aureus* N315. Control-cultured *S. aureus* at 10,000× (**a**) magnification demonstrated that individual staphylococcal cells were clearly visible. *S. aureus* cultured under LSMMG conditions at 10,000× (**b**) confirmed that the cells were much less visible and completely encased in an EPS matrix [5]. Copyright © American Society for Microbiology, Applied and Environmental Microbiology, 77(18):6368–6378, 2011

Growth profiles of *S. aureus* in LSMMG-culture differ among strains. Rosado et al. reported that three clinical isolates (designated RF1, RF6, and RF11) displayed comparable growth profiles in LSMMG and control conditions [25]. In contrast, Castro et al. reported LSMMG culture resulted in a 2.9-fold and 5.6-fold lower total cell concentration for N315 and 8325, respectively, as compared to controls [5]. An increased final cell density has also been reported for *S. aureus* 25923; however, the increase was not noted until deep into stationary growth phase after 40 h of culture in the RWV [27]. Therefore, it is possible that the increase noted by the investigators may not be solely based on growth characteristics in LSMMG culture, but instead may be based on the enhanced survival characteristics associated with changes in nutrient depletion and/or waste build-up. Interestingly, the induction of EPS from *S. aureus* cultures is also dependent on the specific strain [5].

While alterations in *S. aureus* characteristics, such as EPS production and growth profiles, in response to LSMMG culture are strain dependent, common phenotypic changes have been reported in multiple strains in response to this environment. For example, a characteristic of most *S. aureus* strains is their golden yellow color arising from the production of the primary carotenoid pigment, staphyloxanthin, expressed during the stationary phase of growth. Visual inspection of LSMMG-cultured *S. aureus* N315 revealed a decrease in the pigmentation of bacterial pellets (Fig. 14.2a) [5]. Extraction, measurement, and comparison of the carotenoids from LSMMG-cultured *S. aureus* N315 to control cultures and unpigmented *S. aureus* 8325 quanti-

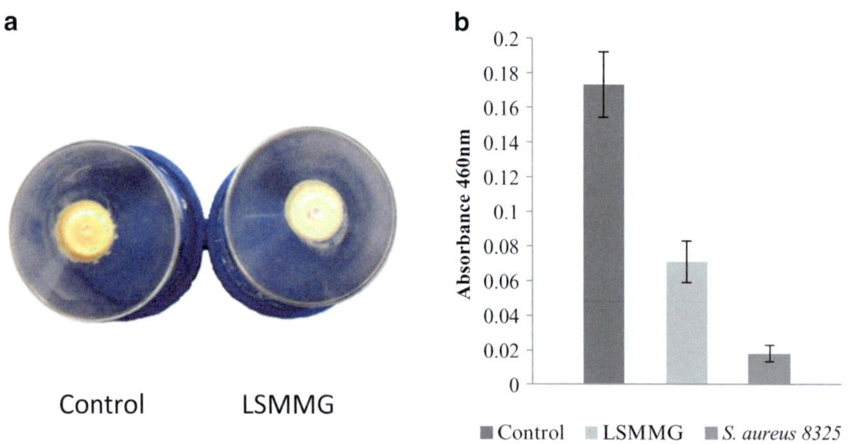

Fig. 14.2 Decreased carotenoid production of *Staphylococcus aureus* N315 in response to LSMMG culture. (**a**) Pellets of control- and LSMMG-cultured *S. aureus* revealing a visual difference in pigmentation of the cells. (**b**) The quantitative difference in pigmentation between control- and LSMMG-cultured *S. aureus* determined by carotenoid extraction and measurement spectrophotometrically at 460 nm. There was a significant reduction in the absorbance of low-shear-cultured bacteria as compared to the control (*, $P<0.0001$). *S. aureus* 8325, which does not produce carotenoids, was used as a negative control for comparison [5]. Copyright © American Society for Microbiology, Applied and Environmental Microbiology, 77(18):6368–6378, 2011

fied the significant decrease in pigmentation (Fig. 14.2b) [5]. Rosado et al. also reported a reduction in *S. aureus* carotenoid production for the clinical strains RF1, RF6, and RF11 in response to LSMMG culture [25]. Collectively, all pigmented *S. aureus* strains that have been assessed for alterations in carotenoid expression in response to LSMMG have displayed a reduction in carotenoid production, indicating a common response among strains to this environmental parameter [5, 25].

The LSMMG environment has been shown to predispose certain Gram negative bacteria with an increased ability to withstand environmental stressors such as high heat and acidic conditions [29]. As the antioxidant properties of carotenoids serve as a shield protecting *S. aureus* from the toxic effects of reactive oxygen species associated with host immunity [30], and as carotenoid production is impacted by the LSMMG environment, the susceptibility of *S. aureus* to oxidative stress was assessed by Castro et al. [5]. After 60 min of exposure to oxidative stress, 50 % of the LSMMG-cultured *S. aureus* had succumbed to damage, whereas the control culture survival rates did not fall below 90 % [5]. Interestingly, by removing the LSMMG- and control-cultured bacteria from the bioreactor vessels and allowing them to sit statically for a period of time and then repeating the oxidative stress assay, it was found that, at a time greater than 1 h but less than 1.5 h, there was no longer a significant difference between the percent survival of LSMMG- and control-cultured *S. aureus* [5]. Based on this data, Castro et al. estimated the half-life of the observed oxidative sensitivity to be 68.3 ± 1.3 min. As opposed to findings with certain Gram negative organisms, neither *S. aureus* N315 nor 8325 cultured in LSMMG revealed a difference in survival to thermal or acid stress.

S. aureus is associated with bacterial sepsis, and thus its ability to survive in the blood stream impacts its dissemination throughout the body [19]. As a result, the vast majority of *S. aureus* strains have an array of mechanisms to procure nutrients and avoid host immunity upon entering the blood stream [19, 31]. To determine if LSMMG culture impacted the ability of *S. aureus* to survive in human blood, the hemolytic ability of *S. aureus* strains RF1, RF6, and RF11 were assessed [24]. All strains demonstrated a reduction in the ability to lyse sheep or rabbit erythrocytes, with the hemolytic capability of RF6 being almost absent in response to LSMMG-culture conditions. In a separate study, upon challenge with freshly drawn human whole blood, LSMMG-cultured *S. aureus* N315 was approximately 30 % more susceptible to being killed as compared to control cultures [5]. Collectively, these studies demonstrate that the survival of *S. aureus* upon interactions with blood may be impaired immediately following growth in the LSMMG environment.

Analyses of the gene expression of *S. aureus* in LSMMG culture have also provided advances in our understanding of the mechanism(s) behind the microbial response in this environment. Depending upon the strain, the differential regulation of 4–25 *S. aureus* genes has been reported in response to the LSMMG environment [5, 25]. A common change in strains RF1, RF6, and RF11 was the down-regulation of VraX, which is proposed to be involved in the cell wall stress stimulon of *S. aureus* [25]. However, upon creating a *vraX* deletion mutation, no differences were noted in growth, viability, the MIC of multiple antibiotics, pigmentation, hemolytic ability, or the differential expression of any genes when compared to the wild-type

[25]. Examination of the *S. aureus* N315 genes that were up-regulated after LSMMG culture suggested an altered metabolic profile, as many of the protein products of the genes were associated with a fermentative metabolism [5]. Moreover, alignment of the LSMMG-responsive genes revealed the conserved consensus sequences for regulatory proteins Rex and SigB [5]. The expression of these genes, in addition to *hfq,* which has been directly associated with the molecular mechanism governing the LSMMG response in both *S.* Typhimurium [9] and *P. aeruginosa* [14] (see Chaps. 11 and 12), was assessed with quantitative real-time PCR. In stationary phase, no changes in expression levels of Rex or SigB were noted; however, a 2.68-fold decrease in *hfq* expression was observed [5]. While the contribution of Hfq to the molecular regulation of *S. aureus* is unclear [32, 33]; Liu et al. speculated that Hfq was a global regulator in *S. aureus* based upon their investigations of an *hfq* mutant [32]. Interestingly, a side-by-side comparison of the *S. aureus* LSMMG-cultured microarray data from Castro et al. [5] with the work by Liu et al. revealed that 7 of the 17 LSMMG-up-regulated genes were also up-regulated in response to an *hfq* mutation. Additionally, 9 of the 17 LSMMG-induced genes were found by Liu et al. to bind Hfq [32]. Moreover, the work by Liu et al. also reported that the mutation of *hfq* impacted pigment production [32]. Taken together, the decreased expression level of *hfq* in response to LSMMG culture, combined with the significant correlations of this work with that of Liu et al., strongly suggests that Hfq is involved in the LSMMG response of *S. aureus*. More importantly, this is the first description associating an Hfq response to the LSMMG environment in a Gram-positive bacterium [5]. The correlation of Hfq to the LSMMG response of *S. aureus*, in addition to the previously documented responses in Gram-negatives, strongly suggests that the ability to sense and respond to mechanical stimuli is evolutionarily conserved among structurally diverse prokaryotes.

The potentially evolutionarily conserved responses of *S. aureus* to LSMMG culture share common mechanistic characteristics with Gram negative organisms, such as the involvement of Hfq; however, a fundamental difference in how these organisms alter virulence characteristics exists. This difference consistently appears to be based on the benefit toward each microorganism's proliferation and perseverance capabilities in this environment. Contrary to previous reports of enhanced virulence and/or virulence properties of *S.* Typhimurium [12] and *P. aeruginosa* [13], *S. aureus* appears to favor a phenotype consistent with colonization, in which it forms a biofilm and down-regulates virulence characteristics [5, 25]. Collectively, these comparisons may afford a unique opportunity to examine the role of environmental parameters serving as cues directing the balance between infection and colonization by *S. aureus* during the initial host–pathogen interaction.

14.3 *Streptococcus pnuemoniae*

***Streptococcus* is non-motile, non-spore forming catalase-negative, Gram positive bacteria commonly arranged in pairs or chains**. S. *pneumoniae* (pneumococcus) grows commonly in chains or pairs, is optochin sensitive, and is alpha hemolytic on blood agar plates. *S. pneumoniae* resides primarily in the nasopharynx and is capable of causing a diverse spectrum of disease ranging from otitis media to bacteremia and meningitis [34–36]. It is the leading cause of community-acquired pneumonia, and a principal cause of meningitis and otitis media [36–40]. Normally an opportunistic pathogen, *S. pneumoniae* affects mainly the very young, the elderly, and the immunocompromised. Carriage rates in healthy individuals vary from 5 to 70 % depending on age, environment, and season (*pneumoniae*). Estimates are that 5–10 % of adults without children are colonized with *S. pneumoniae*. At military installations, *S. pneumoniae* carriage rates can be as high as 50–60 % [41]. *S. pneumoniae* can be a commensal inhabitant of the nasopharyx in humans [41, 42], an opportunistic pathogen in individuals with impaired immune systems, and a mediator of serious disease which can be easily spread via aerosols. In the USA, pneumococcal disease is responsible for an about 3000–6000 cases of meningitis, more than 50,000 cases of bacteremia, estimated 175,000 hospitalizations for pneumococcal pneumonia, and 5,000,000 cases of otitis media annually [41]. *S. pneumoniae* is transmitted by aerosols, and is carried in the upper respiratory tract by healthy individuals [41].

The response of *S. pneumoniae* to LSMMG has been extensively examined [43, 44]. In these experiments, controls with RWVs with an axis parallel to the gravity vector, which were rotating or static were compared to LSMMG conditions (axis perpendicular to the gravity vector). This allowed examination of the effect of modeled microgravity on *S. pneumoniae* gene expression as well as the definition of the contribution of rotation in RWV controls as compared to the static controls. Bacterial growth under LSMMG or static or rotating controls showed no differences in the kinetics of growth or density during logarithmic growth [44]. These studies were performed in THY broth; it is not known if minimal or defined media would yield different results as reported for other bacterial species [10].

The effect of LSMMG on *S. pneumoniae* global gene expression was examined. Microarray analysis comparing RWV static controls and LSMMG conditions revealed 101 genes whose expression was altered [44]. These genes represented a broad range of functional groups including those involved in adhesion, cell envelope, cofactor biosynthesis, iron acquisition, metabolism, proteases, stress response genes, toxin production, transcriptional regulation, transporters, and those in other or unknown groups [44]. For static controls versus $1 \times g$ conditions, 63 genes were up-regulated and 38 were down-regulated [44]. This included genes whose encoded proteins were involved those involved in adherence, proteases, stress proteins, and transport proteins among others [44]. Interestingly, 46 of these same genes were also shown to be differentially expressed in the rotating control versus $1 \times g$ [44]. For these experiments, 4 genes were observed to be up-regulated and 42 down-regulated [44]. Shear forces likely explain the changes seen in the 43 genes altered in LSMMG versus either of the controls. Additionally, six genes (including 2 affecting transporter functions and 2 that are regulatory), were shown to be differentially expressed between the static and the rotating control and LSMMG conditions [44]. These observed changes are likely a specific effect of RWV rotation on gene expression. A comparison of static versus rotating RWV controls has not been extensively reported for other bacterial species but could reveal a conserved repertoire of genes and subsequent properties that arise from the rotational component of RWVs.

Hierarchical clustering comparing the static controls and $1 \times g$ identified a group of 19 genes that had similar expression patterns with 15 showing similar expression differences using four different analysis methods (GenePix Pro 6.0, Spotfire DecisionSite 7.3, Significance Analysis of Microarray, and ANOVA) [44]. Overall, adaptation to LSMMG altered gene expression that could help explain virulence differences observed after spaceflight [44]. Additionally, for *S. pneumoniae*, comparisons of static controls to LSMMG and $1 \times g$ conditions have been found to be exceptionally relevant and should be rigorously evaluated in future experiments [44].

More recently, *S. pneumoniae* was cultured on the ISS (STS-118) and Space Shuttle mid-deck (STS-129) in the SPEGIS (**S**treptococcus **E**xpression of **G**enes **I**n **S**pace) flight experiment (Unpublished Data). The transcriptome of these cultures was compared to ground controls and that obtained for LSMMG cultures. Principal components analysis (Gene Spring after Lowess normalization) also establishes the similarity and distinct qualities of flight and model microgravity studies as shown in Fig. 14.3. This indicates the experimental replicates with the treatment types clustered together and apart from the other experimental treatment type replicates (LSMMG, $1 \times g$, Flight, 25 °C, flight 37 °C).

For two separate analyses comparing the direct analysis of flight versus ground samples and the comparison between spaceflight and LSMMG experiments there were a total of 297 genes (108 up-regulated and 189 down-regulated) that described the overall gene expression alterations occurring during spaceflight (Unpublished

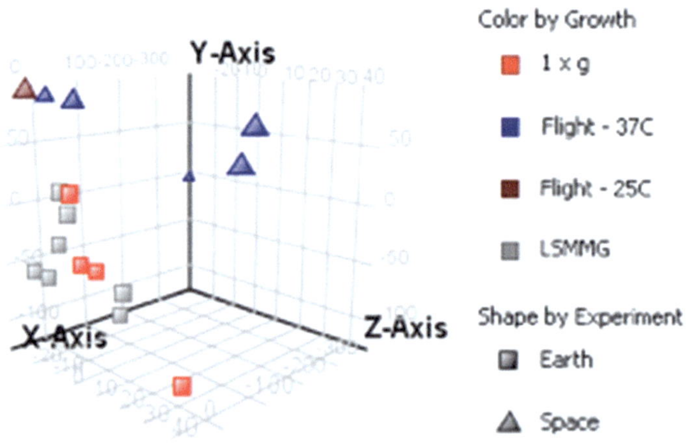

Fig. 14.3 Principal components analysis generated by GeneSpring after Lowess normalization. The colors represent the different experimental replicates (*gray*=*S. pneumoniae* cultured under LSMMG, compared to RWVs gravity controls; *blue*=*S. pneumoniae* grown in flight compared to Earth controls), and shapes represent the two different experiments (*squares*=LSMMG experiments conducted entirely on Earth; *triangles*=shuttle/ISS experiments), as indicated in the legend to the right. The *X*-axis, *Y*-axis, and *Z*-axis components represent 50.75 %, 31.5 %, and 10.42 %, respectively, of the total variability between samples

Data). These genes were identified based on their reproducibility across Space Shuttle flights and their specificity to spaceflight as opposed to those seen under modeled microgravity conditions. These genes belonged to diverse functional groups as shown in Fig. 14.4. Electron transport components were down-regulated in space culture conditions compared to ground-based controls. Genes which function in amino acid and nucleotide synthesis and carbohydrate and cell wall synthesis were also altered. Detailed analysis of individual genes is under review for publication elsewhere. In *Salmonella* [9], *Pseudomonas* [11, 14], and *Staphylococcus* [5], the *hfq* gene is implicated in the regulation of gene expression in response to

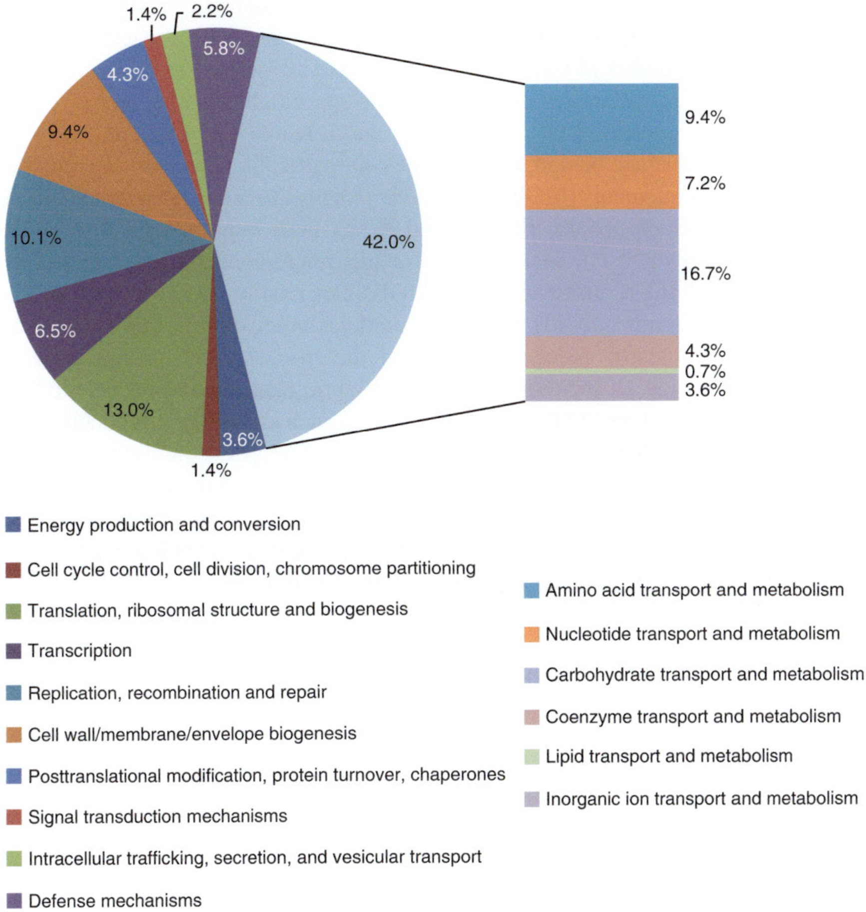

Fig. 14.4 Functions of genes differentially expressed by *S. pneumoniae* grown in space, compared to the bacteria cultured on Earth are shown. Percentages indicate the proportion of altered genes included in the functional category shown, as indicated by color in the legend. Genes with unknown functions were excluded

modeled microgravity and/or true microgravity conditions. Interestingly, *hfq* or an equivalent is not found in the genome of *S. pneumoniae*, and, therefore, the regulation of bacterial gene expression in this microorganism in response to the space environment will be unique from that observed for these other species. The nature of microgravity-induced alterations in the molecular regulation of *S. pneumoniae* will require additional investigation.

14.4 Other Gram Positive Species

Several Gram positive species have been the focus of previous spaceflight and spaceflight analogue studies. Many of these studies provided information into phenotypic changes that could impact disease and its treatment during spaceflight. For example, a small number of spaceflight studies noted an increase in cell concentrations of *Bacillus subtilis* when cultivated in the spaceflight environment as compared to ground controls [3, 45], although this research did not lend itself to identifying a mechanism behind these changes. Of particular relevance, the transfer of mobile genetic elements (MGE) in *Bacillus thuringiensis* has been documented in a study performed aboard the ISS, and it has been hypothesized that MGE exchange may be more efficient in-flight for this microorganism as compared to ground cultures [46]. Confirmation that MGEs can readily be exchanged in the microgravity environment holds value in our understanding of crew health risk as it reveals the potential for endogenous flora of the spacecraft to undergo genetic exchange, presenting an increased risk for antibiotic resistance to occur. Studies using the RWV have focused on bacterial function, such as antibiotic production by *Bacillus brevis*, which was found to be unaffected by LSMMG conditions [47].

Gram positive microorganisms have also been investigated to better understand changes in cellular growth kinetics and culture densities in spaceflight cultures. *B. subtilis* displayed a shortened lag growth phase and an extended exponential growth phase in response to microgravity culture, yielding an increase in final cell density as compared to ground cultures [3]. Kacena et al. also investigated the growth of *B. subtilis* during spaceflight culture and also noted a decreased lag growth phase, that could be temperature dependent, and greater final cell concentrations [45]. Interestingly, Kacena et al. had also investigated the growth of *B. subtilis* in a liquid medium compared to culture of *B. subtilis* on a semi-solid agar [4]. On the semi-solid agar, no differences in growth were observed compared to control cultures suggesting the differences in growth were the result of external physical forces, such as fluid dynamics or extracellular transport, rather than cellular responses directly reacting to gravity [4].

Many spaceflight and spaceflight analogue studies with Gram positive microorganisms have focused on secondary metabolites, especially antibiotic production [48]. In a series of RWV experiments Fang, et al. reported that production of cephamycin C from *Streptomyces clavuligerus* and rapamycin from *Streptomyces hygroscopicus* were inhibited by culture in LSMMG [49, 50]. This inhibition was similar to repressed production of microcin B17 observed in the Gram negative *Escherichia coli* when cultured in LSMMG [6]. Conversely, production of gramicidin S by *Bacillus brevis* cultured in LSMMG was unaffected compared to control cultures [7]. Interestingly, the sites of accumulation for microcin B17 and rapamycin differed depending on whether the bacteria was cultured in the RWV (extracellular localization) or in standard shaking flasks (intracellular localization), and that this localization could be altered by increasing the fluid shear stress in the RWV by the addition of a single glass bead. Investigation of secondary metabolite production in true spaceflight has also provided interesting findings. Benoit et al. reported initial increased production of actinomycin D by *Streptomyces plicatus* in response to spaceflight culture when compared to ground controls. However, this increase was only observed early in the experiment (days 8 and 12), and spaceflight production was consistently below ground-based control for the remainder of the 72-day experiment [51]. Recent spaceflight experiments investigating *Nocardia mediterranei* aboard Shenzhou III [52] and *Streptomyces avermitilis* aboard Shenzhou VII [48] also showed the potential for altered secondary metabolite production during spaceflight culture.

14.5 Conclusion

Gram positive microorganisms are commonly isolated aboard spacecraft [17, 53] **and many species pose a clear risk to the crew of these vehicles**. As current research provides evidence that these organisms are able to sense and respond to the environment of spaceflight and spaceflight analogues, understanding the stimulus/stimuli that triggers these altered responses and the impact on their phenotype is important in protecting the health of the crew on long duration missions. Evidence from spaceflight experiments has also shown us that Gram positive microorganisms respond in unique ways compared to their terrestrial counterparts. These novel responses provide the opportunity to not only benefit space travelers, but also enable the discovery of previously unknown microbial disease mechanisms that could lead to improved health care to the general public on Earth.

Questions for Future Research

- How do the structural differences between Gram positive and Gram negative bacteria affect their response to microgravity?
- As each microorganism can have markedly different phenotypic responses to spaceflight analogue (and possibly spaceflight) environments, do their biochemical and transcriptional responses also vary?
- As many Gram positive pathogens can be carried by crewmembers during spaceflight missions (either as commensals or subclinical pathogens), what is the crew health risk during long term missions (e.g., to Mars)?

References

1. Juergensmeyer, M. A., Juergensmeyer, E. A., & Guikema, J. A. (1999). Long-term exposure to spaceflight conditions affects bacterial response to antibiotics. *Microgravity Science and Technology, 12*, 41–47.
2. Tixador, R., Richoilley, G., Gasset, G., Templier, J., Bes, J. C., et al. (1985). Study of minimal inhibitory concentration of antibiotics on bacteria cultivated in vitro in space (Cytos 2 experiment). *Aviation, Space and Environmental Medicine, 56*, 748–751.
3. Mennigmann, H. D., & Lange, M. (1986). Growth and differentiation of Bacillus subtilis under microgravity. *Naturwissenschaften, 73*, 415–417.
4. Kacena, M. A., Leonard, P. E., Todd, P., & Luttges, M. W. (1997). Low gravity and inertial effects on the growth of E. coli and B. subtilis in semi-solid media. *Aviation, Space and Environmental Medicine, 68*, 1104–1108.
5. Castro, S. L., Nelman-Gonzalez, M., Nickerson, C. A., & Ott, C. M. (2011). Induction of attachment-independent biofilm formation and repression of Hfq expression by low-fluid-shear culture of Staphylococcus aureus. *Applied and Environmental Microbiology, 77*, 6368–6378.
6. Fang, A., Pierson, D. L., Koenig, D. W., Mishra, S. K., & Demain, A. L. (1997). Effect of simulated microgravity and shear stress on microcin B17 production by Escherichia coli and on its excretion into the medium. *Applied and Environmental Microbiology, 63*, 4090–4092.
7. Fang, A., Pierson, D. L., Mishra, S. K., Koenig, D. W., & Demain, A. L. (1997). Gramicidin S production by Bacillus brevis in simulated microgravity. *Current Microbiology, 34*, 199–204.
8. Fang, A., Pierson, D. L., Mishra, S. K., Koenig, D. W., & Demain, A. L. (1997). Secondary metabolism in simulated microgravity: β-Lactam production by *Streptomyces clavuligerus. Journal of Industrial Microbiology, 18*, 22–25.
9. Wilson, J. W., Ott, C. M., Honer Zu Bentrup, K., Ramamurthy, R., Quick, L., et al. (2007). Space flight alters bacterial gene expression and virulence and reveals a role for global regulator Hfq. *Proceedings of the National Academy of Sciences of the United States of America, 104*, 16299.
10. Wilson, J. W., Ott, C. M., Quick, L., Davis, R., Honer Zu Bentrup, K., et al. (2008). Media ion composition controls regulatory and virulence response of Salmonella in spaceflight. *PLoS One, 3*, e3923.
11. Crabbe, A., Schurr, M. J., Monsieurs, P., Morici, L., Schurr, J., et al. (2011). Transcriptional and proteomic response of Pseudomonas aeruginosa PAO1 to spaceflight conditions involves Hfq regulation and reveals a role for oxygen. *Applied and Environmental Microbiology, 77*, 1221–1230.
12. Nickerson, C. A., Ott, C. M., Mister, S. J., Morrow, B. J., Burns-Keliher, L., et al. (2000). Microgravity as a novel environmental signal affecting *Salmonella* enterica serovar Typhimurium virulence. *Infection and Immunity, 68*, 3147–3152.
13. Crabbe, A., De Boever, P., Van Houdt, R., Moors, H., Mergeay, M., et al. (2008). Use of the rotating wall vessel technology to study the effect of shear stress on growth behaviour of Pseudomonas aeruginosa PA01. *Environmental Microbiology, 10*, 2098–2110.
14. Crabbe, A., Pycke, B., Van Houdt, R., Monsieurs, P., Nickerson, C., et al. (2010). Response of Pseudomonas aeruginosa PAO1 to low shear modelled microgravity involves AlgU regulation. *Environmental Microbiology, 12*, 1545–1564.
15. Allen, C. A., Niesel, D. W., & Torres, A. G. (2008). The effects of low-shear stress on adherent-invasive Escherichia coli. *Environmental Microbiology, 10*, 1512.
16. Pierson, D. L. (2001). Microbial contamination of spacecraft. *Gravitational and Space Biology Bulletin, 14*, 1–6.
17. Castro, V. A., Thrasher, A. N., Healy, M., Ott, C. M., & Pierson, D. L. (2004). Microbial characterization during the early habitation of the International Space Station. *Microbial Ecology, 47*, 119–126.

18. Bassinger, V. A., Fontenot, S. L., Castro, V. A., Ott, C. M., & Pierson, D. L. (2004). A survey of *Staphylococcus aureus* and its methicillin resistance aboard the International Space Station. Washington.
19. Lowy, F. D. (1998). Staphylococcus aureus infections. *The New England Journal of Medicine, 339*, 520–532.
20. Pierson, D. L., Chidambaram, M., Heath, J. D., Mallary, L., Mishra, S. K., et al. (1996). Epidemiology of Staphylococcus aureus during space flight. *FEMS Immunology and Medical Microbiology, 16*, 273–281.
21. Pierson, D. L., McGinnis, M. R., & Viktorov, A. N. (1994). Microbiological contamination. In A. E. Nicogossian, S. R. Mohler, O. G. Gazenko, & A. I. Grigoryev (Eds.), *Space biology and medicine* (pp. 77–93). Washington: American Institute of Aeronautics and Astronautics, Inc.
22. Cioletti, L. A., Mills, A. L., & Mishra, S. K. (1991). Microbial growth and physiology in space: A review. SAE Technical Paper Series. San Francisco.
23. Puleo, J. R., Oxborrow, G. S., Fields, N. D., & Hall, H. E. (1970). Quantitative and qualitative microbiological profiles of the Apollo 10 and 11 spacecraft. *Applied Microbiology, 20*, 384–389.
24. Rosado, H., Doyle, M., Hinds, J., & Taylor, P. W. (2010). Low-shear modelled microgravity alters expression of virulence determinants of *Staphylococcus aureus. Acta Astronautica, 66*, 408–416.
25. Rosado, H., O'Neill, A. J., Blake, K. L., Walther, M., Long, P. F., et al. (2012). Rotating wall vessel exposure alters protein secretion and global gene expression in *Staphylococcus aureus. International Journal of Astrobiology, 11*, 71–81.
26. Rosado, H., Stapleton, P. D., & Taylor, P. W. (2006). Effect of simulated microgravity on the virulence properties of the opportunistic bacterial pathogen *Staphylococcus aureus*.
27. Vukanti, R., Model, M. A., & Leff, L. G. (2012). Effect of modeled reduced gravity conditions on bacterial morphology and physiology. *BMC Microbiology, 12*, 4.
28. Shirtliff, M. E., Mader, J. T., & Camper, A. K. (2002). Molecular interactions in biofilms. *Chemistry and Biology, 9*, 859–871.
29. Nickerson, C. A., Ott, C. M., Wilson, J. W., Ramamurthy, R., & Pierson, D. L. (2004). Microbial responses to microgravity and other low-shear environments. *Microbiology and Molecular Biology Reviews, 68*, 345–361.
30. Liu, G. Y., Essex, A., Buchanan, J. T., Datta, V., Hoffman, H. M., et al. (2005). Staphylococcus aureus golden pigment impairs neutrophil killing and promotes virulence through its antioxidant activity. *Journal of Experimental Medicine, 202*, 209–215.
31. Nizet, V. (2007). Understanding how leading bacterial pathogens subvert innate immunity to reveal novel therapeutic targets. *Journal of Allergy and Clinical Immunology, 120*, 13–22.
32. Liu, Y., Wu, N., Dong, J., Gao, Y., Zhang, X., et al. (2010). Hfq is a global regulator that controls the pathogenicity of Staphylococcus aureus. *PLoS One, 5*.
33. Bohn, C., Rigoulay, C., & Bouloc, P. (2007). No detectable effect of RNA-binding protein Hfq absence in Staphylococcus aureus. *BMC Microbiology, 7*, 10.
34. Murray, P. R., Kobayashi, G. S., Pfaller, M. A., & Rosenthal, K. S. (1994). *Streptococcus* and related gram-positive bacteria. In R. Farrel (Ed.), *Medical microbiology* (2nd ed., pp. 180–198). St. Louis: Mosby-Year Book, Inc.
35. Watson, P., Voss, L., Barber, C., Aickin, R., Bremner, D., et al. (1996). The microbiology of chronic otitis media with effusion in a group of Auckland children. *The New Zealand Medical Journal, 109*, 182–184.
36. Jacobs, M. R. (2004). Streptococcus pneumoniae: Epidemiology and patterns of resistance. *American Journal of Medicine, 117*(Suppl 3A), 3S–15S.
37. van der Poll, T., & Opal, S. M. (2009). Pathogenesis, treatment, and prevention of pneumococcal pneumonia. *Lancet, 374*, 1543–1556.
38. Dery, M. A., & Hasbun, R. (2007). Changing epidemiology of bacterial meningitis. *Current Infectious Disease Reports, 9*, 301–307.

39. Kadioglu, A., Weiser, J. N., Paton, J. C., & Andrew, P. W. (2008). The role of Streptococcus pneumoniae virulence factors in host respiratory colonization and disease. *Nature Reviews Microbiology, 6*, 288–301.
40. O'Brien, K. L., Wolfson, L. J., Watt, J. P., Henkle, E., Deloria-Knoll, M., et al. (2009). Burden of disease caused by Streptococcus pneumoniae in children younger than 5 years: Global estimates. *Lancet, 374*, 893–902.
41. CDC. (2012). Pneumococcal disease. epidemiology and prevention of vaccine-preventable diseases. The pink book: Course textbook. 12th ed. http://www.cdc.gov/vaccines/pubs/pinkbook/pneumo.html
42. de Velasco, E. A., Merkus, D., Anderton, S., Verheul, A. F., Lizzio, E. F., et al. (1995). Synthetic peptides representing T-cell epitopes act as carriers in pneumococcal polysaccharide conjugate vaccines. *Infection and Immunity, 63*, 961–968.
43. Allen, C. A. (2007). *Doctoral dissertation*. Galveston: University of Texas Medical Branch.
44. Allen, C. A., Galindo, C. L., Pandya, U., Watson, D. A., Chopra, A. K., et al. (2006). Transcription profiles of Streptococcus pneumoniae grown under different conditions of normal gravitation. *Acta Astronautica, 60*, 433–444.
45. Kacena, M. A., Merrell, G. A., Manfredi, B., Smith, E. E., Klaus, D. M., et al. (1999). Bacterial growth in space flight: Logistic growth curve parameters for Escherichia coli and Bacillus subtilis. *Applied Microbiology and Biotechnology, 51*, 229–234.
46. De Boever, P., Mergeay, M., Ilyin, V., Forget-Hanus, D., Van der Auwera, G., & Mahillon, J. (2007). Conjugation-mediated plasmid exchange between bacteria grown under space flight conditions. *Microgravity Science and Technology, 19*, 138–144.
47. Demain, A. L., & Fang, A. (2001). Secondary metabolism in simulated microgravity. *Chemical Record, 1*, 333–346.
48. Gao, H., Liu, Z., & Zhang, L. (2011). Secondary metabolism in simulated microgravity and space flight. *Protein & Cell, 2*, 858–861.
49. Fang, A., Pierson, D. L., Mishra, S. K., Koenig, D. W., & Demain, A. L. (1997). Secondary metabolism in simulated microgravity: Beta-lactam production by Streptomyces clavuligerus. *Journal of Industrial Microbiology and Biotechnology, 18*, 22–25.
50. Fang, A., Pierson, D. L., Mishra, S. K., & Demain, A. L. (2000). Growth of Streptomyces hygroscopicus in rotating-wall bioreactor under simulated microgravity inhibits rapamycin production. *Applied Microbiology and Biotechnology, 54*, 33–36.
51. Benoit, M. R., Li, W., Stodieck, L. S., Lam, K. S., Winther, C. L., et al. (2006). Microbial antibiotic production aboard the International Space Station. *Applied Microbiology and Biotechnology, 70*, 403–411.
52. Zhou, J., Sun, C., Wang, N., Gao, R., Bai, S., et al. (2006). Preliminary report on the biological effects of space flight on the producing strain of a new immunosuppressant, Kanglemycin C. *Journal of Industrial Microbiology and Biotechnology, 33*, 707–712.
53. Pierson, D., Botkin, D. J., Bruce, R. J., Castro, V. A., Smith, M. J., et al. (2012). Microbial monitoring of the International Space Station. In J. Moldenhauer (Ed.), *Environmental monitoring: A comprehensive handbook* (pp. 1–27). River Grove: DHI Publishing. pp.

Index

C.A. Nickerson et al. (eds.), *Effect of Spaceflight and Spaceflight Analogue Culture on Human and Microbial Cells*, DOI 10.1007/978-1-4939-3277-1

MIX
Papier aus verantwortungsvollen Quellen
Paper from responsible sources
FSC® C105338

If you have any concerns about our products,
you can contact us on
ProductSafety@springernature.com

In case Publisher is established outside the EU,
the EU authorized representative is:
Springer Nature Customer Service Center GmbH
Europaplatz 3, 69115 Heidelberg, Germany

Printed by Libri Plureos GmbH
in Hamburg, Germany